INTRODUCTION TO BIO-ONTOLOGIES

CHAPMAN & HALL/CRC
Mathematical and Computational Biology Series

Aims and scope:

This series aims to capture new developments and summarize what is known over the entire spectrum of mathematical and computational biology and medicine. It seeks to encourage the integration of mathematical, statistical, and computational methods into biology by publishing a broad range of textbooks, reference works, and handbooks. The titles included in the series are meant to appeal to students, researchers, and professionals in the mathematical, statistical and computational sciences, fundamental biology and bioengineering, as well as interdisciplinary researchers involved in the field. The inclusion of concrete examples and applications, and programming techniques and examples, is highly encouraged.

Series Editors

N. F. Britton
Department of Mathematical Sciences
University of Bath

Xihong Lin
Department of Biostatistics
Harvard University

Hershel M. Safer

Maria Victoria Schneider
European Bioinformatics Institute

Mona Singh
Department of Computer Science
Princeton University

Anna Tramontano
Department of Biochemical Sciences
University of Rome La Sapienza

Proposals for the series should be submitted to one of the series editors above or directly to:
CRC Press, Taylor & Francis Group
4th, Floor, Albert House
1-4 Singer Street
London EC2A 4BQ
UK

Published Titles

Chapman & Hall/CRC Mathematical and Computational Biology Series

INTRODUCTION TO BIO-ONTOLOGIES

PETER N. ROBINSON
SEBASTIAN BAUER

CRC Press
Taylor & Francis Group
Boca Raton London New York

CRC Press is an imprint of the
Taylor & Francis Group, an **informa** business
A CHAPMAN & HALL BOOK

CRC Press
Taylor & Francis Group
6000 Broken Sound Parkway NW, Suite 300
Boca Raton, FL 33487-2742

First issued in paperback 2020

© 2011 by Taylor & Francis Group, LLC
CRC Press is an imprint of Taylor & Francis Group, an Informa business

No claim to original U.S. Government works

ISBN-13: 978-1-4398-3665-1 (hbk)
ISBN-13: 978-0-367-65927-1 (pbk)

Visit the Taylor & Francis Web site at
http://www.taylorandfrancis.com

and the CRC Press Web site at
http://www.crcpress.com

Dedication

To my father. And of course to Elisabeth, Sophie, and Anna.

-PNR

To my parents and Denise.

-SB

Preface

Computer scientists recognized the need to capture knowledge in a standardized, computer-readable fashion in order to build large and powerful artificial intelligence systems in the 1970s. This led to the development of ontologies as a tool to allow computer programs to be able to interact with the information produced by other computer programs [183]. Subsequently, these ideas were further developed by Tim Berners-Lee for the World Wide Web as a mechanism by which Internet servers can interoperate with one another and build upon each other's data – the Semantic Web [99].

The still young field of bioinformatics has tended to concentrate on algorithms for the analysis of molecular data such as DNA and protein sequences, gene-expression data, and protein structure, as well as the efficient storage and retrieval of this data from databases. As the amount and complexity of data being produced in modern biomedical research increases, including next-generation sequencing, biomedical imaging, protein interactions, and many other technologies unimaginable only a decade ago, the need for bioinformatics to integrate, manage, and understand the meaning of this data has grown enormously [21].

The Gene Ontology (GO), one of the first major ontologies in the biomedical research community, was first published in 2000 [14] and has since had an enormous influence on the fields of bioinformatics and molecular genetics and has inspired the creation of a large number of other biomedical ontologies. The GO was initially developed to address the need of model organism databases to integrate and exchange genome annotation data, but the GO has since inspired the development of numerous algorithms for the analysis of high-throughput data, and has spurred the development of ontologies for many fields of biomedical research [189, 236], and accordingly bioinformaticians increasingly need to know how to use ontologies.

The aim of this book therefore is to provide an introduction to ontologies for bioinformaticians, computer scientists, and others working in the field of biomedical research who need to know something about the computational background of ontologies. The field of bio-ontologies has become very broad, and people working in bio-ontologies come from many disciplines, including philosophy, statistics, computer science, bioinformatics, molecular biology, and others. Our research group has had experience in many of the fields of bio-ontologies of most relevance to bioinformaticians, including the development of software and algorithms for GO analysis [24, 23, 97, 98, 212], the application of GO analysis to molecular genetics datasets [110, 109, 194, 210], and the development of the Human Phenotype Ontology[1] (HPO) and of a number of algorithms to use the HPO for medical diagnostics and computational phenotype analysis [92, 145, 209, 211]. This book will endeavor to provide a self-contained introduction to bio-ontologies for bioinformaticians or computer scientists who need to understand ontological algorithms and their applications. Therefore, although we will touch upon many subjects that we believe will help bioinformaticians to understand the broader uses of ontologies, our emphasis will clearly be upon computational and algorithmic issues surrounding bio-ontologies.

Introduction to Bio-Ontologies is a self-contained treatment of bio-ontologies that can be used for a one-semester course on bio-ontologies in bioinformatics or computer science curricula or for self-study. The book is divided into four parts.

Part I: Basic Concepts offers an overview of what an ontology is and what is special about bio-ontologies (Chapter 1). Mathematical logic, which is extremely important for understanding concepts of inference in bio-ontologies, is treated in Chapter 2. Statistics is indispensable for most specialties in bioinformatics, and bio-ontologies do not represent an exception to this rule. Chapter 3 treats several topics in the field of probability and statistics that are required to understand the algorithms that will be presented in the subsequent Chapters. Chapter 4 offers an overview of OBO, which is the preeminent ontology language for bio-ontologies, and of the languages of the Semantic Web, RDF, RDFS, and OWL, which are quickly gaining in importance in the field of bio-ontology.

Part II: Bio-Ontologies deals with a number of important bio-ontologies and their applications. The Gene Ontology (GO) has been the most well-known and widely used bio-ontology since its introduction in 2000 [14], and has served as a catalyst for many of the current developments in the field of bio-ontologies. The GO is covered in detail in Chapter 5. Upper-level ontologies are relatively small ontologies that define the general ways that objects and properties, parts and wholes, relations and collectives interact. Upper-level ontologies provide a sort of philosophical framework for the development of bio-ontologies, which in turn provides a logical basis for computational inference. Upper-level ontologies including especially the Basic Formal Ontology and the Relation Ontology are covered in Chapter 6. Chapter 7 offers a survey of several current bio-ontologies and illustrates a number of important concepts along the way.

The final two parts of the book deal with a number of algorithms and applications of ontologies. At the risk of oversimplification, we have divided the algorithms into two categories: graph algorithms and inference algorithms.

---

[1] http://www.human-phenotype-ontology.org/.

As we will see, the graph algorithms make use of inference and vice versa, but the main emphasis of the graph algorithms is more on the use of an analysis of the structure of graphs representing an ontology and annotations to it, and the main emphasis of the inference algorithms is on the discovery of new knowledge based on facts that have been explicitly asserted in a knowledge base. While the graph algorithms have been developed primarily in the specific field of bio-ontologies, the inference algorithms have been developed in the general field of ontologies and have been especially influenced by the advent of the Semantic Web.

Part III: Graph Algorithms for Bio-Ontologies introduces the major graph-based algorithms for bio-ontologies. Probably the single most well-known application of bio-ontologies for molecular biologists and other researchers in biomedicine is overrepresentation analysis, in which the GO and annotations to it are used to characterize the results of high-throughput experiments such as microarray hybridizations. Chapter 8 explains how overrepresentation analysis works and demonstrates a number of improvements to the basic procedure. Chapter 9 introduces model-based procedures for using GO and similar knowledge bases in a similar way. The main algorithm presented in this chapter, MGSA, involves a kind of statistical analysis of a Bayesian network into which the genes and GO terms are embedded. This chapter may be skipped by readers with no prior knowledge of Bayesian statistics, although we have provided some background material on this topic in Chapter 3 for interested readers. Chapter 10 provides an introduction to semantic similarity analysis in ontologies, which represents a fascinating extension to the results of Claude Shannon about information theory that led to the concept of entropy. Chapter 11 describes an algorithm recently developed by the authors of this book that again uses a Bayesian network, this time to improve the results of searching databases by means of semantic similarity.

Part IV: Inference in Bio-Ontologies introduces a number of topics surrounding computational reasoning. Chapter 12 discusses the ways that inference is being used to maintain the structure of the GO ontology and to link it with other bio-ontologies. The remainder of the chapters in this part deal with the ontology languages of the Semantic Web and their applications for inference. The Semantic Web is a proposal of the World Wide Web Consortium (W3C) that is designed to enable richer integration and interoperability of data from multiple sources on the Web. The W3C has developed a number of ontology languages for this purpose, including mainly RDF and OWL. Chapter 13 introduces the formal semantics of RDF and RDFS. This chapter intends to provide readers with an intuition of how formal logical concepts can be used to define reliable computational entailment rules that can be used by actual programs to perform computational inference. Chapter 14 provides a tour of OWL inference rules. Chapter 15 illustrates how inference is actually performed computationally and explains an important inference algorithm in detail. The final chapter of the book, Chapter 16, explains an important query language for RDF knowledge bases called SPARQL, and discusses the current state of the art for querying OWL ontologies.

The book additionally has four appendices. Many of the exercises ask readers to use R to solve problems. **Appendix A** provides an introduction to the R programming language, which will be sufficient for readers with experience in programming languages such as Java or C to complete the exercises. **Appendix B** explains the mathematical background of the concept of information content used in ontological semantic similarity calculations. **Appendix C** explains the RDF/XML format that is commonly used to serialize RDF and OWL ontologies, and covers the main constructs used in RDF and RDFS ontologies. Finally, **Appendix D** provides a detailed introduction to the language constructs most commonly encountered in OWL ontologies and explains how they are represented using the most popular serialization formats of the W3C.

The book should be suitable for a graduate or advanced undergraduate course in a bioinformatics or computer science program but can also be used for self-study. A short course in bio-ontologies with an emphasis on graph algorithms used to support high-throughput molecular biology experiments could comprise Chapters 1 and 3-10. A short course with emphasis on inference and the Semantic Web for biology could comprise Chapters 1-5, and 12-15. For a complete single-semester course, each of the chapters can be covered in about a week depending on the goals and focus of the individual course.

Most of the chapters include theoretical and practical exercises, and some of the exercises represent small tutorials on the use of a particular software. Some of the exercises represent problems in the R programming language. Other exercises are provided using open-source software from our research group.[2]

After finishing this book, readers should be able to use ontologies such as GO to support molecular genetics research projects or as a starting point for their own bioinformatics research projects.

The Book Website

The book Web site at `http://bio-ontologies-book.org` offers software and data designed to complement the material presented in the book. Some of the exercises make use of an R package called `Robo` that was developed for this book. The `Robo` package is available together with a compressed archive containing data and ontology files that are used in some of the exercises from the Web site. Slides that the authors have used in teaching or presentations

---

[2]http://compbio.charite.de/.

are made available, and pointers are made to other Web sites with useful material.

To Our Colleagues

The field of biomedical ontologies has experienced dramatic growth in the last decade, and is already much too large and diverse to be discussed in a single book. We have supplied extensive references to the primary literature as well as pointers to important books and articles and the end of most chapters. Due to space constraints, it has been impossible to include many interesting algorithms and methods, and the choice of algorithms included in this book necessarily reflects the interests and prejudices of the authors.

It is our hope that this book will be able to contribute to bridging the gap that has grown up between OBO ontologies and the Semantic Web by providing a self-contained introduction to both fields and hopefully by helping readers understand what both fields can contribute to the analysis and understanding of biomedical data with the help of ontologies.

Acknowledgments

The authors would like to offer thanks to Chris Mungall, Francesco Couto, Catia Pesquita, Dietrich Rebholz-Schumann, Georgios V Gkoutos, Barry Smith, and Paul Schofield for insightful comments about a draft version of this book. Thanks to Jana Hastings from ChEBI, Darren Natale from the Protein Ontology, Alexander Diehl from the Cell Ontology, Karen Eilbeck from the Sequence Ontology, and José L. V. Mejino Jr. from the FMA for assistance with the material on these ontologies in Chapter 7.

The authors would like to thank other members of the Institute of Medical Genetics of the Charité-Universitätsmedizin Berlin and the Max Planck Institute for Molecular Genetics in Berlin for their contributions, suggestions, and comments, including especially Patrick Booms, Gabriele Carletti, Sandra Dölken, Begoñia Muñoz-García, Johannes Grünhagen, Gao Guo, Denise Horn, Marten Jäger, Stefanie Katzke, Sebastian Köhler, Peter Krawitz, Stefan Mundlos, Claus-Eric Ott, Monika Palz, Angelika Pletschacher, Christian Rödelsperger, Marcel H. Schulz, and Frank Tiecke.

-Peter N. Robinson & Sebastian Bauer
-Berlin

Contents

List of Figures

List of Tables

Symbol Description

α Used to represent the type I error in hypothesis testing

β Use to represent the type II error in hypothesis testing-

H_0 Null hypothesis

H_A Alternative hypothesis

$\vee$ **or**, a logical operator

$\sqcup$ Symbol for **or** in Description Logics

$\neg$ **not**, a logical operator

$\wedge$ **and**, a logical operator

$\sqcap$ Symbol for **and** in Description Logics

$\cup$ **union**, a set operator

$\cap$ **intersection**, a set operator

$\subseteq$ **subset**. If $A \subseteq B$ then A is a subset of, and possibly equivalent to, B.

$\sqsubseteq$ symbol for **subset** in Description Logics

$\equiv$ equivalent to (used for classes in Description Logics and elsewhere)

$\Rightarrow$ implies

$\vDash$ Symbol for entailment

$\Leftrightarrow$ iff, if and only if

$\bot$ **false** (logical constant). In Description Logic, $\bot$ is used to denote the empty concept.

$\top$ **true** (logical constant). In Description Logic, $\top$ is used to denote the set of all concept names.

$n!$ n factorial: $n! = n \cdot (n-1) \cdot (n-2) \cdot \ldots \cdot 2 \cdot 1$.

Σ Summation: $\sum_{i=1}^{n} x_i = x_1 + x_2 + \ldots + x_n$.

Π Product: $\prod_{i=1}^{n} x_i = x_1 \cdot x_2 \cdot \ldots \cdot x_n$.

$:$ Concept or role assertion in Description Logics. $\mathsf{a : C}$ states that individual a is an instance of class C. $\mathsf{(a,b) : R}$ states that individuals a and b are related to one another by role R.

Part I

Basic Concepts

Chapter 1

Ontologies and Applications of Ontologies in Biomedicine

"An ontology is a specification of a conceptualization."
Thomas R. Gruber, 1993 [100]

1.1 What Is an Ontology?

The word *ontology* has been in use for many years, and was originally used to describe a branch of metaphysics concerned with the nature and relations of being. More recently, the word ontology has been used to refer to a common, controlled knowledge representation designed to help knowledge sharing and computer reasoning. In this book, we will cover certain kinds of ontologies in common use in biomedical research called *bio-ontologies*, which have become an essential feature of many fields of biomedical research in which large amounts of data are handled. Bio-ontologies represent a challenging and rewarding field in computational biology and bioinformatics, because practitioners need to know not only about a wide range of algorithms and computational algorithms and to integrate multiple kinds of data from multiple sources, but must also understand the logical and philosophical underpinnings of ontologies to perform computer reasoning and to express and analyze the *meaning* of the data under study. To build an ontology is an intellectually stimulating exercise that amounts to constructing a philosophical and computational model of the world, or at least of some small part of it.

The discipline of ontology goes back at least as far as Aristotle (384 BC–322 BC), a Greek philosopher and a student of Plato, who defined ontology as the science of being as such in the Metaphysics: *There is a science which studies Being qua Being, and the properties inherent in it in virtue of its own nature.* In other works, Aristotle developed conceptual taxonomies that are in some ways similar to modern bio-ontologies (cf. Chapter 6). In modern times, the word ontology was adopted by computer scientists in the field of Knowledge Representation to denote computational systems for describing specific domains of reality. The statement by Thomas R. Gruber quoted above is widely taken to be the first modern definition of *ontology* for computer science. The definition provides a maximally concise description of the two essential components of any ontology: a set of *concepts* or terms that name and define the entities in a domain, and a set of semantic links between the terms that *specify* and also constrain the relations between the entities in the domain.

Ontology: History of a Word

Although the field of Ontology goes back over two millennia to the Greek philosopher Aristotle, the word *ontology* is much younger. Jean le Clerc coined the word in 1692 (ontologie) on the basis of the Greek word for *essence* or *being* (On/Ontos) and the suffix *-logy* for *study of*. According to the *Oxford English Dictionary*, the first recorded usage of the word in English was a definition by Nathan Bailey in his book *An Universal Etymological English Dictionary* published in 1721: "*Ontology*, an Account of being in the Abstract."

Ontologies are thus semantic models for domains of reality. A model is a way of describing the important aspects of a domain while simplifying or omitting less important or irrelevant aspects. Models can be tools for communication, for analyzing or explaining observations, for predicting future developments, and can provide a framework for integrating data from different sources. Models, and ontologies, have different degrees of formality. A free text description using natural language, for instance, is a very informal model. Informal models are mainly intended for human communication. While humans can usually tell from context whether, say, the word *football* in a sentence refers to American football or European football (soccer), such tasks are significantly more difficult for a computer. By increasing the formality of a model, for instance by providing computational rules about syntax and semantics, we can make data more amenable to machine processing.

A big factor responsible for the increasing importance of ontologies in the late 1990s and the early years of this millennium was the growth of the World Wide Web (WWW). Many search engines such as that of Yahoo! make use of ontologies [147], and websites involved in e-commerce and many other kinds of applications use ontologies and ontology algorithms of different degrees of complexity and power [163, 216]. In May 2002, a Google search for the term *ontology* returned about 310,000 websites. At this writing, over 32 million hits are returned. Probably the main factor driving the continued expansion of the field of ontologies in the last several years has been the development of the Semantic Web.

The WWW grew out of proposals by Tim Berners-Lee in 1989 and 1990 to create a Web of hypertext documents using hyperlinks to allow users to browse between documents located on different Web servers. The language for the documents, HyperText Markup Language (HTML), enables authors to create a structure for the presentation of the contents of the document that makes it easy for humans to read and understand. This means that current Web browsers can easily parse Web pages in order to create a nice visual layout and links to other Web pages, but they have no reliable way of recognizing the meaning of the Web pages they present, much less of roaming the Net to harvest the information needed to answer a question. The Semantic Web was proposed by Berners-Lee in 1999 as a framework in which computers are capable of analyzing the meaning – the semantic content – of the data on the Web in order to act as "intelligent agents." The power of the WWW is its universality. Anyone can link to anything. The Web is decentralized, and any kind of document can be published with links to any servers on the Web without asking permission of any central authority. What was required was an agreement about protocols for displaying Web pages and handling HTML links and other protocols. The Semantic Web is designed to be just as decentralized as the WWW. *Anyone can say anything about anything.* What is required is an agreement about how to express and reason over data [32]. As we will see in Chapter 4, the World Wide Web Consortium (W3C) has created languages for the Semantic Web that are specifically designed as a sort of semantic markup language for data: RDF, RDFS, and OWL. Especially OWL has also come to play an important role as a language for bio-ontologies.

1.2 Ontologies and Bio-Ontologies

Ontologies are used in the WWW, in e-commerce, in libraries, and in many other contexts for organizing, annotating, analyzing, searching, and integrating data, as we will see in the course of this book. What, then, makes bio-ontologies special? On the one hand, of course, biomedical ontologies describe biological research or medical data, and one of the most important bio-ontology languages, OBO, was designed specifically for the needs of biological research. A more important difference is the way that bio-ontologies are used to analyze biomedical data, which is often completely different from methodologies commonly used in e-commerce or the Semantic Web. Bio-ontologies are tools for annotation and data integration that provide bench researchers with a common vocabulary to describe and communicate results and give bioinformaticians tools for the functional analysis of microarray data, network modeling, semantic similarity for biological analysis and clinical diagnostics, as well as many other applications.

Nonetheless, bio-ontologies are also used for data integration in much the same way as an ontology about engineering might be used for data integration in an automobile factory, and the Semantic Web for bioinformatics is likely to increase in importance in coming years [48]. This book therefore will provide both an introduction to several classes of algorithms that have been developed specifically for bio-ontologies as well as an introduction to inference algorithms based on the languages of the Semantic Web.

1.3 Ontologies for Data Organization, Integration, and Searching

Semantically aware searching is an extremely important application of ontologies. We will illustrate the topic using biomedical examples, but our comments apply to many other domains of knowledge as well.

One of the biggest challenges in current biomedical research is how to access and analyze ever-increasing amounts of data and harness it to derive novel insights about biology or medicine or to suggest hypotheses for further research. There are currently about 20 million articles on biomedical topics in the PubMed databases, and novel high-throughput technologies including microarrays and massively parallel DNA sequencing are producing previously unimaginable amounts of data.

Perhaps the simplest kind of search in computer science is a search for a string in a document. A string is simply a sequence of characters such as "Traf2 interacts with Smad4 and regulates BMP signaling pathway in MC3T3-E1 osteoblasts" or "Don't panic!". A search on the latter string in most word processing programs will quickly identify all instances of the phrase "Don't panic!" in the document, assuming they are spelled correctly. However, searches that rely on matching a given string, or even sets of words, do not perform well in the biomedical domain. The language of biomedicine contains numerous synonyms, abbreviations, and acronyms that can refer to the same concept. For instance, *TRAF2* is the recommended symbol for the gene TNF receptor-associated factor 2. However, alternate names were (and are) used in the medical literature: TRAP, TRAP3, MGC:45012, and TRAF2. Searches with strings representing the alternative ways of naming the same concept often return wildly different results.[1] One of the purposes of ontologies is to provide a single identifier for describing and disambiguating each concept or entity in a domain and for storing a list of alternative names or synonyms. An ontology can thus be used as a controlled terminology, and search engines can be programmed to return all documents that match either the search term itself or any of its synonyms.

---

[1] At the time of this writing, a search on TRAF2 in PubMed retrieved 976 abstracts, a search on "TNF receptor-associated factor 2" returned 586 abstracts, a search on TRAP returned 19362 abstracts (most of which are not related to the gene), and a search on TRAP3 returned only two abstracts.

Ontologies as Controlled Terminologies

An important, if straightforward, use of ontologies is to provide lists of synonyms for the entities in a domain so that search engines can be designed to retrieve all documents in which a given entity is represented. Here is an example from the Gene Ontology [14] showing part of the information provided for one of the terms.

Name	erythrocyte development
Accession	GO:0048821
Synonyms	RBC development
	red blood cell development
Definition	The process aimed at the progression of an erythrocyte over time, from initial commitment of the cell to a specific fate, to the fully functional differentiated cell.

There are many important medical terminologies that have been used to provide a standard set of terms and synonyms for larger or smaller domains in medicine, including the Medical Subject Headings (MeSH) created by the National Library of Medicine that are used to index each reference to the medical literature in PubMed [54], SNOMED, a systemized nomenclature of medicine design to enable a consistent way of indexing, storing, retrieving and aggregating clinical data [58], and the National Institute of Cancer Thesaurus, which was developed to provide a unified terminology for molecular and clinical cancer research [80].

Ontologies are also used to connect concepts with related meanings. Search engines can use ontologies to retrieve not only documents that contain a string that matches the query, but also documents containing semantically related terms. One straightforward way of using the structure of an ontology to make a search more comprehensive is to expand query terms to include children of the term (also see Section 3.3). In this way, documents that do not explicitly mention the query term or any of its synonyms, but do discuss concepts that represent specialized instances of the query term are returned to the user (Figure 1.1). A number of search engines based on this strategy have been developed to help users find relevant literature. For instance, GoPubMed allows users to search the PubMed database by indexing PubMed abstracts with Gene Ontology terms, which allows users to retrieve abstracts according to GO categories and to exploit the hierarchical nature of GO for the search [73].

If different research labs and clinics agree on a common vocabulary (by using the same set of ontologies to annotate their data), then it becomes possible to integrate data from the different centers. A number of smaller ontologies have been developed in recent years to enable exchange of experimental data generated by new technologies such as microarrays. For example, the Microar-

transcription   gene expression

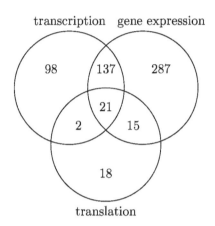

translation

FIGURE 1.1: Extending Searches with Semantically Related Concepts. In GO, the term *gene expression* is used to describe the process by which a gene's sequence is converted into a mature gene product (proteins or RNA). Searching for review articles on gene expression in the zebrafish with the query "gene expression and zebrafish" returned 460 documents. The search engine of PubMed is not designed to return documents that deal with specialized aspects of a given concept. The GO terms *transcription* and *translation* are children of gene expression because they represent specializations of *gene expression*. Searching for "transcription and zebrafish" returned 258 documents, and searching on "translation and zebrafish" return 56 documents, only some of which had been identified by the initial search. The Venn diagram in the figure shows the amount of overlap between the results to the three queries.

ray Gene Expression Database Group (MGED) has developed an ontology to provide standard terms for the annotation of microarray experiments with the goal of defining a common vocabulary for the exchange of microarray data within the biological community [244].

1.4 Computer Reasoning with Ontologies

As we will see in several chapters of this book, ontologies provide a framework for computer reasoning. Although the phrase "computer reasoning" may evoke visions of robots with positronic brains that are learning to "think" like humans, what is actually meant is substantially more modest. The *specification* of an ontology includes simple rules that can be automatically applied to items in a database in order to generate new knowledge. This process is referred to as *inference*. For instance, consider the rule, *All inhabitants of Manhattan are also inhabitants of New York City*.[2] We can apply this rule to a database entry stating that *John Smith is an inhabitant of Manhattan* to infer that John Smith is an inhabitant of New York City, even though there was no database entry to that extent. Although this inference may seem obvious to a human, the only way a computer can "know" this is if it has a framework for doing so. This is an extremely important application of bio-ontologies as well as ontologies in general.

Ontology languages such as OBO and OWL provide a number of constructs for specifying such rules which in combination allow surprisingly powerful inference to be performed on data that has been annotated using the ontologies. Several chapters of this book are concerned with applications of inference in bio-ontologies.

Ontologies provide a framework for integrating data of possibly inhomogeneous sources. Therefore, one important application of reasoning is quality control in such aggregated databases. In particular modern biomedical databases can have hundreds of thousands of entries or more, and it is nearly inevitable that mistakes or inconsistencies will creep into such databases. Ontologies can be used to help identify erroneous or contradictory entries in databases or other large-scale models. Ontologies can also be used for classification of entries, which is essential if thousands of items are being automatically processed. Finally, reasoning based on ontologies can be used to discover new facts about data or new relationships between items in a database in a way that can suggest new directions for research or new explanations for observed phenomena.

---

[2] A rule like this makes sense because Manhattan is one of the five boroughs of New York City.

1.5   Typical Applications of Bio-Ontologies

In addition to the above-described applications, which are important not only for bio-ontologies but for any kinds of ontologies, there are a number of algorithms that have been developed especially for biomedical ontologies. One of the major applications which helped to bring bio-ontologies to prominence was the use of Gene Ontology for the analysis of high-throughput data including especially microarray data. Experiments such as transcriptional profiling with microarrays involve measuring the expression of all genes in the genome under multiple conditions and comparing lists of genes showing differential responses in order to better understand the biology of the system. This kind of analysis tries to identify the ontology terms that best characterize the biological functions of the differential genes, and has come to be known as *overrepresentation analysis* or *gene category analysis*. Chapters 8 and 9 deal with this kind of analysis and related approaches.

Another topic with special importance for bio-ontologies is semantic similarity analysis. This deals with methods to measure similarity between terms in an ontology or between items annotated by those terms on the basis of the meaning of the terms, the semantic structure of the ontology, and the patterns of annotations. This kind of analysis has been used for comparison of proteins and for clinical diagnostics in medical genetics. These methods and their applications will be discussed in Chapters 10 and 11.

Finally, although the basic rules of inference are the same or at least similar for all ontologies, there are characteristic ways and situations in which bio-ontologies such as Gene Ontology are used for inference. Part IV of this book will cover inference in ontologies as well as several special uses of inference for bio-ontologies.

Chapter 2

Mathematical Logic and Inference

We will see that logic, the study of reasoning, and especially a form of mathematical logic termed *predicate logic* is of high importance for understanding computer reasoning and inference in ontologies, which in a word is the process by which a computer program uses a set of known facts expressed as formulas and variables in order to derive new facts that were not explicitly given as input to the program.

2.1 Representation and Logic

Three basic elements are required for computer reasoning: a formal language for knowledge representation, a means for reasoning in the language, and a set of representations of facts about a domain expressed in the formal language. This has long been a major topic in the field of artificial intelligence (AI), and forms a basis of inference in ontologies as well.

George Boole[1] (1815–1864) was one of the first mathematicians to develop a way of expressing the fundamental laws of logic in mathematical form. Let X, Y, and Z represent classes, say X=Mouse, Y=Rodent, and Z=Mammal. It seems intuitively clear that if all X are Y and all Y are Z, then all X are Z as well. Boole denoted the assertion that all X are Y as $X = XY$. He developed a system of purely symbolic manipulation of class representations to prove assertions in logics with relatively simple algebraic manipulations. By hypothesis, all X are Y ($X = XY$) and all Y are Z ($Y = YZ$). With

$$X = XY = X(YZ) = (XY)Z = XZ$$

we have deduced that $X = XZ$, i.e., all X are Z [106]. Based on the work of Boole and other subsequent authors, a number of systems of mathematical logic have been developed. The field of mathematical logic is particularly important for inference in ontologies, and will be discussed in some detail in the chapter.

In computer science parlance, logics deal with sentences and knowledge

---

[1]Boolean variables, i.e., variables that can take on the values TRUE or FALSE, are named in honor of George Boole.

bases. In this context, a *sentence* refers to a representation of some individual fact. A knowledge base is a collection of such sentences and might be implemented computationally in the form of a database or otherwise. The minimal requirement for an AI system is to follow the tell-ask-paradigm, which provides at least two operations:

- *Tell(K,s)* adds a new sentence s to the knowledge base K.

- *Ask(K,s)* queries whether the sentence s is contained in K.

The *Ask* operation is the more interesting one, as its result depends on the already-known facts as well as on the logic and the inference algorithm used. For instance, if a knowledge base contains sentences to the effect that *Mickey* is a mouse, *Mouses* are *Rodents*, and *Rodents* are *Mammals*, then we can query the knowledge base whether the following sentences are true: *Mickey* is a rodent, and *Mickey* is a mammal. Logics, and indeed ontologies, are ways of formalizing the ways we represent information and reason over it. That is, if an appropriate logic is used, then *Ask* would indeed return the truth of both queried statements although the knowledge was not made explicit. Essentially, this paradigm leads to new sentences on the basis of logical analysis of the existing sentences.

As we use the word today, a formal logic consists of two main components [217]:

- A formal language with a syntax (rules for constructing valid sentences) and a semantics (a specification of how the sentences in the language relate to the domain being modeled).

- A proof theory, that is a methodology for deducing the entailments of a set of sentences (inferring new sentences based on the explicitly asserted sentences in the knowledge base).

Entailment and Inference

In English, to *entail* something means to involve it as a necessary outcome, for instance, "Investing in subprime mortgage funds entails a high risk." In logic, we say that a set of one or more sentences $S = \{s_1, s_2, \ldots\}$ **entails** a sentence t if t is true whenever every sentence in S is true. In that case, we write $S \vDash t$. **Inference**, on the other hand, is the process by which we determine the sentences that are entailed by the existing sentences in the knowledge base.

2.2 Propositional Logic

A *mathematical logic* consists of a *syntax*, or a set of rules for forming sentences, as well as the *semantics* of the logic, which are another set of rules for assigning truth values to the sentences of the logics. In this chapter, we will discuss three kinds of mathematical logic that are relevant for understanding inference in ontologies. In this section, we will discuss *propositional logic*, which is a relatively simple logic in which the only objects are *propositions* represented by symbols such as X, Y, or Z. The proposition symbols can be combined with logical connectives to create *sentences* with more complex meanings. Logical connectives are truth functions defined via truth tables that are represented by dedicated symbols and are modeled on the meaning of natural language (Table 2.1).

Symbol	Meaning in natural language
$x \wedge y$	x and y
$x \vee y$	x or y (inclusive)
$\neg x$	not x
$x \otimes y$	x or y but not both (exclusive or)
$x \Rightarrow y$	if x then y
$x \Leftrightarrow y$	if and only if (x implies y and y implies x)

TABLE 2.1: Logical connectives.

Now consider the following propositions:

$X :$ *Mary is an ontologist.*

$Y :$ *Mary is rich.*

$Z :$ *Mary is an investment banker.*

Using logical connectives, we can express more complex sentences such as:

$X \wedge Y :$ *Mary is an ontologist* **and** *Mary is rich.*

$Z \Rightarrow Y :$ **If** *Mary is an investment banker* **then** *Mary is rich.*

Let's say we know that it is **true** that *Mary is an ontologist* and, further, we know that it is **false** that *Mary is an ontologist* **and** *Mary is rich* and we know that is **true** that **if** *Mary is an investment banker* **then** *Mary is rich.* These facts entail the fact that it is **false** that *Mary is an investment banker.* This type of reasoning can be expressed in the language of propositional logic as follows:

$$X \wedge \neg(X \wedge Y) \wedge (Z \Rightarrow Y) \Rightarrow \neg Z. \qquad (2.1)$$

It is possible to create combined propositions by using the $\Leftrightarrow$ connective. For instance, the statement

$$A \Leftrightarrow X \wedge \neg (X \wedge Y) \wedge (Z \Rightarrow Y)$$

creates a kind of "abbreviation" for $X \wedge \neg (X \wedge Y) \wedge (Z \Rightarrow Y)$, and we can now write Equation (2.1) concisely as

$$A \Rightarrow \neg Z.$$

In order to formalize the structure of a syntactically correct formula, a form of grammar is specified. For propositional logics (and all of the other formal languages that we consider in this book), the grammar is context-free, which means that expressions can be nested but the generation of an expression does not depend on its neighborhood (i.e., the context). A context-free grammar can be compactly given via the Backus Naur Form. For propositional logics, the grammar can be specified as:

$$\phi ::= p \mid \bot \mid \top \mid (\phi) \mid (\neg \phi) \mid (\phi \wedge \phi) \mid (\phi \vee \phi) \mid (\phi \Rightarrow \phi) \mid (\phi \Leftrightarrow \phi) \qquad (2.2)$$

Here, ϕ denotes a non-terminal (a symbol that needs to be replaced and does not occur in a syntactically correct formula) and $\mid$ denotes alternatives for how any occurrence of ϕ can be replaced while the other symbols are terminals (and hence may appear in a syntactically correct formula). In particular, p denotes any propositional variable, $\top$ is TRUE and $\bot$ is FALSE and $\neg, \wedge, \vee, \Rightarrow$, and $\Leftrightarrow$ are the logical connectives. For instance, consider the following sentence:

$$((x \vee y) \wedge (\neg z))$$

We can check whether this sentence is syntactically correct by successively applying the rules given in the grammar; if we can generate the sentence from the grammar, then it is syntactically correct.

$$
\begin{array}{ll}
\phi \quad \text{yields} & (\phi \wedge \phi) \\
\quad \text{yields} & ((\phi \vee \phi) \wedge \phi) \\
\quad \text{yields} & ((\phi \vee \phi) \wedge (\neg \phi)) \\
\quad \text{yields} & ((x \vee y) \wedge (\neg z))
\end{array}
$$

Note that in the last line, we have applied the rule $\phi ::= p$ three times and substituted ϕ with the propositional variables x, y, and z.

In FOP formulas, parentheses can be required to avoid ambiguities. For instance, does $x \vee y \wedge \neg z$ mean "x or (y and not z)" or "(x or y) and not z"? However, propositional sentences are usually written in a simplified form using only those parentheses that are needed to avoid ambiguity. The order of precedence of the logical operators and connectives is $\neg, \wedge, \vee, \Rightarrow, \Leftrightarrow$ (from highest to lowest). Thus, $((x \vee y) \wedge (\neg z))$ can be written $(x \vee y) \wedge \neg z$.

x	y	$\neg x$	$x \wedge y$	$x \vee y$	$x \Rightarrow y$	$x \Leftrightarrow y$
⊤	⊤	⊥	⊤	⊤	⊤	⊤
⊤	⊥	⊥	⊥	⊤	⊥	⊥
⊥	⊤	⊤	⊥	⊤	⊤	⊥
⊥	⊥	⊤	⊥	⊥	⊤	⊤

TABLE 2.2: **Truth table for five logical connectives in propositional logic.** The four lines of the truth table show the truth values for the sentences given that x and y take on the values true (⊤) or false (⊥).

The implication of a proposition is determined by the truth values of the propositional variables. Table 2.2 displays the truth values for all possible combinations of two propositional variables using the five logical connectives in a *truth table*. Thus, $x \wedge y$ is true if x is true and y is true, but false if x is true and y is false. Note that the interpretation for $x \Rightarrow y$ is "if x is true then y is true." No claim about y is made if x if false, so that a statement such as $\perp \Rightarrow p$ is always true. The interpretation of $x \Leftrightarrow y$ is that if x is true then y is true and if y is true then x is true (if and only if).

In propositional logic, sentences are *valid* if they are true with all possible combinations of truth values for the propositional variables. For instance, the sentence $(p \vee \neg p)$ is true whether p is true or false because $(\top \vee \perp) \Rightarrow \top$. A sentence is *satisfiable* if there is at least one combination of truth values for the propositional variables for which the sentence is true. $(p \wedge \neg p)$ is thus *unsatisfiable*.

If a computer performs inference or reasoning, it does not "know" the meaning of the sentences it is reasoning on, but it merely is manipulating symbols according to rules. In order for humans to use the results of computer reasoning, we need to be sure that the computer will draw the correct conclusion for any given input data. If only few variables are used in the formula, a reasonable approach to ensure this is to employ truth tables. For example, we can test the validity of Equation (2.1) on page 13 by constructing a truth table with all 2^3 possible combinations of truth values of the three propositional variables X, Y, and Z. The result of this procedure is given in Table 2.3.

The analysis shown in this truth table demonstrates that $A \Rightarrow Z$ is *not* a valid sentence, although it is satisfiable for five of eight combinations of values for X, Y, and Z. For valid sentences, all of the values are true, which is the case for $A \Rightarrow \neg Z$ as was claimed in the beginning of this section. It is essential that all sentences used for computational reasoning are valid if we expect to draw correct conclusions about the input data. The exercises present several problems that ask the reader to use truth tables to check the validity of logical propositions.

X	Y	Z	$\neg(X \wedge Y)$	$Z \Rightarrow Y$	A	$\neg Z$	$A \Rightarrow Z$	$A \Rightarrow \neg Z$
$\bot$	$\bot$	$\bot$	T	T	T	T	$\bot$	T
$\bot$	$\bot$	T	T	$\bot$	$\bot$	$\bot$	T	T
$\bot$	T	$\bot$	T	T	T	T	$\bot$	T
$\bot$	T	T	T	T	$\bot$	$\bot$	T	T
T	$\bot$	$\bot$	T	T	T	T	$\bot$	T
T	$\bot$	T	T	$\bot$	$\bot$	$\bot$	T	T
T	T	$\bot$	$\bot$	T	$\bot$	T	T	T
T	T	T	$\bot$	T	$\bot$	$\bot$	T	T

TABLE 2.3: **Checking validity of sentences with truth table.** For clarity, $X \wedge \neg(X \wedge Y) \wedge (Z \Rightarrow Y)$ has been abbreviated as A. It can be seen that the sentence $A \Rightarrow Z$ is not true for each possible combination of X, Y, and Z. Thus, it is *not* a valid sentence. In contrast, $A \Rightarrow \neg Z$ is true for each possible combination and hence it is a valid sentence.

Rules of Inference in Propositional Logic

It is possible to use truth tables to check the validity of arbitrary statements in propositional logic, although this can be a tedious procedure if a large number of propositional variables are used, because the number of rows is 2^n for n variables. Over time, a number of sentences have been identified that are commonly used and whose validity has been proven. These statements are known as *inference rules*, because valid inference is possible using these rules for arbitrary values of the input variables. We will use the notation $\phi \vDash \psi$ to mean that ϕ entails ψ (another way of saying this is that ψ can be derived from ϕ by inference). As we will see in Chapter 15, a computer program for reasoning can be developed such that it can infer a new statement m on the basis of statements $s_1, s_2, \ldots, s_n$, if there is an inference rule for some s_i such that $s_i \vDash m$.

An inference rule is correct if its conclusion is true in all cases in which the premise is true. That is, if ϕ is true, and our inference rule is $\phi \vDash \psi$, then the rule is correct if ψ is true in all cases in which ϕ is true. No statement is made about cases in which ϕ is false. Table 2.4 summarizes most of the common inference rules of propositional logic.

Of particular importance are the DeMorgan's rules, in which the logical connective $\wedge$ is expressed using $\vee$ and $\neg$, and $\vee$ is expressed using $\wedge$ and $\neg$, i.e.,

$$\neg(x \wedge y) \vDash \neg x \vee \neg y \tag{2.3}$$

$$\neg(x \vee y) \vDash \neg x \wedge \neg y \tag{2.4}$$

Name	Inference Rule	Explanation
Modus ponens	$(x = \top) \land (x \Rightarrow y) \models y$	If x is true and x implies y, then y is true.
Conjunction elimination	$x_1 \land x_2 \land \ldots \land x_n \models x_i$	From a conjunction, infer any of the conjuncts.
Conjunction introduction	$(x = \top) \land (y = \top) \models (x \land y)$	If x and y are true, then so is their conjunction.
Disjunction elimination	$(x \Rightarrow z) \land (y \Rightarrow z) \land (x \lor y) \models z$	If x implies z and y implies z and x or y is true, then z is true.
Disjunction introduction	$(x = \top) \models (x \lor y)$	If a sentence is true, its disjunction with anything at all is true.
Double negation elimination	$\neg(\neg x) \models x$	x can be inferred from not (not x).
Biconditional introduction	$(x \Rightarrow y) \land (y \Rightarrow x) \models (x \Leftrightarrow y)$	If x implies y and y implies x, one can conclude x iff y.
Biconditional elimination	$(x \Leftrightarrow y) \models (x \Rightarrow y)$	x iff y implies that x implies y (and also that y implies x, of course).
Unit resolution	$(x \lor y) \land (y = \bot) \models x = \top$	If one of the disjuncts of a true disjunction is false, the other one can be inferred to be true.
Resolution	$(x \lor y) \land (\neg y \lor z) \models (x \lor z)$	If x is true then $(x \lor y)$ is true by the disjunction introduction rule. If x is false, then the premise that $(x \lor y)$ is true implies that y is true. The premise that $(\neg y \lor z)$ is true together with the fact that y is true implies that z is true. Thus, if the premise (left-hand side) of the rule is true, either x or z must be true.

TABLE 2.4: Inference rules in propositional logic. If the premise ϕ is true and the premise implies a conclusion ψ (i.e., $\phi \models \psi$), then the conclusion ψ must be true. Note that if the premise of an inference rule is false, no conclusion is made. These inference rules can be proven with the use of truth tables.

DeMorgan's rules can be proven using truth tables, as the reader will be asked to do in the exercises.

2.3 First-Order Logic

Propositional logic is very limited as to the types of statements it can make about the world. Essentially, this class of logic models the world as consisting of facts that can be combined using logical connectors. For instance, the statement *All Klingons love Klingon opera* could be represented using the proposition P. However, using this single proposition is not possible to conclude that the individual *Worf* who is a Klingon loves Klingon opera, which however seems to be a natural conclusion. If one is interested in the conclusion that Worf loves Klingon opera one could introduce two propositions: *Worf is a Klingon* and *Worf loves Klingon opera* represented by Q and R, respectively. Then, by considering the sentences Q and $Q \Rightarrow R$, we can conclude that R is true by applying the modus ponens inference rule.

Although this approach leads to the desired result in the specific case of Worf, it is difficult to generalize the statement using the syntax of propositional logic. We know for instance that *Martok is a Klingon* as well. According to the initial statement P, we intuitively conclude that Martok also loves Klingon opera. In propositional logic, we would need to add two new propositions and extend the knowledge base by two new sentences, which is cumbersome.

In contrast to propositional logic, first-order logic (FOL), which is also known as *predicate logic*, makes use of a richer syntax that enables it to make stronger statements about the world. It assumes that the world contains objects that are related to one another. It allows the quantification of objects and thus is much closer to natural language than propositional logic is. For instance, the initial statement can be written as a single formula

$$(\forall x)(\texttt{Klingon}(x) \Rightarrow \texttt{OperaLover}(x)).$$

Translated literally, it means: *For any object x, if x is a Klingon then x is an opera lover.* As we will see later in more detail, if we assert that individual *Worf* is a Klingon by $\texttt{Klingon}(Worf)$ the semantic rules entail the truth of $\texttt{OperaLover}(Worf)$ as well. Due to the usage of the universal quantifier $\forall$ we can replace the name Worf by any other Klingon name and still obtain the desired conclusion.

Syntax and Semantics of FOL

First-order logic is a formal language that is also often used for the description of sets and the relationship of individual elements of the sets to one another and the sets they belong to. The formulas of FOL are constructed from a number of elements that extend the expressive power of propositional logic. FOL uses the logical operators $\wedge, \vee, \neg, \Rightarrow$, and $\Leftrightarrow$ with the same meaning as in propositional logic (see Table 2.1 on page 13), and additionally defines further symbols for quantification of objects. The grammar of FOL is given in Table 2.5.

Formula	$\rightarrow$	Atom
	$\mid$	(Formula Connective Formula)
	$\mid$	(Quantifier Variable) Formula
	$\mid$	$\neg$ Formula
Atom	$\rightarrow$	$P(\text{Term}, \ldots) \mid \text{Term} = \text{Term}$
Term	$\rightarrow$	$f(\text{Term}, \ldots) \mid \text{Constant} \mid \text{Variable}$
Connective	$\rightarrow$	$\wedge \mid \vee \mid \Rightarrow \mid \Leftrightarrow$
Quantifier	$\rightarrow$	$\exists \mid \forall$

TABLE 2.5: FOL syntax. The table shows FOL syntax expressed in abbreviated Backus-Naur form. $P(t_1, t_2, \ldots, t_n)$ is an n-ary predicate on $n \geq 1$ terms. $f(t_1, t_2, \ldots)$ is an n-ary function on $n \geq 1$ terms. Constants and variables are defined as in the text. P is a predicate.

The ability to quantify over objects gave rise to the name *first-order logic*, since objects are considered as the first-order entities that exist in the world. There are two kinds of quantifiers in FOL, the universal quantifier: $\forall$ and the existential quantifier $\exists$. The first quantifier $\forall$ is pronounced "for all." If F is a valid formula of FOL, then $(\forall x)F$ is also valid. Semantically, the formula means that F is true for all objects in the domain under consideration. The other quantifier $\exists$ can be read "there exists," and $(\exists x)F$ means that there exists an object in the domain such that F holds.

Predicates are a further fundamental construct of FOL, which is the reason that FOL is also commonly referred to as predicate logic. As in natural language, the predicate is one of the "sense-bearing" parts of a sentence. They describe characteristics of an object and are sometimes also referred to as *properties*.[2] Predicates are used to refer to a particular *relation*, whereby a relation is a function of one or more arguments that can take on the values {**true**, **false**}. For instance, Klingon(x) is a relation that returns **true** if the object x is a Klingon and **false** otherwise. Similarly, a motherOf(x, y) returns a value of **true** if x is the mother of y and **false** otherwise.

---

[2]Confusingly, different fields have used different words to refer to such entities. In predicate logic, usually the term *predicate* is favored over the term *property*. In other disciplines of logic and mathematics, the word *relation* is more commonly used. In this book we will use the term that is commonly used in the context (predicate is commonly used in OWL, and relation in many other ontology-related contexts).

Predicate

In English grammar, the *subject* and *predicate* are the two main parts of a sentence. The *subject* is what the sentence is about, while the *predicate* tells something about the subject. For instance, in the following sentences, the subject is underlined and the predicate is italicized.

- Elvis *lives.*

- Elvis *is the undisputed king of Rock 'n' Roll.*

- Elvis and Marilyn *were abducted by Martians and are living on a hidden base on the dark side of the moon.*

While statements of facts (propositions) were the basic building blocks of propositional logic, predicates are the building blocks of first-order logic. Predicates are a means of defining properties and relations among the things in a domain.

FOL also uses *function symbols*, which are often written as lower case letters such as f or g. A function can be used for a relation which assigns to objects of the domain of the function exactly one object in the range. Thus, if we know that x is the mother of y and m is a function that represents the relation "is mother of," then $m(y)$ will return x. For instance, assuming x stands for *Marie Curie* and y for her daughter *Irène Joliot-Curie*, the predicate $\text{motherOf}(x, y)$ returns a value of true but the function $\text{motherOf}(y)$ would return *Marie Curie*. Note that although $\text{daughterOf}(y, x)$ is a valid predicate, a "daughter" function $d(x)$ is not valid because a mother can have more than one daughter. Thus, the value of $d(x)$ is not uniquely defined and d is not a valid function. Functions and predicates can be nested in FOL. For instance, if pet is a unary predicate, hasOwner is a binary predicate, and owner is a unary function that returns the name of the owner of a pet, then the following statement is true.

$$(\forall x)(\text{pet}(x) \Rightarrow \text{hasOwner}(x, \text{owner}(x)))$$

A *term* is a logical expression that refers to an object. A term can be either a constant, such as "John" or 7, or a variable, or a n-ary function applied to n arguments, each of which is a term.

An *atom* is an n-ary predicate symbol applied to n arguments, each of which is a term. An atom is used in FOL to state facts, and is thus roughly comparable to a propositional variable in propositional logic.

A *formula* refers to a statement in FOL. The syntax of FOL can be defined recursively, whereby a formula is constructed from atoms, combinations of other formulas using logical connectives, and any variable x in the expressions $(\exists x)\ F$ or $(\forall x)\ F$, where F is a formula (Table 2.5).

Any syntactically valid statement in FOL can be derived from these rules.

For instance, noting that `Klingon` and `OperaLover` are predicates (relations) and x is a variable, we can derive the statement about Klingon opera fans as follows.

Formula	$\rightarrow$	Quantifier Variable Formula
	$\rightarrow$	$\forall x$ Formula
	$\rightarrow$	$\forall x$ Formula Connective Formula
	$\rightarrow$	$\forall x$ Atom $\Rightarrow$ Atom
	$\rightarrow$	$\forall x$ $Klingon(Term)$ $\Rightarrow$ $OperaLover(Term)$
	$\rightarrow$	$\forall x$ $Klingon(x)$ $\Rightarrow$ $OperaLover(x)$

In the first line, we have used the rule that a Formula can be replaced by *Quantifier Variable Formula*, and in the second line we used the rule that *Quantifier* can be replaced by $\forall$, and so on. Readers unfamiliar with formal grammar and Backus-Naur forms can consult the sources listed at the end of the chapter for further information. Using the syntactical rules of Table 2.5, complex formulas can be constructed with logical connectives, universal and existential quantifiers, predicates, and functions.

Intuitively, $(\forall x)P$ can be thought as a logical conjunction. If there are n individuals, then $(\forall x)P$ is equivalent to

$$P(x_1) \wedge P(x_2) \wedge \ldots \wedge P(x_n).$$

On the other hand, $(\exists x)P$ means that $P(x_i)$ must be true for some i, and thus $(\exists x)P$ is equivalent to

$$P(x_1) \vee P(x_2) \vee \ldots \vee P(x_n).$$

According to DeMorgan's rules (Equations 2.3 and 2.4) $\wedge$ and $\vee$ are dual to one another; thus, it is possible to express a universal quantifier using only the existential quantifier and vice versa:

$$\neg((\forall x)P) = (\exists x)\neg P$$
$$\neg((\exists x)P) = (\forall x)\neg P$$

However, usually both types of quantifier are used for clarity's sake. Often, the $\exists$ quantifier is used in patterns such as $(\exists x)(P(x) \wedge Q(x))$, which expresses that there is at least one individual x that has both properties P and Q. In contrast, the $\forall$ quantifier is commonly used with the pattern $(\forall x)(P(x) \Rightarrow Q(x))$, which states that for all objects, if they are members of P then there are members of Q as well. This particular form implies that if there is no x with P, then the expression is regarded as true following the definition of implication. Consequently, the universal quantifier over all members of a set that is empty is also defined to be true for an arbitrary logical statement P. For example, the statement *all pink zebras can fly* is true in the sense that there do not exist any pink zebras that cannot fly (because pink zebras do not exist at all). The ontology language OWL uses an analogous definition for the `owl:allValuesFrom` constraint, as we will see in Chapter 4. Thus, it might be clearer to define the formula $(\forall x)P$ as meaning, "if an x exists, then it must satisfy P."

One of the most important features of FOL consists of the ability to quantify variables using the $\forall$ or $\exists$ quantifiers. The meaning of FOL sentences in which a variable has not been defined is not well defined. For instance, does the following sentence mean that all mice like cheese (which would be the interpretation if the variable x were quantified with $\forall$) or that there exists one mouse (and potentially more than one mouse) that likes cheese (which would be the interpretation if the variable x were quantified with $\exists$)?[3]

$$\texttt{mouse}(x) \Rightarrow \texttt{likesCheese}(x)$$

Therefore, a well-formed formula (wff) in FOL requires that all variables are quantified. For instance, adding a universal quantifier to the previous formula makes it into a wff.

$$(\forall x)(\texttt{mouse}(x) \Rightarrow \texttt{likesCheese}(x))$$

It is possible to combine different quantifiers in a single wff. The order of the quantifiers matters. Compare the following two sentences.

1) $(\forall x)((\exists y)(\texttt{loves}(y, x)))$
2) $(\exists x)((\forall y)(\texttt{loves}(y, x)))$

The first statement can be interpreted as *everybody has someone who loves him (or her)*. The second statement, in contrast, can be taken to mean *there exists somebody whom everybody loves*.

Inference in FOL

Many of the concepts such as *model* and *interpretation* that are important for inference in the Semantic Web can be presented in the setting of FOL in a relatively simple form which will provide a useful introduction for similar concepts used in inference for the Semantic Web (Chapters 13 and 14).

An *interpretation* provides a framework to decide whether FOL statements are true or not depending on the values of the atomic propositions, predicates, and functions. For example, how does one know whether the following wff is true or not?

$$\texttt{worksFor}(\textit{ClarkKent}, \textit{DailyPlanet})$$

The truth or falsity of this statement depends on the interpretation of

---

[3]Note that in some texts, a variable is assumed to be universally quantified unless otherwise specified. This is a convention that we will not follow in this book.

the constants *ClarkKent* and *DailyPlanet* as well as on the definition of the predicate `worksFor`. Note that although it seems obvious to us at first glance that Clark Kent[4] works at the *Daily Planet*, remember that the strings in the sentence such as `worksFor` have no a priori meaning in FOL or for the computer, and could equivalently be written as

`10110011100110111(00010110101010100,11110000000000110)`

Intuitively speaking, an interpretation maps the names (symbols or variables) used in FOL statements to elements of a domain. Formally, an interpretation $\mathcal{I}$ is defined over an alphabet (V, C, F, P) consisting of a set of variables V, and set of constants C, a set of functions F, and a set of predicates P. $\mathcal{I}$ consists of a set Δ, known as the *domain*, along with the following functions.

- $\mathcal{I}_C : C \to \Delta$ ($\mathcal{I}_C$ maps each constant c to an element of the domain $c^{\mathcal{I}} \in \Delta$).

- $\mathcal{I}_V : V \to \Delta$ ($\mathcal{I}_V$ maps each variable x to an element of the domain $x^{\mathcal{I}} \in \Delta$).

- $\mathcal{I}_P : P \to P(\Delta^n)$ ($\mathcal{I}_P$ maps an n-ary predicate symbol p to a relation $(p^{\mathcal{I}} \subseteq \Delta^n) \to \{\text{true}, \text{false}\}$).

- $\mathcal{I}_F : F \to F(\Delta^n)$ ($\mathcal{I}_F$ maps n-ary function symbols to a function $f^{\mathcal{I}} : \Delta^n \to \Delta$).

Consider the atomic formula $p(c)$, where c is a constant and p is a unary predicate. Then $p(c)$ is *satisfied* if c is in the interpretation of p.

$$\mathcal{I} \vDash p(c) \iff c^{\mathcal{I}} \in \mathcal{I}_P(p)$$

Thus, the relation $\mathcal{I}_P(p)$ returns true if $p(c)$ is true.[5] Similarly, if q is a binary predicate and c and d are constants, $q(c,d)$ is satisfied if $<c,d>$ is in the interpretation of q.

$$\mathcal{I} \vDash q(c,d) \iff <c^{\mathcal{I}}, d^{\mathcal{I}}> \in \mathcal{I}_P(q)$$

The interpretation of an existential quantification over a variable x is true if there is (at least) one element e in the domain Δ such that a statement is true.

$$\mathcal{I} \vDash \exists x \, p(x) \iff \text{There exists an element } e \in \Delta \text{ such that } p(e) \text{ is true}$$

---

[4] Kent was born as Kal-El on the planet Krypton and came to be known as Superman.

[5] For example, if the interpretation mapped c to *Mickey* and p to *mouse* in the domain of Disney cartoons, then $p(c)$ would return true.

The interpretation for universal quantifiers is similar.

$$\mathcal{I} \vDash \forall x\, p(x) \iff p(e) \text{ is true for all elements } e \in \Delta$$

An interpretation combines the atoms of an FOL formula in the expected ways (Table 2.6).

$\mathcal{I} \vDash A$	$\mathcal{I} \nvDash \neg A$
$\mathcal{I} \vDash A \wedge B$	$\mathcal{I} \vDash A$ and $\mathcal{I} \vDash B$
$\mathcal{I} \vDash A \vee B$	$\mathcal{I} \vDash A$ or $\mathcal{I} \vDash B$
$\mathcal{I} \vDash A \Rightarrow B$	If $\mathcal{I} \vDash A$ then $\mathcal{I} \vDash B$
$\mathcal{I} \vDash A \Leftrightarrow B$	$\mathcal{I} \vDash A$ iff $\mathcal{I} \vDash B$

TABLE 2.6: Interpretations and connectives in FOL. The left column is true if and only if the right column is true. $\vDash$: entails; $\nvDash$: does not entail.

In FOL, a formula is said to be *valid* if it is true under every interpretation.[6] A formula is said to be *satisfiable* if it is true under at least one interpretation. Otherwise, it is said to be *unsatisfiable* (or *inconsistent*).

Interpretations form the background of proof procedures in FOL, whereby certain formulas called *axioms* are assumed to be true, and a proof is generated by applying a series of formulas derived from axioms, definitions, or previous formulas by applying *inference rules*. We have seen a number of inference rules for propositional logic (Table 2.4) that are also valid in FOL. There are a number of other rules commonly used in FOL, of which perhaps the two most important are:

- Universal instantiation. $(\forall x)(p(x) \vDash p(c))$, where c is any constant (intuitively, if p is true for all elements, it is true for any particular one).

- Existential generalization. $p(c) \vDash (\exists x)(p(x))$ (If p is true for some c, then there is at least one element for which p is true).

We will present a small example proof to give a flavor of how inference rules can be chained together to generate a proof. Later chapters will present more involved (and interesting) examples of inference. Suppose we know that a predicate p is true for some element e' and we also know that for all elements, p(x) implies q(x).

Then, by the universal instantiation rule, we can infer the following:

$$p(x) \Rightarrow q(x) \vDash p(e') \Rightarrow q(e')$$

Then we can use the modus ponens rule (see Table 2.4) from propositional logic to infer that q(e') = **true**:

$$(p(e') = \top) \wedge (p(e') \Rightarrow q(e')) \vDash q(e')$$

---

[6]Sentences that are valid under every interpretation, such as $(x \vee \neg x)$ are also called *tautologies*.

Higher-Order Logic

In higher-order logic, it is additionally possible to quantify over relations and functions as well as over objects. A classic higher-order formula that is not valid FOL is the axiom of induction.

$$(\forall P)(P(0) \wedge ((\forall n \in \mathbb{N})(P(n) \Rightarrow P(n+1))) \Rightarrow (\forall n)(P(n))$$

The quantifier $\forall P$ is over predicates rather than over objects as in FOL.

As we will see in later chapters, the more expressive a logic or ontology language is, the more difficult it is to reason over them. In fact, even FOL can be shown to be *undecidable*, that is, there is no procedure that can be guaranteed to determine whether an arbitrary FOL sentence is valid. However, it is possible to restrict the expressivity of FOL in such a way as to ensure decidability. Description Logics, as we will see below, represent a fragment of FOL that is *decidable*, and provides the basis for the ontology language OWL DL (Chapter 4).

2.4 Sets

Set theory refers to the branch of mathematics that describes collections of arbitrary elements (also known as subjects, items, or objects) of any domain, whereby the order of the elements is of no particular interest. Sets and relations that hold among members of sets can be described using the language of FOL.

The number of elements that a set S contains is referred to as the *cardinality* of S and is denoted as $|S|$. If $|S| = n$, then S contains n distinct items. Note that many sets have a unlimited number of elements, such as the set of the natural numbers denoted by $\mathbb{N}$.

If we know that S contains an item or an element x, then we write $x \in S$ and say that x is in S or x is an element of S or that x *is a member of* S. On the other hand, we write $x \notin S$ to denote that S doesn't contain x, that is, x is not a member of S. Note that x has been used a placeholder for an arbitrary element, which is commonly referred to as a *variable*. As we will see later, being (or not being) a member of a set S is a *property* of the element x that is assigned to a set. There are many ways of defining sets. The easiest way is to enumerate the elements of which the set consists. Another possibility is to name the properties of the elements. If a set S contains no element at all, we write $S = \emptyset$ or alternatively $S = \{\}$. The cardinality of empty sets is 0.

Example 2.1 *The following examples illustrate some of the different ways of defining a set.*

1. $S = \{a, b, c\}$ defines a set that contains elements a, b, and c. Note that the definition $S = \{c, b, a\}$ is equivalent to the first definition, since the order of the elements is irrelevant for sets.

2. $S = \{a, f, b, q, b\}$ defines a set consisting of the elements a, b, f, q.

3. $S = \{\{1\}, \{1, 2\}\}$ defines a set that consists of the sets $\{1\}$ and $\{1, 2\}$.

4. $S = \{n|n$ is a natural number that is divisible by 3$\}$ defines a set that contains all natural numbers that are divisible by 3. This is an intensional definition that describes the defining characteristic of the elements of the set.

5. $S = \{n|n$ is Mersenne prime that is divisible by 2$\}$. We also could have specified $S = \varnothing$ as by definition the only prime that is divisible by 2 is 2 and the smallest Mersenne prime[7] is 3.

6. $S = \{p|p$ is a planet of our solar system$\}$. Note that the cardinality of the set depends on the definition of what is a planet.

Set Extension and Intension

The phrase *extension* is used to denote the definition of a set by enumerating all of its members. For instance, $A = \{A, B, C, D\}$ defines a set A by enumerating its four members. In contrast, *intension* is used to denote the definition of a set by a defining property. For instance, "the set A consists of the first four capital letters of the English alphabet."

As was shown in Example 2.1, sets can themselves be elements of sets. Therefore, it is important to consider special relations over sets. A set A is a subset of B, denoted by $A \subseteq B$, if each member of A is also a member of B. Using the above notation for FOL we can write this as:

$$A \subseteq B \Leftrightarrow (\forall x)(x \in A \Rightarrow x \in B)$$

Using this formula, we can now conclude several facts about sets. For instance, for each set S we can state at that $\varnothing \subseteq S$ and $S \subseteq S$, which literally means that the empty set is a subset of each set, but also that each set is a subset of itself. We can now prove that this statement is indeed true based upon the things which we already know to be true.

In order to verify that $S \subseteq S$, we suppose first that S is a non-empty set. The definition becomes:

$$S \subseteq S \Leftrightarrow (\forall x)(x \in S \Rightarrow x \in S)$$

This obviously still holds, since if $x \in S$ then $x \in S$. Now assume that $S = \varnothing$.

---

[7]Mersenne primes are all prime numbers M_p for which $2^p - 1$ for $p > 1$ holds.

Then $x \in S$ is false and $x \in S \Rightarrow x \in S$ is true because $X \Rightarrow Y$ is always true if X is false (see Section 2.2 on page 15).

If we want to prove that $\emptyset \subseteq S$ is true, we note that if the set being quantified is empty, then the quantification always yields a true statement. It also follows that $\emptyset \subseteq \emptyset$. Thus we have proved that $S \subseteq S \leftrightarrow (\forall x)(x \in S \Rightarrow x \in S)$ is true for empty or non-empty sets (i.e., all sets).

Set A is equal to set B, denoted as $A = B$, if each element of A is also in B and if each element of B is also in A. We can express this in FOL as

$$A = B \leftrightarrow (\forall x)(x \in A \Rightarrow x \in B) \wedge (x \in B \Rightarrow x \in A),$$

which is equivalent to:

$$A = B \leftrightarrow A \subseteq B \wedge B \subseteq A.$$

Ordered Sets

Sometimes, it is desirable to impose an order when specifying a collection. Such ordered sets are often referred to as *tuples*. By convention, tuples are written using parentheses. A tuple $(a_1, a_2, \ldots, a_n)$ is called an n-tuple. The 3-tuple (a, b, c) is different from the 3-tuple (c, b, a). A *pair* is an important n-tuple with $n = 2$.

Operations on Sets

The *union* of two sets A and B, denoted by $A \cup B$ is the set that contains the elements that are in A in addition to the elements in B.[8] The *intersection* of A and B, denoted by $A \cap B$ contains the elements that are both in set A and in set B. The formal definitions are:

$$A \cup B \leftrightarrow \{x | x \in A \vee x \in B\}$$
$$A \cap B \leftrightarrow \{x | x \in A \wedge x \in B\}$$

Another important operation is the *set difference*, which is defined as the set containing each element of A that is not in B:

$$A - B \leftrightarrow \{x \in A | x \notin B\}.$$

In contrast to that, the *symmetric set difference* of two sets A and B contains the elements which are in A but not in B together with the elements that are in B but not in A. It can be defined in terms of the set difference as:

$$A \triangle B \leftrightarrow (A - B) \cup (B - A).$$

The *Cartesian product*, which is also called a cross-product, is an operation on $n \geq 2$ sets and yields a set of n-tuples. For $n = 2$ it is defined as:

$$A \times B \leftrightarrow \{(a, b) | a \in A \wedge b \in B\}$$

---

[8] Elements in both sets are just counted once.

Example 2.2 *Consider the sets* $A = \{a, b, c, d, e\}, B = \{d, e, f\}$, *and* $C = \{1, 2\}$. *Then*

$$A \cup B = \{a, b, c, d, e, f\} \qquad \text{(union)}$$
$$A \cap B = \{d, e\} \qquad \text{(intersection)}$$
$$A - B = \{a, b, c\} \qquad \text{(difference)}$$
$$A \vartriangle B = \{a, b, c, f\} \qquad \text{(symmetric difference)}$$
$$B \times C = \{(d, 1), (e, 1), (f, 1), (d, 2), (e, 2), (f, 2)\} \qquad \text{(cross-product)}$$

Set relation and set operations as well as specific set configurations are commonly depicted using Venn diagrams (Figure 2.1).

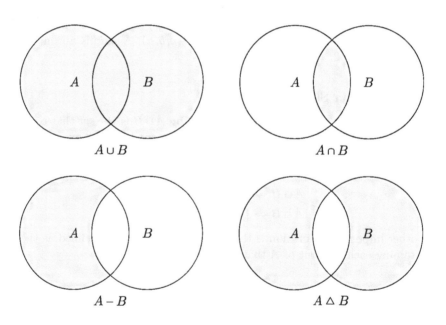

FIGURE 2.1: Set Operations. Venn diagrams are used to illustrate the operations of set union ($A \cup B$), set intersection ($A \cap B$), set subtraction ($A - B$), and symmetric set difference ($A \vartriangle B$).

2.5 Description Logic

A description logic is a mathematical formalism that provides the foundation for most of the ontology languages, including the Web Ontology Language OWL-DL and the OBO language developed especially for bio-ontologies (see Chapter 4). Description logics, which are sometimes also referred to as terminological logics, comprise a family of related knowledge representation languages. The name *description logic* suggests that we want to describe *concepts* of the world by using a formal semantics based on mathematical logic. Many of the elements of descriptions logic are directly inherited from FOL. In many cases, description logics and FOL use a similar but not identical notation to describe identical concepts.

Similar to FOL, description logics model *concepts*, *roles* (i.e., relations), and *individuals*, and their relationships to one another. Readers may thus ask why there is a need for another logic. One reason is that FOL is in a sense too flexible, which leads to certain undesirable computational characteristics including undecidability. That is, it has been proved by Church and Turing that is impossible to verify whether arbitrary statements in FOL are true or false by computational means. In contrast, the expressiveness and hence the generality of description logics are constrained in comparison to those of FOL in favor of computational decidability and efficiency of inference algorithms.

In this chapter, we will present the basic syntax and semantics of description logics. Algorithms that operate on DL to perform inference will be presented in Chapter 15.

2.5.1 Description Language $\mathcal{AL}$

One of the most simple description logic languages is the Attributive Concept Language, which is denoted as $\mathcal{AL}$ for short. The $\mathcal{AL}$ syntax of specifying classes is defined by a context-free grammar $G = (V, \Sigma, R, C)$ where

- the non-terminal characters are $V = \{A, C, R\}$;

- the terminal characters are $\Sigma = \{alphanum, (,), \neg, \sqcap, \bot, \top, \forall, \exists, .\}$; and

- the production rules R are described as

$$A \rightarrow alphanum^+ (\text{literal, class name}) \mid \top \mid \bot$$
$$R \rightarrow alphanum^+ (\text{literal, role name})$$
$$C \rightarrow A \mid \neg A \mid (C) \mid C \sqcap C \mid \forall R.C \mid \exists R.\top$$

- the start symbol is C.

The axioms that define a knowledge base following the $\mathcal{AL}$ description logics can be classified as *assertional axioms*, which is commonly referred to as the

ABox, and *terminological axioms*, which form the so-called *TBox*. An ABox contains assertions of the form $C(a)$ and $R(a, b)$. The former states that a is a member of a constructed class C, while the latter states that a and b are related via role or relation R. The TBox contains general concept inclusions, which are written as $C \sqsubseteq D$. Statements following this syntax mean that each member of class C is also in class D, and thus encode subsumption relations. If in addition, each member of D is also a member of C, then we write $C \equiv D$ for short, and say that C is defined by D. A TBox therefore represents the schema of the knowledge base and thus the ontology of the concepts. Note that writing $C \sqsubseteq D$ is the same as writing $\neg C \sqcup D$.

ABox, TBox, and RBox

A collection of assertional axioms has come to be known as an **ABox** (assertional box). An ABox comprises assertions about individuals that state what class or classes an individual belongs to, or relate individuals to one another by roles.

woman(Elisabeth), ophthalmologist(Elisabeth)
hasChild(Elisabeth,Sophie), hasChild(Elisabeth,Anna)

The **TBox** (terminological box) describes the taxonomy of the classes.

mother $\sqsubseteq$ woman

The **RBox** describes relations of roles.

hasParent $\equiv$ hasChild⁻

Due to the similarity of the symbols used by description logics and first-order logic and set theory, the semantics of $\mathcal{AL}$ can be intuitively understood. More formally, the semantics of $\mathcal{AL}$ are defined by the following recursive function FOL, in which the $\mathcal{AL}$ description logics terminological expression is mapped to an equivalent first-order logic expression:

$$\text{FOL}(C \sqsubseteq D) = (\forall x)(\text{FOL}(C(x)) \Rightarrow \text{FOL}(D(x))) \qquad (2.5)$$

$$\text{FOL}(C \equiv D) = (\forall x)(\text{FOL}(C(x)) \Leftrightarrow \text{FOL}(D(x))) \qquad (2.6)$$

$$\text{FOL}(A(x)) = A(x) \qquad (2.7)$$

$$\text{FOL}(\neg A(x)) = \neg A(x) \qquad (2.8)$$

$$\text{FOL}((C \sqcap D)(x)) = \text{FOL}(C(x)) \wedge \text{FOL}(D(x)) \qquad (2.9)$$

$$\text{FOL}((\forall R.C)(x)) = (\forall y)(R(x, y) \Rightarrow \text{FOL}(C(y))) \qquad (2.10)$$

$$\text{FOL}((\exists R.\top)(x)) = (\exists y)(R(x, y)) \qquad (2.11)$$

Note that in these expressions, A is used to represent an atomic class, C and D are used to represent classes that may be either atomic or compound, and R is used to represent a role.

In description logics, the symbol ⊤ is defined as the class that contains all members, i.e., $⊤ ≡ C ⊔ ¬C$ and therefore denotes the most general class of an ontology. In contrast, ⊥ denotes the most specific class that contains no members, i.e., $⊥ ≡ C ⊓ ¬C$.

As can be seen, in contrast to FOL, $\mathcal{AL}$ allows only for unary and binary predicates. Additionally, there is no support for functions. Both of these restrictions apply for all members of the common DL family. Quantification is also available only in a limited form, although other common DL languages allow more sophisticated patterns. Other restrictions are relaxed in more expressive DL languages, as we will see in the next section.

2.5.2   Description Language $\mathcal{ALC}$

The foundation of many description languages is $\mathcal{AL}$ extended to also support complex concept negation, which is a language feature of description logic that is commonly abbreviated as $\mathcal{C}$. This means that the negation symbol can now be also used before a constructed class C, while in $\mathcal{AL}$ it could only be used for atomic classes. The introduction of complex concept negation brings in two further features that can be declared explicitly as well:

Concept union $(\mathcal{U})$. Means that it is possible to take the union of two complex concepts. This is denoted as $C ⊔ D$. For $\mathcal{ALC}$ it follows from concept negation as $C ⊔ D ≡ ¬(¬C ⊓ ¬D)$.

Full existential quantification $(\mathcal{E})$. Extends the limited existential quantification to take an arbitrary concept C as the filler. It is denoted as $∃R.C$ and follows from concept negation as $∃R.C ≡ ¬∀R.¬C$.

Strictly speaking, $\mathcal{AL}$-family language supporting $\mathcal{C}$ but lacking $\mathcal{U}$ and $\mathcal{E}$ are syntactically different from languages supporting $\mathcal{U}$ and $\mathcal{E}$ but lacking $\mathcal{C}$. But as they are equally expressive and thus semantically equivalent, the constructors for $\mathcal{UE}$ are commonly also used when referring to $\mathcal{ALC}$. The semantics of the new constructors can be again defined using a mapping to predicate logic as follows:

$$\text{FOL}(¬C(x)) = ¬C(x) \tag{2.12}$$

$$\text{FOL}((C ⊔ D)(x)) = \text{FOL}(C(x)) ∨ \text{FOL}(D(x)) \tag{2.13}$$

$$\text{FOL}((∃R.C)(x)) = (∃y)(R(x,y) ∧ \text{FOL}(C(y))) \tag{2.14}$$

Example 2.3 *Transcription factors bind to specific DNA sequences. For the purpose of this example, we will assume that transcription factors contain one or more DNA-binding domains. Therefore, if a protein has at least one DNA-binding domain, then it is a transcription factor. We use the following classes* *and relations in the constructor to express this knowledge in $\mathcal{ALC}$:*

T *denotes the class of transcription factors*
D *subsumes all classes of DNA-binding domains*
hasDomain *relates proteins to domains*

The class of transcription factors is then defined by

$$T \equiv \exists \text{hasDomain}.D.$$

By applying FOL *for this expression we derive*

$\text{FOL}(T \equiv \exists\text{hasDomain}.D)$
$\quad = (\forall x)(\text{FOL}(T(x)) \leftrightarrow \text{FOL}((\exists\text{hasDomain}.D)(x)))$ *by (2.6)*
$\quad = (\forall x)(T(x) \leftrightarrow \text{FOL}((\exists\text{hasDomain}.D)(x)))$ *by (2.7)*
$\quad = (\forall x)(T(x) \leftrightarrow (\exists y)(\text{hasDomain}(x,y) \wedge \text{FOL}(D(y))))$ *by (2.11)*
$\quad = (\forall x)(T(x) \leftrightarrow (\exists y)(\text{hasDomain}(x,y) \wedge D(y)))$ *by (2.7)*

as the corresponding expression in first-order logic.

2.5.3   Further Description Logic Constructors

In addition to $\mathcal{ALC}$, there are many other members of description logics. The most widely used description logics extend $\mathcal{ALC}$ by new constructors in order to increase their expressiveness, which in turn also often raise the complexity of corresponding inference algorithms. Depending on what features are added, the feature is commonly abbreviated using a calligraphic letter. The remainder of this section gives a brief overview about the syntax and semantics of some features that are commonly used in description logic (see above for $\mathcal{U}$ and $\mathcal{E}$).

Inverse Roles: $\mathcal{I}$

$\mathcal{I}$ adds support for inverse properties. It expressed as R^- for a relation R. In order to define the semantics in FOL, we need to introduce the function FOL that maps complex role constructors to FOL. In previous mapping definitions, we replace all occurrences of $R(x,y)$ by $\text{FOL}(R(x,y))$. Then we add

$$\text{FOL}(R(x,y)) = R(x,y) \qquad\qquad (2.15)$$
$$\text{FOL}(R^-(x,y)) = \text{FOL}(R(y,x)) \qquad\qquad (2.16)$$

to the mapping.

Example 2.4 *Sara is a parent of Alan. Implicitly, we know that Alan is a* *child of Sara as well. We can use the notion of inverse roles to express this in DL as follows:*

$$\text{isParentOf}(\text{Sara}, \text{Alan})$$

$$\text{isChildOf} \equiv \text{isParentOf}^-$$

Using inference rules (2.15) and (2.16), we obtain that

$$\text{isChildOf}(\text{Alan}, \text{Sara}).$$

Transitive Properties: Trans

If a role R is transitive, it means that if a member x is related by R to y and y is related by R to z, then x is also related by R to z. In DL, this is commonly denoted as *Trans*(R). Statements about roles are often called *role axioms* and part of the so-called *RBox*. The mapping of *Trans*(R) to FOL is defined as:

$$\text{FOL}(\textit{Trans}(R)) = (\forall x, y, z)(R(x,y) \land R(y,z) \Rightarrow R(x,z)). \tag{2.17}$$

No special letter is defined to indicate that a particular DL supports transitive roles. However, $\mathcal{S}$ is considered to be an abbreviation for $\mathcal{ALC}$ plus transitive roles.

Example 2.5 *The* nucleoplasm is_part_of *the* nucleus. *The* nucleus is_part_of *the cell. Intuitively, we expect that* nucleoplasm is_part_of *the cell as well. In DL, this can be accomplished by specifying that* is_part_of *is transitive, i.e.,*

$$\text{partOf}(\text{nucleoplasm}, \text{nucleus})$$

$$\text{partOf}(\text{nucleus}, \text{cell})$$

$$\text{Trans}(\text{partOf})$$

yields

$$\text{partOf}(\text{nucleoplasm}, \text{cell}).$$

This special domain of knowledge, i.e., the domain of how the cell is structured, is represented by cellular component ontology of Gene Ontology (see Chapter 5).

Example 2.6 *The town* Kahla *is located in the federal state* Thuringia. *The federal state* Thuringia *is located in* Germany, *which in turn is* located in Europe. *By declaring the role* is located in *as transitive, the knowledge that* Kahla *is located in* Germany *and that* Kahla *is located in* Europe *does not need to be explicitly stated.*

Transitivity is a very important property in mathematics for concrete domains. Transitivity is typical for so-called *order relations*, such as "larger than" and "smaller than."

Role Hierarchies: $\mathcal{H}$

$\mathcal{H}$ allows hierarchies for roles, i.e., $R \sqsubseteq S$ where R and S are both roles. Statements of this kind are called *role axioms* and as such belong to the RBox. The mapping to FOL is as follows:

$$\text{FOL}(R \sqsubseteq S) = (\forall x, y)(R(x, y) \Rightarrow S(x, y)). \qquad (2.18)$$

Example 2.7 (Continuation of Example 2.4 on page 32) Sara *is not only a* **parent** *of* Alan, *but she* is the mother of *him. Being the mother of someone is more specific than being a parent of someone. In DL, we express this by relationships between roles, i.e.,*

isMotherOf $\sqsubseteq$ isParentOf.

Composition-Based Regular Role Inclusion: $\mathcal{R}$

$\mathcal{R}$ adds support for regular role inclusions axioms (regular RIAs), which allows for the propagation of roles along other roles. A RIA is written as $R_1 \circ R_2 \circ \ldots \circ R_n \sqsubseteq S$ for roles $R_1, \ldots, R_n$ and S, where $\circ$ denotes the composition. A set of RIAs is *regular* if there are no cycles between two RIAs. For instance, RIAs $R \circ S \sqsubseteq S$ and $S \circ R \sqsubseteq R$ are regular if they are considered separately. However, a set consisting both of them is not regular.[9] As such, RIAs generalize the concept of transitivity, i.e., $R \circ R \sqsubseteq R$ and role hierarchies, i.e., $R \sqsubseteq S$.

Example 2.8 *That the owner of something is also an owner of its parts can be expressed as:*

owns $\circ$ hasPart $\sqsubseteq$ owns

If Joe owns a bike that has a part called handlebar then he also owns the handlebar.

Example 2.9 *Gene Ontology consists of terms that describe features of genes or gene products. Terms can be connected using* **partOf** *relation. Genes or their products are related to terms using* **isAnnotated** *relations. That those propagate along the* **partOf** *relations is expressed using complex role inclusions as follows:*

isAnnotated $\circ$ partOf $\sqsubseteq$ isAnnotated

Of course, Gene Ontology consists of many more relationships that can be combined with one another. These are explained in detail in Chapter 12.

Depending on the logic, the attribute $\mathcal{R}$ also adds some other features. Among them are *disjoint roles, reflexive, irreflexive* and *antisymmetric* roles, a universal role U and so-called *local reflexivity*. See [126] for more details on syntax and semantics.

---

[9]The distinction between regular and non-regular RIAs is made because regular RIAs are decidable, whereas non-regular RIAs lead to undecidability. See [128] for more details.

Nominals: $\mathcal{O}$

$\mathcal{O}$ means that nominals are supported. Nominals are denoted by enumerating the individuals in a set, e.g., $\{i_1, \ldots, i_n\}$ describes a class that consists of n specific individuals.

Example 2.10

$$\text{Good} \equiv \{\text{Elaine}, \text{Guybrush}\}$$
$$\text{Bad} \equiv \{\text{LeChuck}\}$$

Unqualified Cardinality Restrictions: $\mathcal{N}$

$\mathcal{N}$ adds support for unqualified cardinality restrictions $\leq nR$ and $\geq nR$, which stand for at most and at least, respectively. The semantics can be defined via predicate logics as follows:

$$\text{FOL}((\geq nR)(x)) = (\exists y_1, \ldots, y_n)\left(\bigwedge_{1 \leq i \leq n} R(x, y_i) \wedge \bigwedge_{1 \leq i < j \leq n} y_i \neq y_j\right) \quad (2.19)$$

$$\text{FOL}((\leq nR)(x)) = (\forall y_1, \ldots, y_{n+1})\left(\bigwedge_{1 \leq i \leq n+1} R(x, y_i) \Rightarrow \bigvee_{1 \leq i < j \leq n+1} y_i = y_j\right)$$
$$(2.20)$$

That is, if $(\geq nR)(x)$ then there are at least n other distinct elements, to which x is related by R. On the other hand, the construct $(\leq nR)(x)$ states that there are at most n elements to which x is related by R. Note that the construct $= nR$ is also commonly used. It is the exact unqualified restriction, which can be expressed by $\leq nR \sqcap \geq nR$.

Example 2.11 *A lucky person has at least 3 friends. We express this using unqualified cardinality restrictions as:*

$$\text{Lucky} \sqsubseteq (\geq 3\text{hasFriend})$$

Example 2.12 *A soccer team is a sports team that consists of 10 field players and a goalkeeper. This can be written as:*

$$\text{SoccerTeam} \sqsubseteq \text{SportsTeam} \sqcap (= 10\text{hasFieldPlayer}) \sqcap (= 1\text{hasGoalKeeper})$$

Functional Roles: $\mathcal{F}$

A special case of unqualified cardinality restrictions is the feature of functional roles, whose presence is indicated via letter $\mathcal{F}$. If a role R is declared as functional then any instance can be related to at most one member by R. In that sense, functional roles have the same characteristics as functions in analysis. Functional roles are denoted as $\top \leq 1R$.

Qualified Cardinality Restrictions

Q adds support for cardinality restrictions that can be qualified by an arbitrary constructed class. They are denoted as $\leq nR.C$ and $\geq nR.C$. In particular, $(\leq nR.C)(x)$ means that x is related to at most n other elements of class C. Similarly, $(\geq nR.C)(x)$ means that x is related to at least n other elements of class C.

Example 2.13 *Using qualified cardinality restrictions we can refine our definition of class* Lucky *from Example 2.11. A lucky person is someone who has at least three friends who are true friends. This can be expressed as:*

$$\text{Lucky} \sqsubseteq (\geq 3\textit{hasFriend}.\text{True}),$$

whereby True *describes the class of true friends.*

Concrete Domains: $(\mathcal{D})$

Support of concrete domains is denoted by $(\mathcal{D})$.[156] A concrete domain $\mathcal{D}$ consists of a set and associated predicates. For instance, the set could be the the set of natural numbers, i.e., $\mathbb{N}$, and the predicate defines relations over the set, e.g., $<, \leq, =, \geq, >$.

Example 2.14 *With appropriate domain-based constructors, it is possible to express that a teenager is a human whose age is between 13 and 19, i.e.,* Teenager $\equiv$ Human $\sqcap \exists \textit{age}. \geq 13 \sqcap \exists \textit{age}. \leq 19$.

Key Boxes

$\mathcal{K}$ adds support for a key box that contains key definitions. A key definition can be used to assert functional dependencies as known from relational databases. For instance, one can state that a social security number (ssn) is a key for U.S. citizens; hence, two individuals that share the same ssn must be the same individuals.

2.6   Exercises and Further Reading

Artificial Intelligence: A Modern Approach by S. Russell and P. Norvig [217] is a classic work on many aspects of the field of Artificial Intelligence including excellent introductions to propositional and first-order predicate logic that go beyond the concise treatments of these topics presented here. *The Description Logic Handbook: Theory, Implementation, and Applications*, edited by Franz Baader and colleagues, covers many aspects of the field of Description Logic and includes several introductory chapters [15].

In this chapter, we have written slightly simplified grammars in Backus Naur Form (BNF) in order to define the syntax of the various kinds of mathematical logic we presented. In brief, there are four components in a BNF grammar.

1. A set of *terminal symbols*, which are symbols and words that are not further transformed by the productions of the grammar. For instance, symbols such as ⇒ or Klingon are terminals.

2. A set of *non-terminal symbols*, which are symbols that stand for subphrases of the language. For instance, in FOL, *Connective* is a nonterminal symbol that stands for any of the terminal symbols ∨, ∧, ⇒, or ⇔.

3. A *start symbol*, which is a non-terminal that stands for complete strings in the language. For instance, *Formula* is used as a start symbol in the BNF grammar for FOR presented in Table 2.5.

4. A set of *productions*, which are rules for rewriting non-terminal symbols on one side of a production in terms of zero or more terminal and nonterminal symbols on the other side of the production.

BNF grammars can be used to determine if a string expressing some concept in a mathematical logic is syntactically valid. An example of this was given at Section 2.2 on page 14. A deeper understanding of BNF will not be important for the material in later chapters, but for readers who would like to know more about the topic, many books, including *Artificial Intelligence* by Russel and Norvig [217] provide more detailed treatments of BNF than we do here.

Exercises

2.1 Express the following sentences using statements from propositional logic:

- The Yankees are the best baseball team of all time, or anabolic steroids improve athletic performance.
- The Yankees are the best baseball team of all time, and it is not true that anabolic steroids improve athletic performance.
- It is not true that the Yankees are the best baseball team of all time.
- If anabolic steroids improve athletic performance, then the Yankees are the best baseball team of all time.
- The Yankees are the best baseball team of all time if and only if anabolic steroids improve athletic performance.

Try to use as few propositions as possible to express all sentences.

2.1 Create a truth table similar to Table 2.3 for the proposition $((x \vee y) \wedge y) \Rightarrow x$. Is this proposition valid?

2.2 Create a truth table to check the validity of the statement $x \wedge \neg x \Leftrightarrow \bot$.

2.3 Create a truth table to test whether DeMorgan's laws are valid. That is, check the two propositions:

$$\neg(x \wedge y) \quad \Leftrightarrow \quad \neg x \vee \neg y$$
$$\neg(x \vee y) \quad \Leftrightarrow \quad \neg x \wedge \neg y$$

2.4 Use truth tables to check the validity of the inference rules given in Table 2.4.

2.5 Although it is common to use both the connectives $\wedge$ (AND) and $\vee$ (OR) for easier legibility, it is possible to formulate the inference rules of propositional logic without making explicit use of $\wedge$. Consider a way of combining parentheses, the NOT operator $\neg$ and $\vee$ that will yield and identical truth table for two propositional variables as the AND connector $\wedge$ does.

2.6 Translate the following facts and rules that are given in common speech into their Description Logic counterparts.

- Dogs are a kind of pet.
- Dogs are the same thing as canines.
- Fido is a dog.
- It is not true that Cleopatra is a dog.
- All cats are fickle.
- Dalmatians are spotted dogs. (Hint: Create a class representing spotted entities.)
- There exists a cat that loves me.

2.7 Translate the following knowledge base into common speech. Let D represent the class of Dolphins, M the class of Mammals, W the class of animals that live in the Water, F the class of Fish, and let d, f, and x be individuals. The string `eats` is a predicate.

- D(d)
- F(f)
- D ⊑ M
- D ⊑ W
- D ⊑ M ⊓ W
- ¬D(f)

- D ⊑ ∀eats.F
- ∃eats.D ⊓ M

2.8 Convert the $\mathcal{ALC}$ expressions $\top \sqsubseteq \forall R.C$ and $\exists R.\top \sqsubseteq C$ to their FOL equivalents.

2.9 Show that $\mathcal{ALC}$ is semantically equivalent to $\mathcal{ALEU}$. (Hint: Convert expression to first-order logic, then show the equivalence in this space using known equivalences such as DeMorgan's rules.)

2.10 Find the FOL equivalent to the qualified number restrictions, i.e., $\leq nR.C$ and $\geq nR.C$.

2.11 Show that the subset relation $\subset$ is transitive.

Chapter 3

Probability Theory and Statistics for Bio-Ontologies

This chapter offers a review on some mathematical and statistical topics that will be essential for understanding the algorithms presented in Parts III and IV of this book.

3.1 Probability Theory

Probability theory deals with the analysis of experiments that have distinct outcomes such as "heads" or "tails" as outcomes of tossing a coin, or the 52 outcomes of drawing a card from a standard deck of playing cards. The set of all possible outcomes of such an experiment is called the *sample space*, which is usually symbolized with the Greek letter Omega (Ω). Denoting the elements (the individual outcomes, also known as the elementary events of the experiment) as e_i, we have that

$$\Omega = \{e_1, e_2, \ldots, e_n\}$$

For instance, if we denote "heads" as e_h and "tails" as e_t, then the sample space of the coin-toss experiment is $\Omega = \{e_h, e_t\}$. For the card draw, $\Omega = \{e_1, e_2, \ldots, e_{52}\}$, where each of the 52 elements e_i symbolizes the drawing of one of the 52 cards.

Definition 3.1 *A **probability measure** P is defined over subsets of the events in Ω, and must fulfill the following three requirements:*

1. *$0 \leq P(\{e_i\}) \leq 1$ for all i.*

2. *$\sum_{i=1}^{n} P(\{e_i\}) = 1$.*

3. *For any compound event defined as $E = \{e_{i_1}, e_{i_2}, \ldots, e_{i_k}\}$, $P(E) = P(\{e_{i_1}\}) + P(\{e_{i_2}\}) + \ldots + P(\{e_{i_k}\})$.*

Intuitively, the last condition states that, say, the probability of drawing an ace is equal to the combined probability of drawing an ace of spades or an ace of clubs, or an ace of hearts, or an ace of diamonds: $P(\text{ace}) = \frac{1}{52} + \frac{1}{52} + \frac{1}{52} + \frac{1}{52} = \frac{1}{13}$.

The *conditional probability* of X given Y is denoted as $P(X|Y)$, and indicates the probability of X occurring given that Y is known to have occurred. If X is the event of drawing a heart from a deck of cards, then $P(X) = \frac{13}{52} = \frac{1}{4}$ if the draw is done from a complete deck of cards. But if Y represents the event that a heart is (previously) drawn from the deck, then $P(X|Y) = \frac{12}{51}$. If $P(X) \neq 0$, then the conditional probability of X given Y is defined as

$$P(X|Y) = \frac{P(X \cap Y)}{P(X)}. \qquad (3.1)$$

Here, $P(X \cap Y)$ denotes the probability of X *and* Y occurring.[1] One can provide a sort of intuitive proof of this equation by counting events. For instance, if we want to know the probability of drawing a queen given that we have drawn a royal card (king, queen, or jack), then clearly $P(Q|R) = \frac{1}{3}$. If we let $n = 52$ be the total number of cards, $n_R = 12$ be the total number of royal cards, and $n_{QR} = 4$ be the total number of cards that are both royal cards and queens, then we have that $P(Q \cap R) = n_{QR}/n$, then we have that

$$P(Q|R) = \frac{P(Q \cap R)}{P(R)} = \frac{n_{QR}/n}{n_R/n} = \frac{n_{QR}}{n_R} = \frac{4}{12} = \frac{1}{3}$$

For any probability space, two events, say A and B, are said to be independent if and only if $P(A \cap B) = P(A)P(B)$. Thus, if A and B are independent, it follows that $P(A) = P(A|B)$.

A *random variable* is a function that maps elements from the *sample space* Ω to a measurable, real-valued space, called *the state space*. A probability distribution is a probability measure over the state space.

If the sample space of a random variable is finite or countable, then the random variable is said to be *discrete*. The probability measure is then described by a *probability mass function* (pmf).

For instance, the pmf of a coin toss with the outcomes "heads" or "tails" is a *Bernoulli distribution*, which, in this particular case, would assign to both elementary events 0.5 if the coin is fair. The *binomial distribution* is the discrete probability distribution of the number of successes in n trials. If the probability of success in a single trial is p, the probability of observing k successes in n trials can be written as

$$P(K = k) = \binom{n}{k} p^k (1 - p)^{n-k} \qquad (3.2)$$

The Bernoulli distribution is thus a special case of the binomial distribution where $n = 1$.

Now consider a random experiment, in which every trial results in one

---

[1] Note that $P(X, Y)$ is an equivalent, alternative notation for the joint probability.

of k possible outcomes, where the probability of observing an outcome i is given by p_i. When repeating this random experiment m times, let the random variable X_i count the number of times outcome i is observed. The pmf is then described by a *multinomial distribution* which is given by

$$P(X_1 = x_1, \ldots, X_k = x_k) = \frac{(x_1 + \ldots + x_k)!}{x_1! \ldots x_k!} p_1^{x_1} \ldots p_k^{x_k}, \qquad (3.3)$$

where $\sum_{i=1}^{k} x_i = m$. It can be easily seen that the binomial distribution is a special case of the multinomial distribution where there are only two outcomes. Thus, for the coin example we would have $k = 2$ and $p_1 = p_2 = 0.5$.

A random variable X is said to be continuous if its probability distribution is continuous, i.e., it is a probability density function $f(x)$, which is $f(x) \geq 0$ for all $x \in \mathbb{R}$ and

$$\int_{-\infty}^{\infty} f(x)dx = 1.$$

The probability of $a \leq X \leq b$ denoted as $P(a \leq X \leq b)$ can be calculated by integrating the density function from a to b. Note that this implies that for continuous random variables $P(X = a) = 0$, for all $a \in \mathbb{R}$.

3.1.1   Hypothesis Testing

"Everyone charged with a penal offense has the right to be presumed innocent until proved guilty according to law in a public trial…"

–Universal Declaration of Human Rights, Article 11

The purpose of statistical inference is to draw a conclusion about a population based on data obtained from a sample of the population. Hypothesis testing provides a kind of standardized procedure for evaluating the strength of evidence provided by the sample, in which the research question is posed as a test between two exhaustive and mutually exclusive hypotheses.

- **Null hypothesis** (H_0): The claim that is initially assumed to be true.

- **Alternative hypothesis** (H_A): The claim that is initially assumed not to be true.

Usually, we are interested in the relationship between a predictor and an outcome in the population. The null hypothesis is generally taken to be a lack of association between the predictor and the outcome, and the alternative hypothesis indicates the existence of an association. In hypothesis testing, the burden of proof is on us to disprove the null hypothesis, analogous to a criminal court case, in which the accused is assumed innocent until otherwise proven.

In addition to defining the two hypotheses, the researcher must select the significance level for the test, which is the probability of rejecting H_0 even though H_0 is true (i.e., a false-positive conclusion). The significance level is usually symbolized by the Greek letter alpha (α) and is often chosen to have a value of 5%, meaning that there is a 1 in 20 chance of a false-positive conclusion that an association between the predictor and the outcome exists when in fact there is no such association. The probability of a false-negative conclusion that there is no association when in fact an association does exist is called beta (β). Therefore, the power of a test to detect an association when one exists is given by $1 - \beta$. Finally, the researcher chooses an appropriate distribution and test statistic to model the data.

Imagine for example a study concerning a treatment for increased blood cholesterol levels. Let us assume that the mean level of low-density lipoprotein cholesterol (LDL-C) is well characterized in the population and is known to be 87.9 mg/dl. The study involves 60 persons who are to receive a new dietary supplement over three months. Before performing the study, however, the investigators would like to know if the 60 persons are representative of the population as a whole, that is, whether they have baseline LDL-C levels of about 87.9 mg/dl. We can state the null and alternative hypotheses as follows:

- H_0: $\mu = 87.9$ mg/dl

- H_A: $\mu \neq 87.9$ mg/dl

Since the sample is reasonably large, we can assume that the sample mean $\bar{x}$ has approximately a normal distribution. We measure LDL-C levels in the 60 persons and obtain $\bar{x} = 91.3$ and the sample standard deviation $s = 9.7$, which leads us to reject the null hypothesis (Figure 3.1).

3.1.2   *p*-Values and Probability Distributions

As described statistical hypothesis tests can be used to decide whether an observed value falls within the range defined by the critical values of a distribution.[2] If not, the null hypothesis is rejected. In practice, hypothesis testing is often combined with the determination of a *p*-value. The *p*-value is defined as the probability of observing a test statistic that is at least as extreme as the one that was observed given that the null hypothesis is true. The null hypothesis is rejected if the *p*-value is equal to or lower than the significance level alpha (α), which by convention is often taken to be 5%. If on the other hand, the *p*-value for the test statistic is larger than this, we fail to reject the null hypothesis.

In traditional statistics and using strict definitions, if a *p*-value is judged to be significant because it is less than or equal to a given significance level

---

[2]The critical value of a distribution defines the value beyond which the null hypothesis must be rejected. For instance, if we are testing the null hypothesis $H_0 : \mu = \mu_0$ against the alternative hypothesis $H_A : \mu < \mu_0$ then the critical value of the random value X would be x_c such that $P(X \leq x_c) = \alpha$.

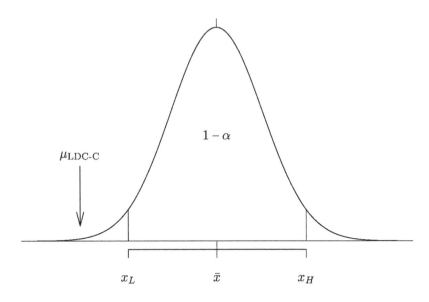

FIGURE 3.1: Hypothesis Testing. The sampling density of $\bar{x}$ is approximately normal because the sample size is relatively large (this is a result of the central limit theorem). Under the null hypothesis, the probability that either $\bar{x} \leq x_L$ or $\bar{x} > x_H$ is α. The fact that the mean value of LDL-C in the general population is less than the value of x_L in our sample, means that it is improbable that the 60 persons forming the test group are representative of the general population with respect to LDL-C levels. Therefore, we reject the null hypothesis that $\mu = 87.9$ mg/dl, and conclude that the 60 persons are not suitable for investigating the expected effect of the new medication in the general population.

that had been set prior to performing the experiment, then the actual value of the p-value is unimportant. For example, if the significance level has been set to 0.05, then it would not be correct to say that a p-value of 0.00001 is more significant than a p-value of 0.04. According to this view, a result is either statistically significant or it is not. However, in science and in bioinformatics, the actual level of the p-value is usually interpreted as meaningful, whereby smaller p-values are taken to be more statistically significant. This is the viewpoint that we will take in this book.

One of the most important uses of statistical tests in bio-ontologies is in term overrepresentation analysis. This topic will be discussed in detail in Chapter 8, but we will briefly describe this kind of analysis here so that readers will know what the statistical tests will be used for. In a typical microarray experiment, expression values for essentially all genes in the genome are obtained under control and test conditions, and a list of genes that respond to the test conditions by increasing or decreasing their expression is obtained. This list, the so-called study set, typically comprises a relatively small fraction of all the genes on the microarray. Now imagine we are interested in whether small GTPases are one of the factors responsible for the biological response observed in the experiment. A total of 418 human genes are annotated to the GO term *small GTPase mediated signal transduction* (GO:0007264). If a total of 20,000 genes, including these 417 genes, are represented on the microarray, then $100\% \times 418/20,000 = 2.1\%$ of all genes on the microarray are annotated to *small GTPase mediated signal transduction*. Under the null hypothesis that *small GTPase mediated signal transduction* is not related to the experiment, we would expect 2.1% of the differentially expressed genes also to be annotated to this term. That is, if our study set comprises 1000 genes, we would expect 2.1% of them, or 21 genes, to be annotated to *small GTPase mediated signal transduction* by pure chance. Let's say we observe that 23 genes are annotated to *small GTPase mediated signal transduction*. Is this more than we expect by chance? What if we observe 39 genes to be annotated to *small GTPase mediated signal transduction*?

The Fisher exact test is the statistical test that is most commonly used to address this issue. Before discussing the Fisher exact test, it will be necessary to explain the binomial and the hypergeometric distributions.

Many statistical tests involve the concept of a Bernoulli trial, that is, an event such as the tossing of a coin that results in one of two outcomes ("heads" vs. "tails" or "success" vs. "failure"). Assuming a fair coin, the probability of tossing "heads" is then simply $p(H) = 0.5$. To calculate the probability of a sequence of events, one multiplies the probabilities of each individual event. For instance, in calculating the probability of tossing "heads" once followed by three "tails" (which we will denote $HTTT$), we obtain $p(HTTT) = 1/2 \times 1/2 \times 1/2 \times 1/2 = (1/2)^4 = 1/16$. If we are interested in the overall probability of seeing "heads" once if we toss a coin four times (rather than the probability of a specific sequence of tosses), we note that there are four ways of this happening, $HTTT$, $THTT$, $TTHT$, and $TTTH$. Therefore, the overall probability of getting "heads" once in four coin tosses is 4/16. Similarly, we can calculate the probabilities of observing a total of 0, 1, 2, 3, or 4 "heads" in four coin tosses (Table 3.1).

The binomial coefficient provides a general way of calculating the number of ways k objects can be chosen from a set of n objects. Recall that $n! = n \times (n-1) \times (n-2) \times \ldots \times 2 \times 1$ is the number of ways of arranging n objects in a series. There are n choices for the first object. Once that object has been chosen, the second object can be chosen from among the $n-1$ remaining objects, and the third object can then be chosen from among the remaining $n-2$ objects, and so on.[3]

In order to calculate the number of ways of observing k "heads" in n coin

---

[3]Recall that $n!$ is spoken as "n factorial."

4 Heads	3 Heads	2 Heads	1 Head	0 Heads
HHHH	*HHHT*	*HHTT*	*HTTT*	*TTTT*
	HHTH	*HTTH*	*THTT*	
	HTHH	*HTHT*	*TTHT*	
	THHH	*THHT*	*TTTH*	
		THTH		
		TTHH		
1/16	4/16	6/16	4/16	1/16

TABLE 3.1: Probabilities of getting 0,1,2,3, or 4 "heads" on four coin tosses.

tosses, we first examine the sequence of tosses consisting of k "heads" followed by $n - k$ "tails":

H	H	H	...	H	H	T	...	T	T	T
1	2	3	...	$k-1$	k	$k+1$	...	$n-2$	$n-1$	n

Each of the $n!$ rearrangements of the numbers $1, 2, \dots, n$ defines a different rearrangement of the letters $HHH\dots HHTT\dots TT$. However, not all of the rearrangements change the order of the H's and the T's. For instance, exchanging the first two positions leaves the order $HHH\dots HHTT\dots TT$ unchanged. Therefore, to calculate the number of rearrangements that lead to different orderings of the H's and the T's (e.g., $HTH\dots HHTT\dots TH$), we need to correct for the reorderings that merely change the H's or the T's among themselves. Noting that there are $k!$ ways of reordering the "heads" and $(n - k)!$ ways of reordering the "tails," it follows that there are

$$\binom{n}{k} = \frac{n!}{(n-k)!k!} \tag{3.4}$$

different ways of rearranging k "heads" and $n - k$ "tails." This quantity (read "n choose k") is known as the binomial coefficient. We can now use this to calculate the probability of obtaining k *successes* in n trials, whereby the probability of *success* in one trial is p, and the probability of *failure* is $q = 1 - p$. To obtain the probability of any one particular order of trials with k successes and $n - k$ failures, we multiply the probabilities of the individual trials to obtain $p^k(1-p)^{n-k}$. In order to find the total probability of obtaining k successes, we simply multiply the probability of one particular trial order with k successes by the number of trial orders resulting in k successes, which is given by the binomial coefficient. This probability distribution is called the *binomial distribution*:

$$P(k, n, p) = \binom{n}{k} k^p (n - k)^{1-p} \tag{3.5}$$

For instance, to calculate the probability of observing two "heads" in four coin tosses, we calculate the probability of "heads" as $p = 1/2$ and that of "tails" as $1 - p = 1/2$, and the number of sequences of four coin tosses with two "heads" as $\binom{4}{2} = 6$, corresponding to a probability of $6 \times (1/2)^2 (1 - 1/2)^2 = 6/16$ (see also Figure 3.2).

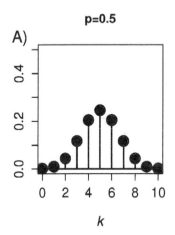
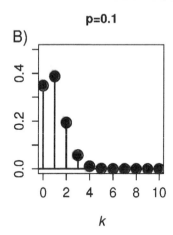

FIGURE 3.2: **Binomial Probability Distribution.** **A)** Probability of $k = 1, \ldots, 10$ "heads" in ten coin tosses using a fair coin ($p = 0.5$). **B)** Probability of $k = 1, \ldots, 10$ "heads" in ten coin tosses using a biased coin ($p = 0.1$).

GO overrepresentation analysis can be modeled as a binomial distribution in which each gene in the study set is considered to be a "trial," and "success" occurs if the gene is annotated to a GO term of interest.[4] The probability of "success" is simply the frequency of the annotation to the GO term among all genes. For instance, if there are 10,000 genes on a microarray chip, 600 of which are annotated to some GO term t, then the probability of t is $p = 600/10,000 = 0.06$. If we perform an experiment using microarray hybridizations and identify 250 differentially expressed genes (the "DE set"), 30 of which are annotated to t, we can use the binomial distribution to find the probability of observing exactly 30 genes in the DE set annotated to t as $p(30, 250, 0.06) = 1.44 \times 10^{-4}$. In order to use this for a statistical test, we now define a *null hypothesis* to be that the biological function described by GO term t is not overrepresented among the differentially expressed genes. According to this null hypothesis we would expect the same proportion of genes (or less) in the DE set to be annotated to t as in the population (if we symbolize the proportion of genes in the population and study set that are annotated to the term as π_p and π_s, then $H_0 : \pi_s \leq \pi_p$). In GO overrepresentation analysis, we are looking

---

[4]We will see shortly that the binomial distribution only approximates the true probability in this setting.

for terms that annotate a higher than expected proportion of the genes in the study set. Therefore, the alternative hypothesis is $H_a : \pi_s > \pi_p$. If the probability of the observed result of our experiment (the observation of 30 genes in the DE set being annotated to t) is less than α, we reject the null hypothesis in favor of the alternative hypothesis, which would imply that the GO term t has something to do with the observed differential expression. Since we have divided the set of all possible results into two classes, it is necessary to collect all outcomes that are as extreme or more so than the observed one to be part of the rejection class. In our example, this means that we need to sum up the probability of observing 30 *or more* genes annotated to GO term t in the DE set of 250 genes. Note that if there were only 100 genes annotated to a term t in the entire population of genes on the microarray chip, the probability of observing 101 or more genes annotated to t is zero. Therefore, in general we can take the sum to the maximum of the number of genes annotated to a GO term t (denoted p_t) or to the size of the DE set (denoted $|DE|$). We thus obtain:

$$p(k \geq k', n, p) = \sum_{k=k'}^{\max(p_t, |DE|)} \binom{n}{k} k^p (n-k)^{1-p}. \tag{3.6}$$

Figure 3.3 displays an approximation of the probability mass function for $k = 0, 1, \ldots, 40$, in which the discrete points are dipicted as a smoothed line. The gray region indicates the values of k which encompass 95% of the probability mass under the null hypothesis. Since the observed value lies outside of this region, we reject the null hypothesis that genes annotated to t are not enriched in the DE set. In order to calculate a p-value, we add up the probabilities for observing $k = 40, 41, \ldots, 250$ genes annotated to t in the DE set. The exercises at the end of this chapter provide some tutorials on this.

Although this approach provides a nearly correct answer for terms with many annotations in the population and large DE sets, it is actually only an approximation. The problem is that the binomial distribution is based on the assumption that the individual trials are independent of one another. While this is clearly true for coin tosses, it is not for GO overrepresentation analysis, which is rather comparable to a lotto game in which labelled balls are taken from an urn and not replaced. For instance, if there are 60 balls numbered 1 to 60, the chance of getting a 7 on the first draw is $1/60$. If a 7 was drawn the first time, the chance of getting it the second time is zero. If another number was drawn the first time, the chance of getting a 7 the second time is $1/59$, because some other ball has been removed. In our microarray experiment, we imagine that the set of all 10,000 genes on the chip is represented by an urn with 10,000 balls. A total of 100 genes are annotated to some GO term t, and thus 600 balls in the urn are colored blue. We then model the observation of 250 differentially expressed genes as the process of randomly removing 250 balls in turn from the urn without replacement. Clearly there are many possible outcomes to this experiment. The statistical analysis basically just counts up

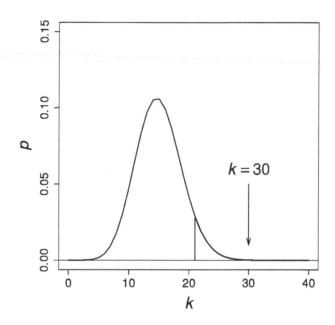

FIGURE 3.3: Smoothed pmf for the Binomial Distribution. The smoothed variant of the probability mass function for the binomial distribution, with $p = 0.06$ and $n = 250$. Values for $k = 0, 1, \ldots, 40$ are shown (the values for $k > 40$ are nearly zero and are not shown). The observed value for k of 30 lies outside of the range containing 95% of the probability mass, and we reject the null hypothesis that the genes in the study set are not enriched in genes annotated by t. The p-value for this observation can be calculated by Equation (3.5) to be 1.68×10^{-8}.

all the ways of getting k genes annotated to t and $n - k$ remaining genes not annotated to t, and compares this number to the number of all possible outcomes. If we let m be the number of genes on the microarray chip, m_t be the number of these genes that are annotated to GO term t, n be the number of DE genes (the study set) and k be the number of DE genes that are annotated to t, we have

$$p(k, n, p) = \frac{\binom{m_t}{k}\binom{m - m_t}{n - k}}{\binom{m}{n}}. \tag{3.7}$$

There are $\binom{m_t}{k}$ ways of choosing k genes annotated to t. There are $\binom{m - m_t}{n - k}$ ways of choosing the remaining $n - k$ genes that are not annotated to t. In total, there are $\binom{m_t}{k}\binom{m - m_t}{n - k}$ ways of choosing the genes in the study set such that k are annotated to t and $n-k$ are not. There are $\binom{m}{n}$ total ways of choosing a study set with n genes from the population of m genes with arbitrary annotations. Equation (3.7) therefore reflects the proportion of all possible study sets that can be chosen from the population such that k of the genes are annotated to t. If this proportion is very small, it is unlikely that such a study set will be chosen at random.

Equation (3.7) is known as the *hypergeometric distribution*. Analogously to the situation with the binomial distribution, we need to calculate the probability of observing some number k or more genes annotated to t in order to have a statistical test. In this case, the sum over the tail of the hypergeometric distribution is known as the *Exact Fisher Test*:

$$p(K \geq k', n, p) = \sum_{k=k'}^{\max(K, |DE|)} \frac{\binom{m_t}{k}\binom{m-m_t}{n-k}}{\binom{m}{n}} \qquad (3.8)$$

Equation (3.8) gives the probability of observing k' or more genes annotated to t in a study set of size n from a population of size m in which m_t genes are annotated to t. This equation is the workhorse of Gene Ontology overrepresentation analysis that is often used in the evaluation of microarray experiments. More details about how this test is used in practice and about other extensions to this test will be presented in Chapter 8. Several computational problems surrounding the exact Fisher test are presented in the exercises at the end of this chapter.

3.1.3 Multiple-Testing Correction

The p-value is the probability, under the null hypothesis, that the test statistic assumes the observed or a more extreme value. It is important to realize that if we go on testing long enough, we will inevitably find something which is "significant" by chance alone. If we test a null hypothesis that is true using a significance level of $\alpha = 0.05$, then there is a probability of $1 - \alpha = 0.95$ of arriving at a correct conclusion of non-significance. If we now test two independent true null hypotheses, the probability that neither test is significant is $0.95 \times 0.95 = 0.90$. If we test 20 independent null hypotheses, the probability that none will be significant is then $(0.95)^{20} = 0.36$. This corresponds to a probability of $1 - 0.36 = 0.64$ of getting at least one spurious significant result, and the expected number of spurious significant results in 20 tests is $20 \times 0.05 = 1$. If we perform 100 such tests, the probability that none will be significant is $(0.95)^{100} = 0.01$ and there is a 99% probability of getting at least one significant result [35]. As we will see in Chapter 8, in GO overrepresentation analysis, we perform a hypothesis test for each individual GO term, meaning that we perform thousands of tests (usually one test is made for each GO term that annotates at least one of the genes in the population of genes represented on the microarray chip). Therefore, unless we take into account the fact that multiple statistical tests are being performed, it is nearly certain that we will obtain one or more false-positive results. A number of multiple-testing correction (MTC) procedures have been developed to limit the probability of false-positive results. The MTC procedures differ in complexity, in their assumptions about the data, and in the type of control they provide. In this section, we will explain those MTC procedures commonly used in the setting of GO overrepresentation analysis and ontological semantic similarity evaluations.

The oldest and simplest MTC procedure is the Bonferroni correction. Since the probability of getting a non-significant result for a single test if the null hypothesis is true is $1 - \alpha$, the probability of getting no significant result for k such tests is $(1-\alpha)^k$. For a single test, α is often chosen to be 0.05, so that $1 - \alpha = 0.95$. If we make α small enough, then we can make $(1-\alpha)^k = 0.95$. If α is very small, then $(1-\alpha)^k \approx 1 - k\alpha$. Therefore, if we set $\alpha = \frac{0.05}{k}$, then the probability that one of the k tests will be called significant if the null hypothesis is true is now 0.05, because $(1-\alpha)^k = (1 - \frac{0.05}{k})^k \approx 1 - k\frac{0.05}{k} = 0.95$. In essence, the Bonferroni procedure relies on the fact that the probability of a union of events (i.e., one or more tests being called significant) is less than or equal to the sum of the probabilities of the individual events. Thus we have

$$p(\text{one or more tests called significant}) \leq \sum_{i=1}^{k} \frac{\alpha}{k} = \alpha.$$

The Bonferroni procedure thus provides control of the family-wise error rate (FWER), which is the probability of at least one Type I error.

There are two ways of implementing a Bonferroni correction. One can either adjust the significance level by dividing α by the number of tests performed as above, or one can multiple the p-value returned by each test (which is call the *nominal* p-value) by the number of tests performed. The second method is more commonly seen in bioinformatics applications. With the second method, it is possible that multiplying a relatively large p-value by the number of tests results in a number that is larger than one. Therefore, what is usually reported is $\max(1, k \times p)$, where k is the number of tests performed and p is the nominal p-value.

The Bonferroni method is simple to implement and can be used in any multiple testing situation. The big disadvantage of the Bonferroni procedure is its low power, meaning that one pays for the simplicity and universality of the Bonferroni procedure by the fact that one is more likely to miss a true effect. For this reason, a number of MTC procedures have been developed with the goal of providing adequate control of the type I error rate without losing too much power. We will review three MTC procedures with different assumptions that are commonly seen in ontology applications. For more details on other MTC procedures, readers are referred to recent reviews [74, 257].

Holm's method, like the Bonferroni method, controls the FWER and is universally applicable. However, Holm's method achieves a greater power.

Holm's Multiple-Testing Correction Procedure

The p-values are first placed in ascending order. We will refer to the p-values in this order as $p_1, p_2, \ldots, p_n$, where $p_1 \leq p_2 \leq \ldots \leq p_n$. We now compare each nominal p-value to $\alpha/(n - i + 1)$. Starting from $i = 1$, we compare each p-value in turn and reject the null hypothesis if $p_i \leq \alpha/(n - i+1)$. We continue in this way until the first non-rejection, after which (since the p-values have been placed in ascending order) all other null hypotheses are also not rejected.

Note that because the divisors are $n-i+1$ instead of n, the Holm procedure never rejects less hypotheses than the Bonferroni procedure. The increased power arises from the fact that the Holm procedure may reject some null hypotheses that are not rejected by Bonferroni. The following theorem [3] shows that the Holm procedure provides adequate control of the FWER.

Theorem 3.1 *The Holm procedure bounds the type I error at α.*

Proof 3.1 *Assume that n tests have been performed and that t of the n null hypotheses are true. Let P denote the minimum p-value associated with a true null hypothesis. A type I error occurs if P causes rejection of a true null hypothesis at step i of the Holm procedure, implying $P \leq \alpha/(n-i+1)$. Because this is the first correct null hypothesis and because of the way the p-values were ordered, the hypotheses $1, 2, \ldots, i-1$ were correctly rejected in favor of the alternative hypothesis. Assuming all hypotheses from $i, \ldots, n$ are true null hypotheses, then $n - t = i - 1$. Since some of the null hypotheses $i, \ldots, n$ could be false, we have $i-1 \leq n-t$ which implies that $\alpha/(n-i+1) \leq \alpha/t$. Therefore, a type I error implies that $P \leq \alpha/t$, and since there are t true null hypotheses, the total probability of a type I error is less than or equal to $t \times \alpha/t = \alpha$ (see Figure 3.4).*

Resampling procedures rely on taking repeated, random samples of the data in order to estimate the distribution of p-values under the null hypothesis [84]. This type of procedure has been used extensively in microarray analysis in experiments that seek to identify genes differentially expressed between test and control conditions, whereby a certain number of control and a certain number of test samples are hybridized and a t test or variant thereof is performed for each of the many thousands of genes on the microarray. Resampling methods use a technique called *bootstrapping*, whereby samples are chosen at random with replacement, such that the original dataset gives rise to many others. Since the labels of the samples are reshuffled at random, the comparisons between two groups no longer reflect the differences between the test and control groups, but the dependencies between gene expression within individual samples are retained. In the context of GO overrepresentation analysis, we are analyzing a list of genes (the study set) for enrichment of annotations to thousands of GO terms. The equivalent permutation here is

A)

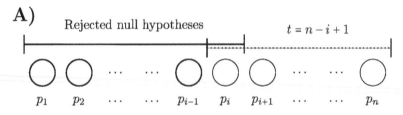

B)

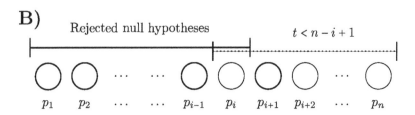

FIGURE 3.4: Holm's Multiple Testing Correction. The proof of the Holm MTC procedure first arranges the p-values in ascending order. False null hypotheses are shown as gray circles, and true null hypotheses as white circles. **A)** Assume that the first correct null hypothesis is associated with the ith p-value and that t of the null hypotheses are true. If all of the null hypotheses after the ith one are true, then $i - 1 = n - t$. **B)** If one or more of the null hypotheses after i are false, then $i - 1 < n - t$. We conclude that $i - 1 \leq n - t$ and thus $n - i + 1 \geq t$ and $\alpha/(n - i - 1) \leq \alpha/t$.

to create random study sets with the same number of annotated genes as the actual one. This maintains the dependency structures of the GO annotations while creating random study sets. Although there are numerous variations upon the theme, the basic algorithm, which is outlined within the box at the next page, is relatively simple.

Another way of dealing with multiple testing involves shifting the focus from control of the family-wise error rate (the chance of making even a single type-I error) to procedures that control the false-discovery rate (FDR), which describes the proportion of significant results that are type I errors (false discoveries). The situation is illustrated in Table 3.2. Control of the FWER limits the probability of even one type I error to α or less: $P(V \geq 1) \leq \alpha$. FDR control implies that the expected proportion of type I errors among all results declared significant is not more than a desired level: $E[V/R] \leq q$. If some of the null hypotheses are false, FDR control is less strict that FWER control, and thus FDR-controlling procedures are potentially more powerful.

Benjamini and Hochberg introduced a simple procedure that achieves FDR-control. The nominal p values are first placed in ascending order. For simplicity, we will number the ordered p values obtained for m statistical

---

Resampling Algorithm to Adjust p-Values

The nominal p-values are first calculated for each GO term, as will be described in more detail in Chapter 8.

For the bth permutation, $b = 1, \ldots, B$,

1. Create a permuted dataset X

2. Compute the nominal p values for each hypothesis (i.e., GO term)

The adjusted p-values can be calculated as the proportion of times in which the p-values for the permuted datasets were lower (more significant) than the observed p-values:

$$p_j^{adj} = \frac{\sum_{b=1}^{B} I(p_j \leq p_j^b)}{B} \tag{3.9}$$

It should be kept in mind that the accuracy of the p-values obtained by resampling procedures is limited by $1/B$. For many purposes, a value of $B = 10,000$ is adequate.

---

tests as $p_1 \leq p_2 \leq \ldots \leq p_m$. The corresponding null hypotheses are denoted $H_0, H_1, \ldots, H_n$. For a desired FDR level q (e.g., $q = 0.05$), the ordered p-value p_i is compared to the critical value $q \times \frac{i}{m}$. Now let $k = \max\left\{i : p_i \leq q \times \frac{i}{m}\right\}$. Reject $H_1, H_2, \ldots, H_k$ if such a k exists. It can be shown that this procedure controls the FDR at the level q if the test statistics are independent or positively correlated [28, 29, 203]. Control of FDR has become widely adopted in genomics research for situations such as microarray hybridization in which thousands of hypothesis tests are performed and a Bonferroni correction is too strict. FDR correction may be desirable for GO term overrepresentation analysis, in which thousands of GO terms are individually tested for overrepresentation.

	Declared non-significant	Declared significant	Total
True null hypotheses	U	V	m_0
Non-true null hypotheses	T	S	$m - m_0$
	$m - R$	R	m

TABLE 3.2: Number of errors committed when testing m hypotheses.

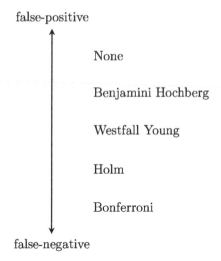

false-positive

None

Benjamini Hochberg

Westfall Young

Holm

Bonferroni

false-negative

FIGURE 3.5: **Multiple-Testing Correction Procedures**. The Bonferroni procedure is the most conservative MTC method, and is associated with a relatively high risk of false negative results. As we go toward the false-discovery rate controlling procedure of Benjamini and Hochberg, more hypotheses are rejected, reducing the risk of false-negative results and increasing the risk of false-positive ones. Using no MTC procedure at all is more likely to result in false-positive results when multiple tests are performed.

3.2   Bayes' Theorem

Bayes' theorem follows from the definition of the conditional probability and relates the conditional probability $P(A|B)$ to $P(B|A)$ for two events A and B such that $P(B) \neq 0$:

$$P(A|B) = \frac{P(B|A)P(A)}{P(B)}. \tag{3.10}$$

In this context, $P(B|A)$ is referred to as the *likelihood*, as it is a probability of parameter B, in contrast to $P(A|B)$, which is called the *posterior*, since it is derived from the knowledge of B. $P(A)$ is referred to as the *prior*, as it represents the knowledge of A prior to the application of knowledge B. $P(B)$ is the *normalization constant*.

Bayes' theorem is often used for a set of n mutually exclusive events $E_1, E_2, \ldots, E_n$ such that $\sum_i P(E_i) = 1$. Then, equation (3.10) can be written as

$$P(E_i|B) = \frac{P(B|E_i)P(E_i)}{\sum_i P(B|E_i)P(E_i)}. \tag{3.11}$$

This form of Bayes' theorem makes it clear why $B = \sum_i P(B|E_i)P(E_i)$ is called the normalization constant, because it forces the sum of all $P(E_i|B)$ to be equal to one, thus making $P(\cdot|B)$ a real probability measure as demanded by Definition 3.1.

There are many ways of interpreting Bayes' theorem. One a basic level, the Bayes' theorem implies that the interpretation of new information (posterior probability) should depend on what we already know (prior probability). This strategy describes a classic application of Bayes' theorem in medical diagnostics. An example that is seen in many textbooks describes the situation in which a chest X-ray (CXR) is obtained as a part of a routine checkup and a shadow is seen in one of the lungs that is suspicious for lung cancer. The referring physician consults a study that documents the following conditional probabilities:

- $P(\text{CXR=positive}|\text{Lung cancer=present})=0.8$

- $P(\text{CXR=positive}|\text{Lung cancer=absent})=0.02$

That is, the CXR has a false-negative rare of 20%, and a false-positive rate of 2%. A logical fallacy that is occasionally committed is the assumption that since the false-positive rate is 2%, the probability of there being a true finding of lung cancer must be 98%. This is known as the *base-rate fallacy* [18]. The assumption is incorrect because it ignores our prior knowledge that lung cancer is relatively uncommon in the general population undergoing a routine checkup. In fact, the probability that is of interest to the patient is not one of the above listed probabilities, but rather, the probability of having lung cancer given that the CXR is positive. This can be calculated using the second form of Bayes' theorem given by equation (3.11) if we know the prior probability of a patient having lung cancer. For the sake of argument, let us say that this probability is 1 in 500. That is, $P(\text{Lung cancer=present})=0.002$ and $P(\text{Lung cancer absent}=0.998)$. Bayes' theorem then implies (using the symbols **CA** for lung cancer present, $\neg$**CA** for lung cancer absent, and **+CXR** for chest X-ray positive):

$$
\begin{aligned}
P(\text{CA}|\text{+CXR}) &= \frac{P(\text{+CXR}|\text{CA})P(\text{CA})}{P(\text{+CXR}|\text{CA})P(\text{CA}) + P(\text{+CXR}|\neg\text{CA})P(\neg\text{CA})} \\
&= \frac{0.8 \times 0.002}{0.8 \times 0.002 + 0.02 \times 0.998} \\
&= 0.074
\end{aligned}
$$

Thus, the adjusted probability of the patient having lung cancer is about 7.4%, rather than 98%.

It is also possible to interpret Bayes' theorem as a way of relating observations to a hypothesis with the goal of inferring whether the hypothesis is likely to be true or not. This is one form of Bayesian inference. In the above example, there are two hypotheses: "The patient has lung cancer" and "The patient does not have lung cancer." We are unable to observe the true state (i.e., the diagnosis) with a chest X-ray (let us say that a lung biopsy is needed to make a definitive diagnosis). However, we do know what the conditional probability of the observable information is given that one or the other hypothesis is true. In the context of bioinformatics, Bayesian inference is often used to identify the most likely model: For instance, we observe a DNA sequence and would like to know if it is a gene (M_1) or not (M_2). We have no way of directly calculating the probability of there being a gene given a DNA sequence, but this is what we would like to know. Often, the model is symbolized by M and the observed data by D. Then, Bayes' theorem can be given as:

$$P(M_1|D) = \frac{P(D|M_1)P(M_1)}{P(D|M_1)P(M_1) + P(D|M_2)P(M_2)} \qquad (3.12)$$

In some cases, we may only be interested in the relative likelihood of two models. Thus, if we know that $P(M_1|D) = \frac{P(D|M_1)P(M_1)}{P(D)}$ and $P(M_2|D) = \frac{P(D|M_2)P(M_2)}{P(D)}$ it may be sufficient for us to know whether M_1 or M_2 is more likely. By taking their ratio, the normalization constant $P(D)$ cancels out:

$$\frac{P(M_1|D)}{P(M_2|D)} = \frac{P(D|M_1)P(M_1)}{P(D|M_2)P(M_2)} \qquad (3.13)$$

This approach is attractive in situations in which it is difficult or impossible to derive an explicit expression for $P(D)$, as we will see in Chapter 9.

3.3   Introduction to Graphs

Graphs are abstract entities of discrete mathematics which are used to encode relationships of interest between objects of the same domain. Formally, a graph is a pair $G = (V, E)$, in which V is finite set of *vertices* (or nodes), representing the objects, and E a set of pairs of distinct elements of V, which is a binary relation over V. Elements of E are called *edges* (or arcs). The pairs may be ordered or not. If the pair is unordered, we say that the edge is *undirected*. If all edges in a graph are undirected, the graph is called an *undirected graph* (Figure 3.6).

Undirected graphs can be used to describe connections between objects where neither of the objects "comes first." Thus, no distinction is made between the source and target vertices of an edge. An undirected graph is *connected* if for any pair of vertices in the graph there is a path between them. Many important graph algorithms, such as depth-first search and breadth-first search, can be implemented for undirected graphs.

Undirected trees are an important special kind of undirected graphs. A tree is a connected graph that has no cycles (Figure 3.7).

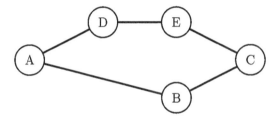

FIGURE 3.6: **Graphical Representation of an Undirected Graph.**
Oftentimes, nodes are represented using shapes (circles in this case) and edges
via connecting lines.

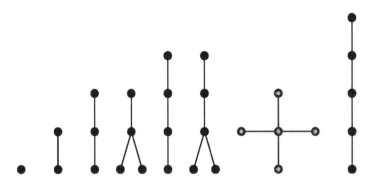

FIGURE 3.7: **The First Eight Undirected Trees.** An undirected graph
is called an *undirected tree* if it is connected and has no cycles. The figure
shows the first eight such trees comprising up to five vertices.

If the pair is ordered, we say that the edge is *directed*. If all edges of *G* are
directed, the graph is *directed* (Figure 3.8). Many applications in computer
science can be represented as directed graphs, including scheduling problems
and graphs representing metabolic networks. We will see further on that ontologies can be represented as directed graphs, whereby the vertices represent
entities in a domain and the edges represent relations between the entities.

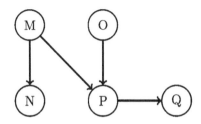

FIGURE 3.8: **Directed Graph.**

An edge such as $M \to N$ in Figure 3.8 is *directed* from M to N; M is

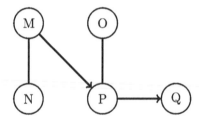

FIGURE 3.9: Partially Directed Graph.

called the head and N is called the tail of the edge. A series of one or more connected edges is called a *path*. The number of edges in a path is referred to as the *length* of the path. For instance, in Figure 3.8, the edges $M \to P$ and $P \to Q$ form a path of length 2. A path with length n is a sequence of vertices $(v_1, ..., v_n)$, which respects the edges, i.e., $(v_i, v_{i+1}) \in E$ for all i. A *cycle* is a special path whose start vertex v_1 is the same as the end vertex v_n. In directed graphs, the *indegree* of a vertex is defined to be the number of edges that are directed to the node, and the *outdegree* of a vertex is defined as the number of edges that originate from the node.

A graph can contain both directed and undirected edges. If at least one edge is directed we call the graph a *partially directed graph* (Figure 3.9).

A *directed path* is a path, in which the edges between the vertices are all directed. A directed graph is *acyclic* if it contains no directed cycle; such graphs are referred to as *directed acyclic graphs* (DAGs). A *partially directed acyclic graph* (PDAG) is a graph that contains directed and undirected edges, but which doesn't contain any directed cycle. See Figure 3.10 for an example of a graph with a cycle.

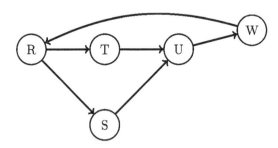

FIGURE 3.10: A Directed Graph with a Cycle. There is a cycle involving nodes R,T,U, and W. Note that nodes R,S,T, and U do not form a cycle because there is no directed path involving just these nodes whose start vertex is also the end vertex.

We say that vertex v_i is a *parent* of vertex v_j, if there is a directed edge (v_i, v_j) in G, in which case vertex v_j is also called a *child* of vertex v_i. The set of all parents of v_i is denoted by $pa(v_i)$. A *family* is defined as the set of a vertex and all of its parents. The set of *descendants* of vertex v_i consists of all vertices to which a directed path that originates from v_i can be constructed. All other vertices are said to be *non-descendants* of v_i.

We will see in Parts III and IV of this book that graphs are commonly used to represent ontologies. By convention, the terms of an ontology are represented as vertices of a graph, and the relations between the terms are represented by arcs. We will see that there are two kinds of relations in ontologies: those that are explicitly stated ("asserted") and those that are inferred. By convention, only the asserted relations are shown in graphical representations of ontologies unless otherwise stated.

Graph algorithms represent an important and vast topic in computer science. Nevertheless, a few relatively algorithms suffice to perform simple forms of inference in graphs representing ontologies. For instance, in Figure 3.11, *forebrain development* is a subclass of *brain development*.

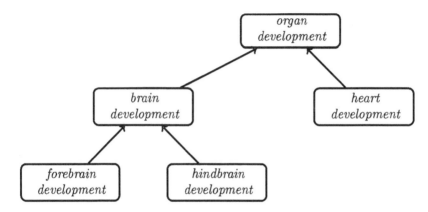

FIGURE 3.11: Subsumption Inference in Ontology Graphs. The Figure shows an imaginary ontology representing developmental processes, whereby the edges of the graph represent subclass relations.

For the sake of example, imagine that the terms of this ontology have been used to annotate a database of scientific articles about development, and that a computational data structure has been created such that the terms of the ontology are associated with the names of all articles that deal with the corresponding topics. In order to perform a search for all articles that deal with *forebrain development*, a computer algorithm could be developed to first find the vertex representing *forebrain development* in the graph and then return the names of all of the articles associated with that vertex. In order to perform a search for all articles that deal with *brain development*, the algorithm would first identify the vertex representing *brain development* in the graph, and then use a graph traversal algorithm such as depth-first search to visit all of the descendents of this term in the graph, and to report all articles associated with either *brain development* or any of its descendent nodes. Because the descendent nodes represent subclasses of brain development, this search procedure will return all articles associated with brain development or any subclass or part of brain development. This algorithm implicitly makes use of subsumption inference that will be explained in more detail in Chapter 5.

3.4 Bayesian Networks

Bayesian networks are a generalization of the Bayes' theorem to more than two random variables. Bayesian networks have very numerous applications in bioinformatics including the analysis of DNA sequences, haplotype inference, pedigree analysis, and inference of genetic network structures. In Chapter 9 we will examine an algorithm that uses Bayesian network analysis to search for an optimal combination of Gene Ontology terms (a *model*) to explain an observed pattern of differential gene expression (the observed *data*). Thus, a major use of Bayesian networks in bioinformatics is to infer the most likely model to explain data, as described above for Bayes' theorem. In fact, Bayesian networks involve interpreting the terms of Equation (3.12) as probabilities of models with multiple variables rather than single variable models such as a variable representing the presence or absence of lung cancer.

Definition 3.2 *A Bayesian Network B is a pair (G, P), in which G is a directed acyclic graph. The vertices of G represent a set of random variables $X = \{X_1, \ldots, X_n\}$, while the edges assert statistical dependency relations between the variables. The set $P = \{P(X_1|pa(X_1)), \ldots, P(X_n|pa(X_n))\}$ defines local probability distributions (LPDs) on the variables.*

Bayesian networks thus can be seen as a mixture of graph theory and probability theory. In addition to direct dependence relations, the DAG of a Bayesian network also encodes independence relations following the Markov condition, which states that a variable given its parents doesn't depend on any other non-descendant. The joint probability distribution thus can be calculated as:

$$P(X_1, \ldots, X_n) = \prod_{i=1}^{n} P(X_i|pa(X_i)).$$

Although any LPD can be used for BN analysis, two are extensively used in practice: the multinomial distribution (MD) for discrete variables and the normal (Gaussian) distribution (GD) for continuous variables.

The MD for a variable X_i with m discrete states is a function of all members of the variable's family which maps all possible configurations to a probability value between 0 and 1 such that for every parent configuration π_j

$$\sum_{i=1}^{m} P(X_i|pa(X_i) = \pi_j) = 1.$$

Usually, the MD is given as a conditional probability table (CPT).

For the GD, the distribution for each variable follows a normal distribution whose mean depends linearly on the configuration of the parents:

$$P(X_i|pa(X_i)) = N(X_i, \mu_i + \sum_{j \in pa(X_i)} b_{ij}(X_j - \mu_j), \sigma_i^2),$$

where b_{ij} defines the strength of the influence of variable X_j on X_i. Note that $b_{ij} \neq 0$, otherwise one would not include X_j in the parent set of X_i. Note that while non-linear relationships can be modeled using the MD, the fact that the mean of the GD is a linear function of the states of the parents means that non-linear relationships cannot be modeled with the GD. Also note that a BN is not required to have either discrete or continuous nodes. Instead one can mix nodes by defining different types of LPDs for the nodes.

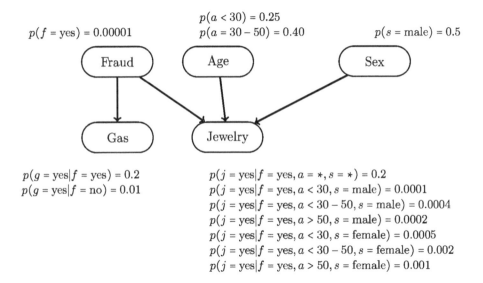

FIGURE 3.12: A Bayesian Network. This network is redrawn from a figure in the excellent tutorial of David Heckerman on Bayesian networks [111]. The network is intended to help detect credit card fraud. Edges are drawn from cause to effect. An asterisk is shorthand for "any state." We are interested in the probability that the current purchase is fraudulent, which is represented by the Fraud vertex. Other relevant variables represented in this network are whether there was a gas purchase within the last 24 hours (Gas), whether there was a jewelry purchase within the last 24 hours (Jewelry), and the age and sex of the card holder. The structure of the Bayesian network entails the following conditional independencies: $P(a|f) = p(a)$, $P(s|f,a) = P(s)$, $P(g|f,a,s) = P(g|f)$, and $P(j|f,a,g,s) = P(j|f,a,s)$.

Figure 3.12 shows an example Bayesian network that encodes the probabilities for credit card fraud. The goal of such a system would be to help detect fraud based on observations of other variables. In our example, we might want to calculate the probability of fraud given observations of the other variables:

$$P(f|a,s,g,j) = \frac{P(f,a,g,s,j)}{P(a,g,s,j)} = \frac{P(f,a,g,s,j)}{\sum_{f'} P(f',a,g,s,j)}.$$

Note that the sum over f' runs over the states **fraudulent** and **not fraudulent**. Note that by the definition of joint and conditional probability, we have that

$$P(f,a,g,s,j) = P(f)P(a|f)P(s|a,f)P(g|f,a,s)P(j|f,a,g,s)$$

Because of the conditional independence relationships encoded in the graph (Figure 3.12), this expression can be simplified as follows:

$$P(f,a,g,s,j) = P(f)P(a)P(s)P(g|f)P(j|f,a,s)$$

We simplify the calculations by using the conditional independencies encoded within the Bayesian network.

$$
\begin{aligned}
P(f|a,s,g,j) &= \frac{P(f)P(a)P(s)P(g|f)P(j|f,a,s)}{\sum_{f'} P(f')P(a)P(s)P(g|f')P(j|f',a,s)} \\
&= \frac{P(f)P(g|f)P(j|f,a,s)}{\sum_{f'} P(f')P(g|f')P(j|f',a,s)}
\end{aligned}
$$

This has been a very concise introduction to the rich and fascinating topic of Bayesian networks. The information should be enough for readers who were previously unfamiliar with the topic to understand the material presented in Chapters 9 and 11, in which Bayesian networks are used to perform analyses in ontologies. Bayesian networks have numerous other applications in bioinformatics, and tips for further reading are given below.

3.5 Exercises and Further Reading

Principles of Statistics by M.G. Bulmer [43] is a highly readable introduction to the mathematical background of statistics. *Statistics and Data with R*, by Y. Cohen and J.Y. Cohen [53] provides an excellent overview of data analysis and statistical evaluation in R. *Learning Bayesian Networks* by Richard Neapolitan [182] provides a comprehensive introduction to many aspects of Bayesian networks. The previously mentioned tutorial on Bayesian networks by David Heckerman is also highly recommended [111].

Exercises

Many of the exercises in this book will ask the reader to use R, a freely available programming language that is designed for statistical computing and data visualization. R is also the basis for Bioconductor, which is a widely used set of tools for the analysis of high-throughput genomic data that comprise many important bioinformatic algorithms [88]. For readers who are not familiar with R, we have provided an introduction to the features of R that will be needed to do the exercises as well as pointers to further sources of information in Appendix A.

3.1 Calculate the p-value for the experiment shown in Figure 3.3 using the binomial distribution. That is, we observe that 30 genes in a study set of 250 genes are annotated to a Gene Ontology term t, whereas 600 of 10,000 terms in the population are. Write an R script to calculate the p-value according to Equation (3.5). Use the R command dbinom(k,n,pr), where k is the number of success, n is the number of trials, and pr is the probability of success of a single trial. Check your answer using the R command pbinom(k,n,pr), which represents the cumulative distribution function (cdf) of the binomial distribution, that is, $p(K \leq k)$, the probability that the random variable K takes on a value less than or equal to k.

3.2 Calculate the probability of drawing 20 blue balls from an urn containing 30 blue balls and 40 red balls if a total of 35 balls are drawn. Use the R command choose(n,k) to calculate $\binom{n}{k}$ and calculate the probability using Equation (3.7). Check your answer using the R command dhyper(x,m,n,k), where x is the number of blue balls drawn, m is the total number of blue balls in the urn, n is the total number of red balls, and k is the number of balls drawn from the urn.

3.3 Consider the following R commands. The second line generates a matrix with 200 rows of 5 numbers each drawn at random from a normal distribution with mean 0 and standard deviation 1. The third line uses the R command apply to apply a t test to each row (to test whether the mean value is significantly different from zero) and store the result in the vector P. The fourth line again uses apply to apply a Bonferroni correction to the p-values.

```
N<-200
dat<-matrix(rnorm(5*N),nrow=N)
P<-apply(dat,1,function(x) t.test(x)$p.value)
P.corr <-apply(matrix(P),1,
        function(x,factor) min(1,x*factor),factor=N)
```

Now use the commands sum(P<= 0.05) and sum(P.corr<= 0.05) to count the number of entries for which the nominal p-values are significant and for which the corrected P-values are significant. Repeat the experiment with other values of N. Explain the results in light of the descriptions of the Bonferroni procedure.

3.4 The higher power of the Holm procedure results from the fact that Holm may reject some null hypotheses not rejected by Bonferroni. This is because $\alpha/(n-i+1)$ is much larger than α/n for large values of i. If $\alpha/(n-i+1) \geq P_i > \alpha/n$, the ith null hypothesis will be rejected by Holm but not by Bonferroni [120]. Use R to perform a multiple testing correction with the Bonferroni and the Holm procedures for the following nominal p-values. 0.0001, 0.0002, 0.0012, 0.0022, 0.0044, 0.0084, 0.016, 0.0273, 0.033, 0.039, 0.047, 0.049. How many p-values are significant after correction for each procedure?

3.5 Calculate the ratio of the probabilities of lung cancer being present or absent in the scenario described in Section 3.2 on page 57 using Equation (3.13).

3.6 Calculate the probability of fraud in the Bayesian network of Figure 3.12 if the credit card holder is 32 years old, male, and has purchased gas but not jewelry in the last 24 hours. Note that not all of the probabilities are shown explicitly in the figure, but they can all be inferred from the values that are shown. For instance, the probability that the credit card holder is female can be inferred from the fact that $p(s = \text{male}) = 0.5$.

Chapter 4

Ontology Languages

Two ontology languages, OBO and OWL, account for most of the predominant bio-ontologies. Good open-source software is available for creating and using ontologies in these languages. As we will see, OWL was developed by the W3C as a general tool for the Semantic Web, and OWL therefore represents an extremely powerful but also relatively complex language that can be used for a number of purposes including the development of bio-ontologies. OBO was developed by the Gene Ontology (GO) consortium for the GO and has been widely adopted by numerous other bio-ontologies. For many, but not all, concepts of importance in the biomedical domain, equivalent ontologies can be constructed with either OBO or OWL.

In the following sections, we present the most important aspects of both ontology languages and present several examples of how a set of concepts and relations can be expressed in both languages.

4.1 OBO

The OBO ontology language models a subset of the semantics available in OWL 2. Using the notation of description logics, most ontologies that are expressed using OBO can be characterized by $\mathcal{EL}++$. Thus, constructors for intersection and existential quantification are used. The two $++$ signs indicates further constructors, mostly affecting the RBox. For instance, inverse roles and composition rules play an important role in the OBO semantics.

An OBO document is presented in a relatively simple and human-readable flat file format. It begins with a header and contains an arbitrary number of stanzas. The header is an unlabeled section at the beginning of the OBO file. Each line of the header contains one tag-label pair with a name, a colon, and a value. The header ends when the first stanza is encountered. The following example header was excerpted from the GO OBO file:

```
format-version: 1.2
date: 29:07:2010 12:36
auto-generated-by: OBO-Edit 2.0
subsetdef: goslim_candida "Candida GO slim"
```

```
subsetdef: goslim_generic "Generic GO slim"
default-namespace: gene_ontology
```

We see that the `format-version` (tag) is 1.2 (value), and so on. The `auto-generated-by` tag shows that the OBO file was generated using the OBO-Edit program, which is used by the GO consortium and many other bio-ontology groups to create and manage OBO ontologies [64]. The `subsetdef` tag is used to provide a label and a description of a term subset, which will be used to indicate the terms (stanzas) that belong to the subset. In the GO, it has proven useful to collect sets of top-level terms that describe important general biological processes, which have come to be known as "GO Slims" (see also Chapter 5). To indicate that a specific GO term is a member of a GO Slim, a line must be included in the corresponding stanza. For instance, to indicate that the GO term *cell wall organization* belongs to the *Candida albicans* (a fungus model organism) GO Slim, the following line is included in the stanza for that GO term.

```
subset: goslim_candida
```

4.1.1 OBO Stanzas

OBO stanzas can refer to universal types [Term], type definitions [Typedef], or instances [Instances]. Each stanza begins with one of these keywords and then contains a number of key-value lines.[1] By convention, stanzas are separated by an empty line, but this is not required and empty lines are ignored by OBO parsers. Anything that follows an exclamation mark (!) up to the end of a line is interpreted as a comment. The following stanza is one of nearly 30,000 stanzas in the GO OBO file:

```
[Term]
id: GO:0000031
name: mannosylphosphate transferase activity
namespace: molecular_function
def: "Catalysis of the transfer of a mannosylphosphate
      group from one compound to another." [GOC:jl]
is_a: GO:0016740 ! transferase activity
```

The stanza begins with the id of the term, which is comparable to the accession number of a nucleotide sequence in the GenBank database, in that the id is a unique identifier for the entity being referred to by the term. Each term also has a name, which is a concise human-readable description of the term. All other tag-value entries are optional, but some are very commonly encountered. In the GO, terms are classified into one of three subontologies,

---

[1] Note that some key-value lines are printed over multiple lines in this book, but only a single line is allowed in the OBO file.

molecular function, biological process, and cellular component (see Chapter 5), and this is indicated by the namespace tag. The def entry provides a concise definition of the term, whereby a reference for the definition must be provided in square brackets (in this case, the GO curator *jl* is shown as the reference for the definition).

There are several ways of indicating the semantic links of the ontology. The is_a (subclass) links are shown using the keyword is_a. In the above example, we see that the GO term *mannosylphosphate transferase activity* is a subclass of the GO term GO:0016740. The comment following the exclamation mark provides the name of the term (*transferase activity*) to make it easier for humans to understand the stanza, but this information is disregarded by OBO parsers.

The following GO stanza shows some other important features, including a synonym representing the Enzyme Commission (EC) entry 3.1.3.21 and external references (xref) to the EC entry as well as to an entry of the MetaCyc database.

```
[Term]
id: GO:0000121
name: glycerol-1-phosphatase activity
namespace: molecular_function
def: "Catalysis of the reaction: glycerol 1-phosphate
     + H2O = glycerol + phosphate." [EC:3.1.3.21]
synonym: "alpha-glycerol phosphatase activity"
         EXACT [EC:3.1.3.21]
xref: EC:3.1.3.21
xref: MetaCyc:GLYCEROL-1-PHOSPHATASE-RXN
is_a: GO:0016791 ! phosphatase activity
```

Terms may have any number of is_a (subclass) relations. Because of the importance of this kind of relation in ontologies, the keyword is_a is used for them. For all other kinds of relations, a line consisting of the keyword relation and the *relation type name* and *target term id* is used to declare the relation. The relation type name must be a relation that is defined in a typedef stanza; for instance, the part of relation is defined in the GO OBO file by means of the following stanza:

```
[Typedef]
id: part_of
name: part_of
xref: OBO_REL:part_of
is_transitive: true
```

There is a cross-reference to the OBO Relation Ontology (see Chapter 6), and the relation is declared to be transitive (i.e., if A part_of B and B part_of C, then we can infer that A part_of C). Any term of the OBO file can use the part-of relation (or any other relation) defined in such a typedef stanza. For instance, the *alpha-1,6-mannosyltransferase complex* is declared to be `part_of` of the *Golgi cis cisterna* in the following stanza.

```
[Term]
id: GO:0000136
name: alpha-1,6-mannosyltransferase complex
namespace: cellular_component
def: "A large, multiprotein complex..." [GOC:mcc]
is_a: GO:0031501 ! mannosyltransferase complex
is_a: GO:0044431 ! Golgi apparatus part
relationship: part_of GO:0000137 ! Golgi cis cisterna
```

Additionally, the term is defined to be a subclass of *mannosyltransferase complex* as well as of *Golgi apparatus part* using the `is_a` keyword. Thus, the term *alpha-1,6-mannosyltransferase complex* has two `is_a` parents and one `part_of` parent.

As we will see in Chapter 6, the semantics of relations needs to be further clarified for term-level relations as no quantification is involved in the definition of a term. Most of the relations in OBO follow the ALL-some pattern, which can be expressed by subclass relations and existential quantification.

4.1.2   Intersections: Computable Definitions

An important feature of OBO ontologies is the option of defining a term using other terms from the same or different ontologies. We will see in Chapter 12 how this feature can be used to perform inference (computer-based reasoning) to check ontologies for consistency and to discover relations between terms that had not been identified by manual curation. The definitions are constructed using the keyword `intersection_of`. Two or more `intersection_of` lines in a stanza indicate that the term is equivalent to the intersection of the terms referred to in the `intersection_of` lines. For instance, the GO term *cytoplasmic ubiquitin ligase complex* is defined to be the intersection of *ubiquitin ligase complex* and the class of all things that are `part_of` the *cytoplasm* by the following stanza:

```
[Term]
id: GO:0000153
name: cytoplasmic ubiquitin ligase complex
namespace: cellular_component
def: "A ubiquitin ligase complex found in the cytoplasm."
    [GOC:mah]
is_a: GO:0000151 ! ubiquitin ligase complex
is_a: GO:0044444 ! cytoplasmic part
intersection_of: GO:0000151 ! ubiquitin ligase complex
intersection_of: part_of GO:0005737 ! cytoplasm
```

This kind of definition, which is also referred to as a *logical definition*, captures the same information for computers as the def tag is intended to capture for humans: they help to "understand" the meaning of the term. Thus, logical definitions can be seen as computer-readable translations of the human readable definition or vice versa.

Although with the term *cytoplasmic ubiquitin ligase complex* it would be reasonably easy to use text mining techniques to conclude that the term is equivalent to a *ubiquitin ligase complex* that is cytoplasmic, in general text mining techniques cannot perform this kind of decomposition and definition of ontology terms in a reliable fashion. This kind of definition is sometimes also called a *cross-product definition*, and is comparable to definitions constructed in OWL using the intersectionOf keyword (see Appendix D).

Cross-product definitions usually should follow the genus-differentia pattern [177]. Thus, the cross-product definition for *cytoplasmic ubiquitin ligase complex* states that the term is equivalent to a *ubiquitin ligase complex* (genus) that is part of the *cytoplasm* (differentia). These are necessary and sufficient conditions for being a *cytoplasmic ubiquitin ligase complex*.

4.2 OWL and the Semantic Web

The second ontology language with a wide following within the bio-ontology community is OWL. In contrast to OBO, which was specifically designed to fit the perceived needs of ontologies in biological and medical research communities, OWL was developed by the W3C, the international standards organization for the Internet, as a language for defining structured, Web-based ontologies which enable richer integration and interoperability of data from multiple sources on the Web. OWL is thus part of a major effort to create a smarter infrastructure for the Internet, the so-called *Semantic Web*, that will improve the availability and consistency of data on the Web.

The Semantic Web is an alternative to the current World Wide Web, which primarily presents data as HTML pages that are easily interpretable to human eyes. That is, the current structure of the Web is mainly made of a distributed network of Web pages that use hyperlinks (uniform resource locators or URLs) to refer to one another. In contrast, the Semantic Web is moving towards Web sites that function as collections of semantically defined data, which can be read by other servers or be used to dynamically generate a Web page intended for human consumption.

The main idea of the Semantic Web is to connect a network of Web sites at the level of data rather than at the level of the presentation of data as HTML pages. OWL has attracted interest not only as an ontology language for the Semantic Web, but also as a tool for performing inference in one or several ontologies. A number of important bio-ontologies are written in OWL or have OWL versions. In this book, we will not provide a detailed explanation of the Semantic Web; a number of other excellent books are available on this topic, some of which are listed at the end of this chapter. This section will provide an intuitive introduction to RDF, RDFS, and OWL, which will provide a starting point for understanding the role of computational inference in the chapters of Part III. In Part IV we will return to the languages of the Semantic Web and will describe in detail how inference is performed in RDF, RDFS, and OWL. The languages of the Semantic Web can be written down ("serialized") in a number of different formats, which can be confusing for beginners. Appendix C provides a detailed description of the syntax of RDF and RDFS, and Appendix D provides a detailed description of the syntax of OWL.

The OWL language is built upon RDF and RDFS, which we will describe first. RDF is intended to be used to express propositions using precise formal vocabularies, particularly those specified using RDFS, for access and use over the World Wide Web. RDF and RDFS provide a basic foundation for more advanced assertional languages such as OWL [108].

4.2.1   Resource Description Framework

The data model used by the Semantic Web infrastructure and OWL is called the *Resource Description Framework* or RDF. RDF was designed by the W3C as a language for representing information about resources in the World Wide Web in a way that would provide a common framework for computer processing. The basic building block of RDF is the *triple*: a statement about a resource consisting of a subject, predicate, and object which together represent some piece of information. In RDF, the *subject* identifies the thing the statement is about, the *predicate* identifies the property or characteristic of the subject that the statement specifies, and the object identifies the value of that property [160]. Consider the following RDF triplet:

`http://www.example.org/  hasCreator John Smith`

Here, the subject is the Web page at `http://www.example.org/`, the predicate is the property (i.e., "hasCreator") the RDF statement is describing, and the object is the value of that property ("John Smith").

RDF is designed to make computer-processable statements in the context of the Semantic Web. Therefore, it requires a system of machine-processable identifiers for identifying resources that are *unique* across the entire Web as well as a computer language for representing and processing these statements. The existing Web architecture provides one form of identifier, the Uniform Resource Locator (URL). URLs include HTTP and FTP Web addresses and are used to locate resources, by identification of the resource location as well as a protocol for retrieving the resource. For example, for the URL `http://www.example.org`, the resource location is www.example.org, and the protocol for retrieving the resource is http. URLs are one of two specific classes of *Uniform Resource Identifier* (URI), which is a string of characters used to identify a name or a resource on the Internet (see also Appendix C, page 403). The other class of URI is a *Uniform Resource Name* (URN), which represents a name for a resource that is unique and global throughout the Web. For instance, the URI urn:isbn:0691141339 provides a unique identifier (the International Standard Book Number or ISBN) for Julian Havel's book entitled "Gamma: Exploring Euler's Constant." However, although this URI is unique, it cannot be used to retrieve the book from the Web. The W3C has published a detailed clarification of the differences between URIs, URLs, and URNs [256].

RDF itself is an abstract data model. There are several ways of representing RDF as a computer file, a process which is referred to as *serialization*. Many RDF files are written in a form of XML called RDF/XML. The following example is modified from an RDF file created by the Uniprot consortium [255]; the original file contains descriptions of the each entry of the Enzyme Commission (EC) numerical classification scheme for enzymes. The example is a complete RDF file containing just a single such entry (1.14.11.2, which corresponds to Procollagen-proline dioxygenase). Line numbers are shown for clarity and are not a part of the actual file.

```
1.  <?xml version='1.0' encoding='UTF-8'?>
2.  <rdf:RDF xmlns="http://purl.uniprot.org/core/"
3.          xmlns:rdf="http://www.w3.org/1999/02/22-rdf-syntax-ns#"
4.  >
5.
6.    <rdf:Description
7.        rdf:about="http://purl.uniprot.org/enzyme/1.14.11.2">
8.      <rdf:type rdf:resource="http://purl.uniprot.org/core/Enzyme"/>
9.      <name>Procollagen-proline dioxygenase</name>
10.     <name>Procollagen-proline 4-dioxygenase</name>
11.     <name>Prolyl 4-hydroxylase</name>
12.     <activity>L-proline-[procollagen] + 2-oxoglutarate + O(2)
13.            = trans-4-hydroxy-L-proline-[procollagen] +
14.               succinate + CO(2).
15.     </activity>
16.     <cofactor>Iron</cofactor>
17.     <cofactor>L-ascorbic acid</cofactor>
18.   </rdf:Description>
19. </rdf:RDF>
```

Line 1 is the XML declaration, which indicates that the following content is XML version 1.0. Line 2 opens an rdf:RDF element, which indicates that the XML content from line 2 up to the closing tag at line 19 represents RDF. Following the rdf:RDF on line 2 is an XML namespace declaration, represented as an xmlns attribute of the rdf:RDF start-tag. We see that the default name space (xmlns) for this file is http://purl.uniprot.org/core/ of the UniProt consortium. In addition, the name space of the RDF specification of the W3C is indicated. The RDF file located at http://www.w3.org/1999/02/22-rdf-syntax-

Subject	Predicate	Object
:1.14.11.2	rdf:type	:Enzyme
:1.14.11.2	:name	Procollagen-proline dioxygenase
:1.14.11.2	:name	Procollagen-proline 4-dioxygenase
:1.14.11.2	:name	Prolyl 4-hydroxylase
:1.14.11.2	:activity	L-proline-[procollagen] + 2-oxoglutarate + O(2) = trans-4-hydroxy-L-proline-[procollagen] + succinate + CO(2).
:1.14.11.2	:cofactor	Iron
:1.14.11.2	:cofactor	L-ascorbic acid

TABLE 4.1: RDF triples for procollagen-proline dioxygenase. Each of the predicates indicates a different property of the enzyme 1.14.11.2. The triples with the predicate :name should be read as ":1.14.11.2 has the name Procollagen-proline dioxygenase" etc., and the triples with the predicate :cofactor should be read as ":1.14.11.2 has the cofactor Iron" etc.

ns# contains definitions for the RDF vocabulary defined in the RDF namespace. The namespace declaration states that the prefix *rdf* is equivalent to the namespace URI http://www.w3.org/1999/02/22-rdf-syntax-ns#, and allows the prefix rdf to be used in place of the complete URL in the remainder of the RDF file.

Lines 6–18 contain the specific information of the RDF file. RDF statements are descriptions of resources, and line 7 indicates that the *subject* of the RDF statements contained within the Description element is http://purl.uniprot.org/enzyme/1.14.11.2. In the following text, we will abbreviate this as :1.14.11.2. This is done using the rdf:about attribute to specify the URI of the subject resource.

Line 8 contains the first of several triplets contained in the Description element. For each triplet, the subject is :1.14.11.2. The predicate rdf:type means that the subject of the triple is an *instance of* the class represented by the object of the triple. Thus, :1.14.11.2 is an instance of Enzyme. In all, there are seven triples about :1.14.11.2 in this fragment. They are summarized in Table 4.1.

RDF models statements as nodes and edges in a graph. For any given triple, there is a node for the subject, a node for the object, and an edge for the predicate, which is directed from the subject node to the object node. For instance, the triple :1.14.11.2 rdf:type :Enzyme can be represented as the graph shown in Figure 4.1.

Another commonly used serialization format for RDF that is somewhat easier to read in printed text is *Notation 3 RDF* (N3 for short). N3 is more compact than XML/RDF and is probably easier to understand for most people. N3 makes use of qnames ("qualified names"), which essentially represent abbreviations of full-length URIs. Qualified names can comprise

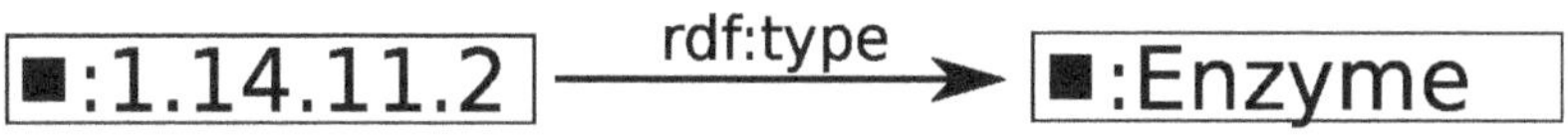

FIGURE 4.1: **Graphical Representation of an RDF Triple.** The subject, :1.14.11.2, and the object, :Enzyme, are shown as nodes, while the predicate is shown as an edge leading from the subject to the object.

a prefix and a local part; for instance in the above RDF document the statement `xmlns:rdf="http://www.w3.org/1999/02/22-rdf-syntax-ns#"` declares `rdf` to be equivalent to (i.e., an abbreviation for):

`http://www.w3.org/1999/02/22-rdf-syntax-ns\#`

With this declaration, `rdf:type` is a valid qname, whereby `rdf` is the prefix and `type` is the local part. Thus, `rdf:type` is equivalent to:

`http://www.w3.org/1999/02/22-rdf-syntax-ns\#type`

   N3 uses a binding between local qnames and global URIs by beginning with a preamble which defines these bindings. For instance, the following line in N3 defines the above-described mapping for rdf:

`@prefix rdf: <http://www.w3.org/1999/02/22-rdf-syntax-ns#> .`

If the prefix of the qname corresponds to the default namespace, this can be indicated by a colon. The qnames can now be used to express the triples by listing the three members of the subject/predicate/object triplet using the qnames instead of the full URIs. Because of the mapping at the beginning of the N3 document, the triples are still uniquely defined. For instance, the first triple of the above RDF/XML document can be expressed in N3 by listing the three resources in the order subject, predicate, and object using the qname abbreviations. The triple is followed by a space and a period.

`:1.14.11.2 rdf:type :Enzyme .`

Since rdf:type is very commonly used in RDF, N3 allows the abbreviation "a" to be used for it. Therefore, the following triplet is equivalent to the previous one:

`:1.14.11.2 a :Enzyme .`

This triplet can thus be read as *1.14.11.2 is an instance of Enzyme*. In case multiple triples share the same subject, only the first triple is shown with all three resources. This triple is not terminated with a period but with a semicolon. Subsequent triples that share the same subject only show the predicate and object, and also are terminated by a semicolon, except for the last triplet sharing the same subject, which is terminated by a period. For instance

```
:1.14.11.2 rdf:type    :Enzyme ;
            :name       "Procollagen-proline dioxygenase" ;
            :cofactor   "Iron" .
```

Finally, if there are triplets that share both the subject and the predicate but have different objects, the triplets can be written together on the same line by separating the different objects with a comma. In the above RDF example, there were three alternative names for the enzyme 1.14.11.2, each of which was expressed in a separate triple. This can be expressed compactly in N3 as follows:

```
:1.14.11.2 :name "Procollagen-proline dioxygenase",
                 "Procollagen-proline 4-dioxygenase",
                 "Prolyl 4-hydroxylase" .
```

The entire RDF/XML document shown above can be expressed in N3 as follows:

```
@prefix : <http://purl.uniprot.org/core/> .
@prefix rdf: <http://www.w3.org/1999/02/22-rdf-syntax-ns#> .

:1.14.11.2 a      :Enzyme ;
            :name "Procollagen-proline dioxygenase",
                  "Procollagen-proline 4-dioxygenase",
                  "Prolyl 4-hydroxylase" ;
        :activity "L-proline-[procollagen] + 2-oxoglutarate
                  + O(2) = trans-4-hydroxy-L-proline-[procollagen]
                  + succinate + CO(2)" ;
        :cofactor "Iron", "L-ascorbic acid" .
```

This is a collection of RDF statements that correspond to a labeled directed graph (Figure 4.2).

RDF provides several syntactic constructs for expressing more complex statements that go beyond what can be expressed by the simple subject/predicate/object form of the triples presented up to now. Suppose we want to represent something that does not have a URI, for instance, a thing that we know to exist but whose exact identity we do not know. This can be done using an RDF construct called a *blank node* (*bnode* for short). We will see that bnodes are used in OWL to represent complex data such as anonymous classes such as restrictions, unions, or intersections of other classes, but for now we will explain a simpler example in RDF. Let us say we want to express the statement *Someone called Douglas wrote a book entitled "Hitchhiker's Guide to the Galaxy."* Clearly, it is impossible to provide a URI for the concept *Someone called Douglas*. Instead, we use a bnode to express the statement about this "someone." There are several ways of expressing this using the N3 notation [31]; here, we will just show the so-called square bracket blank node syntax. In our example,

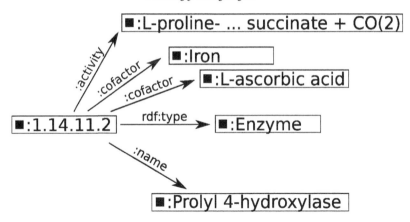

FIGURE 4.2: **Graphical Representation of a Collection of RDF Triples.** Five of the triples of the example RDF document are shown in this graph.

```
[ :firstname "Douglas"]
```

This statement means X, such that there exists some X (i.e., somebody) such that X has a first name Douglas. It is convention to leave a space after the opening square bracket as if to indicate the presence of the "blank" subject of these triples. We can now use this statement as an rdf:subject.

```
[ :firstname "Douglas"] dc:wrote
        [ dc:title "Hitchhiker's Guide to the Galaxy"] .
```

In this statement, the rdf:subject is [:firstname "Douglas"], the rdf:predicate is dc:wrote,[2] and the rdf:object is another bnode, [dc:title "Hitchhiker's Guide to the Galaxy"]. This bnode means Y, such that there exists some Y such that Y has a title "Hitchhiker's Guide to the Galaxy." In English, the entire statement can be given as: *Some person who has a first name Douglas wrote a thing entitled "Hitchhiker's Guide to the Galaxy."* Let's say we then do a search in our favorite Web search engine and find a Wikipedia article saying that Douglas Adams wrote a book entitled "Hitchhiker's Guide to the Galaxy." Furthermore, let's say we are skeptical about this statement and want to use RDF to express the fact that Wikipedia *claims* that Douglas Adams wrote a book entitled "Hitchhiker's Guide to the Galaxy." This can be modeled by means of *reification*,[3] in which a triple is

---

[2]The Dublin Core Metadata Initiative (dc) namespace provides interoperable metadata standards that are broadly used in RDF models and include definitions for classes such as *creator*, *dateCopyrighted*, and *language*.

[3]The word *reification* is derived from the Latin word *res* (thing) and the suffix *-fication* which in turn is derived from the word *facere* (to make) and thus means literally, to make into a thing, to materialize. In the current context, reification describes the fact that we are "making" a triple into a "thing" about which other RDF statements can be made.

assigned an identifier and treated as a resource about which additional statements can be made. RDF represents a reified statement as four statements
using RDF properties and objects: the triplet (S, P, O), reified by resource R,
is represented by:

- R rdf:type rdf:Statement

- R rdf:subject S

- R rdf:predicate P

- R rdf:object O

Assuming that the appropriate qname prefixes have been defined, we can
express our statement about Wikipedia's claim as follows:

```
R rdf:type rdf:Statement
R rdf:subject "Douglas Adams"
R rdf:predicate dc:wrote
R rdf:object "Hitchhiker's Guide to the Galaxy"
web:wikipedia q:says R
```

RDF is the basic framework on which RDFS, OWL, and the Semantic
Web are based. RDF data are often stored in so-called RDF stores, which are
databases designed to work with RDF triples. RDF stores are able to merge
RDF triples from various sources, which is important in Semantic Web applications that combine data from various Web sites. RDF stores are accessed by
RDF query engines, often by means of the query language SPARQL (which
will be discussed in Chapter 16). Thus, SPARQL is to RDF stores that what
SQL is to relational databases. The fact that all data in an RDF store are represented as triples means that querying an RDF database can be substantially
simpler than querying a relational database.

The entire RDF standard is relatively small. Appendix C provides a detailed description of the keywords and syntax of RDF.

4.2.2 RDF Schema

A schema is a formal definition of the syntax of a language, and a schema
language is a language for expressing that definition. In SQL, the schema is
the structure of the database that defines the objects in the database. In the
extended markup language (XML), the Document Type Definition (DTD) is
a schema that defines where particular elements and attributes may appear
in an XML document and what datatypes the contents of the elements may
assume. An XML Schema is a related schema language for XML that is itself
written in XML. Intuitively speaking, schemas provide a framework for interpreting the meaning of data. RDF Schema (RDFS) is intended as a framework

for interpreting the meaning of data expressed in RDF. RDFS provides additional specifications and keywords that allow many kinds of *inference* to be performed.

Asserted and Inferred Triples

The Oxford English Dictionary defines inference as *drawing a conclusion from known or assumed facts or statements*. That is, inference generates conclusions that go beyond what was originally stated. In our context, each RDF triple represents a statement. The process of inference can be used to generate novel RDF triples using logical rules. The RDF triples present in the original RDF store prior to inference are known as *asserted triples*, and those generated by the inference process are known as *inferred triples*.

RDFS builds on the syntax of RDF and extends its vocabulary. The core vocabulary of RDFS is associated with the prefix `rdfs`.[4] By themselves, RDF properties (as identified by the `rdf:predicate` of the triple) indicate the attributes of the resources represented by the subject of the triple. RDF properties can also represent relationships between resources. However, RDF provides no mechanisms for describing these properties or for describing the relationships between these properties and other resources. That is the role of RDFS [39]. RDFS vocabulary descriptions are written in RDF and define classes and properties that may be used for defining custom classes (sets), and of analyzing how classes and individuals are related to one another (Table 4.2). That is, RDFS supports the concepts of resources and properties like RDF, but additionally supports subclassing and superclassing, instantiation of instances of classes, inheritance, and domain/range property restrictions.

Subclasses and Type Propagation

Probably the most important concept for inference in RDFS is that of the class. Classes provide an abstraction mechanism for grouping resources with similar characteristics.

Imagine we have defined an ontology and have provided definitions at `http://www.my-example.org`, which we assign to the qname prefix x as described above. It is already possible in RDF to express the statement *Mickey is a mouse*:

```
x:Mickey rdf:type x:Mouse
```

However, there is no mechanism built into RDF with which we could express the statement that *mouse* is a subclass of *rodent*. The RDFS keywords `rdfs:Class` and `rdfs:subClassOf` can be used for this purpose. We first

---

[4]The complete specification is found at the URI-Reference `http://www.w3.org/2000/01/rdf-schema#`.

RDFS keyword	Explanation
rdfs:Resource	All things described by RDFS are called resources, and are instances of this class.
rdfs:Class	The class of resources that are RDF classes. Particular classes are instances of rdfs:Class.
rdfs:Literal	The class of literal values such as strings and integers.
rdfs:Datatype	The class of RDF datatypes, which are the same as the datatypes defined in XML Schema and include datatypes such as integer, string, and many others.
rdfs:Container	The class of RDF containers which comprises rdf:Bag, rdf:Seq, and rdf:Alt.
rdfs:subClassOf	An RDF property whereby the subject is a subclass of the object.
rdfs:subPropertyOf	An RDF property whereby the subject is a subproperty of the object.
rdfs:domain	An RDF property defining the domain of the subject property.
rdfs:range	An RDF property defining range of the subject property.
rdfs:label	An RDF property providing human-readable name for the subject.
rdfs:comment	An RDF property providing a description of the subject resource.
rdfs:seeAlso	An RDF property providing further information about the subject resource.
rdfs:isDefinedBy	An RDF property providing the definition of the subject resource.

TABLE 4.2: RDFS keywords. This table provides brief explanations of RDFS keywords. For fuller explanations, the reader is referred to Chapter 13 and Appendix C.

declare that x:Mouse and x:Rodent are RDFS classes, and then state that x:Mouse is a subclass of x:Rodent.

```
@prefix x: <http://www.my-example.org> .
x:Mouse rdf:type rdfs:Class .
x:Rodent rdf:type rdfs:Class .
x:Mouse rdfs:subClassOf x:Rodent .
```

One of the RDFS inference rules states that if there are asserted triples of the form

```
X rdfs:subClassOf Y .
b rdf:type X .
```

---

is a ain't always *is a*

It is important to distinguish between two different usages of the phrase "is a" in the English language. When we say *Mickey is a mouse*, we mean that a particular individual called *Mickey* is an instance of the class Mouse. Thus, we have made a statement that relates a particular instance to a class. On the other hand, if we say *A mouse is a rodent*, we mean that the class "mouse" is a subclass of the class "rodent," which implies that every being that is a mouse is also a rodent. The second statement thus relates one class to another class. As we shall see, RDFS has a different syntax for expressing each relation.

---

then the following triple can be inferred:

```
b rdf:type Y .
```

Thus, we can infer that Mickey is an instance of the class *rodent*!

Another inference rule states that the subclass relation is transitive. That is, if there are asserted triples about classes X, Y, and Z:

```
X rdfs:subClassOf Y .
Y rdfs:subClassOf Z .
```

then the following triple can be inferred:

```
X rdfs:subClassOf Z .
```

Thus, if we assert the following triples,

```
x:Mouse rdf:type rdfs:Class .
x:Rodent rdf:type rdfs:Class .
x:Mammal rdf:type rdfs:Class .
x:Mouse rdfs:subClassOf x:Rodent .
x:Rodent rdfs:subClassOf x:Mammal .
```

then we can infer that mice are mammals:

```
x:Mouse rdfs:subClassOf x:Mammal .
```

Similar inference rules apply to properties. For instance, we can define a simple hierarchy of biochemical regulation with the following RDF statements:

```
@prefix r: <http://www.regulation.org> .
r:regulates rdf:type rdfs:Property .
r:positively_regulates rdf:type rdfs:Property .
r:negatively_regulates rdf:type rdfs:Property .
r:positively_regulates rdfs:subPropertyOf r:regulates .
r:negatively_regulates rdfs:subPropertyOf r:regulates .
```

If proteins A and B have been defined in a namespace `pro`, we can state that A positively regulates B by asserting the following triple:

```
pro:A  r:positively_regulates pro:B .
```

There is an inference rule about RDFS properties according to which if the following triples are asserted:

```
P rdfs:subPropertyOf Q .
X P Y .
```

then the following triple can be inferred:

```
X Q Y .
```

Thus we can infer that

```
pro:A  r:regulates pro:B .
```

Similarly to the situation with classes, the subproperty relation is transitive. Thus, if

```
P rdfs:subPropertyOf Q .
Q rdfs:subPropertyOf R .
```

then

```
P rdfs:subPropertyOf R .
```

Property Domains and Ranges

RDFS does not provide the elaborate mechanisms for defining property restrictions that OWL does, but does provide a syntax for specifying the range and domain of properties. Recall that the *domain* of a function is the set of values for which a function is defined. The *range* of the function is the set of all values the function takes for the values in the domain. Although one might be tempted to think that the primary purpose of `rdfs:domain` and `rdfs:range` is to constrain the values of the subject and object of an RDF statement, much as the declaration of a variable as an integer constrains the variable to take on integer values rather than say string values, this is not the case. Rather, it enables inference. The RDFS statement `P rdfs:domain C` for an `rdf:Property` P and an `rdf:Class` C means that the resource denoted by the subject of a triple whose predicate is P is an instance of the class C. That is, if the following two triples are asserted:

```
P rdfs:domain D .
X P Y .
```

then the following triple can be inferred:

```
X rdf:type D .
```

A similar inference can be made about the range of a property; if the following two triples are asserted:

```
P rdfs:range R .
X P Y .
```

then the following triple can be inferred:

```
Y rdf:type R .
```

The National Cancer Institute (NCI) of the United States is a major agency for cancer research and training. The NCI has developed a large-scale ontology, the *NCI Thesaurus*, that is used to enable storage, retrieval, and analysis of data from a number of fields in scientific and clinical research in cancer [65]. The following highly simplified example based on entries from this ontology exemplifies the kind of inference that can be done using domain and range restrictions in RDFS constructs. We first define some classes and a property with domain and range restrictions in the default namespace using these triples:

```
:drug_affects_protein rdf:type rdfs:Property .
:Drug rdf:type rdfs:Class .
:Protein rdf:type rdfs:Class .
:drug_affects_protein rdfs:domain :Drug .
:drug_affects_protein rdfs:range :Protein .
```

If we now additionally assert the following triple

```
:Nilotinib   :drug_affects_protein :BCR-ABL .
```

then we can infer the following two triples:

```
:Nilotinib rdf:type :Drug .
:BCR-ABL rdf:type :Protein .
```

In fact, Nilotinib is a selective Bcr-Abl tyrosine kinase inhibitor that is used in the treatment of chronic myeloid leukemia. Figure 4.3 displays the graph corresponding to these triples. The inferred triples are shown using dashed lines.

Intersections and Unions in RDFS

Yet other forms of inference in RDFS resemble to a certain extent conclusions about set intersections and set unions. If Z is a subclass of both class X and class Y and b is defined as an instance of Z, then b can be inferred to be an instance of X and an instance of Y. That is, the triples

```
Z rdfs:subClassOf X .
Z rdfs:subClassOf Y .
b rdf:type Z .
```

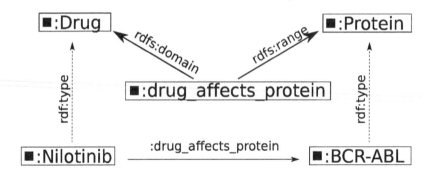

FIGURE 4.3: Example of RDFS Inference in an RDF Graph.
Because of the domain and range restrictions on the `rdf:predicate`
`:drug_affects_protein`, it can be inferred that Nilotinib is a drug and BCR-ABL is a protein.

allow the following two triples to be inferred:

```
b rdf:type X .
b rdf:type Y .
```

For instance, if we state that

```
:Pig rdfs:subClassOf :Mammal .
:Pig rdfs:subClassOf :FarmAnimal .
```

then we can conclude that a particular pig is both a mammal and a farm animal, and thus the class :Pig can be considered to occupy the intersection between the latter two classes. On the other hand, the triples

```
:Cow rdfs:subClassOf :Mammal .
:Cow rdfs:subClassOf :FarmAnimal .
```

do not imply the reverse. That is, even though a cow is a mammal and a farm animal, it is not a pig. The RDFS constructs are not designed to express an exact mathematical relationship between sets such as $A = B \cap C$.

It is also possible to express something similar to set union in RDFS. If we assert the following triples

```
X rdfs:subClassOf Z .
Y rdfs:subClassOf Z .
```

then if either b `rdf:type` X . or b `rdf:type` Y ., it can also be inferred that

```
b rdf:type Z .
```

For instance, if we assert the triples

```
:HairBearingAnimal rdfs:subClassOf :Mammal .
:AnimalWithMammaryGlands rdfs:subClassOf :Mammal .
```

then we can infer that any animal with hair is a mammal, and we can also infer that any animal with mammary glands is a mammal.

More information on RDFS can be found in Appendix C and Chapter 13.

4.2.3 The Web Ontology Language OWL

The acronym OWL stands for **W**eb **O**ntology **L**anguage. Attentive readers will have noted that the most natural acronym for Web Ontology Language would have been WOL, but the acronym OWL was chosen because "owl" is easier to pronounce.[5] In essence, OWL is an ontology language that builds upon RDF and RDFS but has more powerful inference rules and built-in constructs.

OWL can be understood as an extension of RDFS that allows more powerful inference to be performed. We have seen in the last section that inference in RDFS is based on a set of constructs as well as a set of rules about the constructs. To a first approximation, we can say that OWL extends the inference capabilities already present in RDFS by the addition of new constructs and new rules. In this section, we will present many of these constructs and rules and demonstrate them using small examples. In Chapter 14 we will present an in-depth explanation of inference in OWL ontologies.

Any OWL ontology is an RDF graph, that is, a set of RDF triples. Similar to RDF, an OWL ontology can be written in a number of alternative syntactic forms, which can then be stored as a document (file) in the World Wide Web (see Appendices C and D). OWL documents consist of an optional ontology header, and arbitrary numbers of class axioms, property axioms, and facts about individuals expressed using the axioms. The built-in OWL constructs are all defined at http://www.w3.org/2002/07/owl#, which corresponds to the namespace owl.

Defining Classes in OWL

An important aspect of OWL is its ability to defines classes with considerably more expressivity than RDFS. This of course allows us to be more specific about the restrictions pertaining to individuals that are instances of an OWL class as well as about the interrelationships between various classes and individuals. Classes may represent either a specific collection of individuals (e.g., Paul, John, and Mary) or concepts that can have an indeterminate number of instances (e.g., Mitochondrion).

OWL provides six different mechanisms for defining classes [262]:

1. The name of a class indicated by a URI.

---

[5]The claim that the choice of the acronym OWL had something to do with the character Owl from Winnie the Pooh, who spelled his name WOL, is apocryphal.

2. An enumeration of all individuals that make up the class.

3. A property restriction.

4. As an intersection of two or more class descriptions.

5. As the union of two or more class descriptions.

6. As the complement of a class description.

Aside from the first type, which describes a class by merely providing the name of the class in form of a URI (which may or may not point to a human-readable explanation of the "meaning" of the class available on the Web), each of these mechanisms defines classes by placing restrictions on the things that belong to the class. Perhaps the easiest to understand is the enumeration type, which essentially just provides a list of all individuals that belong to the class. For instance, the following class definition expressed using the RDF/XML syntax creates a class of all four Beatles:

```
<owl:Class>
  <owl:oneOf rdf:parseType="Collection">
    <owl:Thing rdf:about="#John"/>
    <owl:Thing rdf:about="#Paul"/>
    <owl:Thing rdf:about="#George"/>
    <owl:Thing rdf:about="#Ringo"/>
  </owl:oneOf>
</owl:Class>
```

Property restrictions are commonly used in OWL. Consider the following class definition, which was modified from an OWL class of the NCI ontology [65]. *A-Microtubule* can be defined as "A complete cylindrical microtubule that is part of a microtubule doublet in cilia." Consider the following OWL class definition:

```
1.  <owl:Class rdf:about="#A-Microtubule">
2.    <rdfs:label>A-Microtubule</rdfs:label>
3.    <rdfs:subClassOf rdf:resource="#Cilium Microtubule"/>
4.    <rdfs:subClassOf>
5.      <owl:Restriction>
6.        <owl:onProperty rdf:resource="#is Physical Part of"/>
7.        <owl:someValuesFrom rdf:resource="#Cytoskeleton"/>
8.      </owl:Restriction>
9.    </rdfs:subClassOf>
10.   <rdfs:subClassOf>
11.     <owl:Restriction>
12.       <owl:onProperty rdf:resource="#is Physical Part of"/>
13.       <owl:someValuesFrom rdf:resource="#Cilium"/>
14.     </owl:Restriction>
```

```
15.    </rdfs:subClassOf>
16. </owl:Class>
```

Line 3 defines *A-microtubule* to be a subclass of *Cilium Microtubule*. The two following subClassOf elements represent restrictions that define what kind of subclass *A-microtubule* represents. Lines 5–9 state that instances of this class must represent a physical part of the *Cytoskeleton*. Lines 10–15 state that they must also represent a physical part of the *Cilium*. A restriction in OWL is defined in terms of existing properties and classes. For the sake of legibility, the RDF/XML has been slightly simplified from that in the original file of the NCI thesaurus.[6]

A property restriction is a special kind of class description that describes an anonymous class, namely a class of all individuals that satisfy the restriction. Thus, the above RDF/XML code should be read as stating that the class *A-Microtubule* is a subclass of an anonymous class that is defined by the restriction that its instances are a physical part of some cytoskeleton as well as a subclass of another anonymous class that is defined by the restriction that its instances are a physical part of some cilium. In plain English, every A-microtubule is part of a cilium and simultaneously part of a cytoskeleton.

OWL distinguishes two kinds of property restrictions: value constraints and cardinality constraints. A property restriction on some property P has the following general form:

```
<rdfs:subClassOf>
  <owl:Restriction>
    <owl:onProperty rdf:resource="P" />
    <owl:allValuesFrom rdf:resource="#V"/>
  </owl:Restriction>
</rdfs:subClassOf>
```

That is, we state that the class that contains this `rdfs:subClassOf` element is a subclass of an anonymous class of the form "all those instances for which all values of the property P are from class V." N3 provides an equivalent and more concise notation:

```
[ rdf:type owl:Restriction;
  owl:onProperty P;
  owl:allValuesFrom V]
```

`owl:allValuesFrom` states that for each instance of the class that is being described, every value for P must fulfill the constraint. Somewhat confusingly, `owl:allValuesFrom` does not necessarily mean that an instance of the class must have the property P, merely that if an instance does have P, then it must have a value from class V. This is analogous to the *universal quantifier* we

---

[6] See Appendix C for more information about the syntax of RDF/XML.

met in Chapter 3. Recall that the formula $(\forall x)P$ means "if an x exists, then it must satisfy P."

The following excerpt from a class definition of the Ontology for Biomedical Investigations (OBI) illustrates a typical use of `owl:allValuesFrom`. The class describes the antigen presentation function as a subclass of *Function* as defined in the BFO (see Chapter 6). Thus, the class refers to the function inhering in a cell that expresses MHC molecules which is realized in the process of antigen processing and presentation of antigen derived parts by MHC molecules on the cell surface. The prefix CL stands for the Cell Type Ontology [19].

```
<owl:Class rdf:about="#antigen presentation function">
  <rdfs:label>antigen presentation function</rdfs:label>
  <rdfs:subClassOf>
    <owl:Restriction>
      <owl:allValuesFrom>
        <owl:Class>
          <owl:intersectionOf rdf:parseType="Collection">
            <owl:Class rdf:about="CL:cell"/>
            <owl:Restriction>
              <owl:onProperty>
                <owl:TransitiveProperty rdf:about="#has_part"/>
              </owl:onProperty>
              <owl:someValuesFrom
                  rdf:resource="#MHC protein complex"/>
            </owl:Restriction>
          </owl:intersectionOf>
        </owl:Class>
      </owl:allValuesFrom>
    <owl:onProperty>
      <owl:FunctionalProperty rdf:about="#inheres_in"/>
    </owl:onProperty>
    </owl:Restriction>
  </rdfs:subClassOf>
  <rdfs:subClassOf rdf:resource="#Function"/>
</owl:Class>
```

The class *antigen presentation function* is declared to be a subclass of the class *Function* by the next-to-last line:

```
<rdfs:subClassOf rdf:resource="#Function"/>
```

The first `rdfs:subClassOf` statement, which takes up the rest of the definition, states that the class is a subclass of an anonymous class that is a restriction on the property *inheres in*.[7] Thus, if the function exists, it must

---

[7] We will see in Chapter 6 that *inheres is* is synonymous with *has bearer*; the inverse relation is *bearer of*. If a quality q inheres in an independent continuant (bearer) B, then B is the bearer of the quality q.

inhere in (`owl:allValuesFrom`) a certain kind of cell that is defined as a cell that has to have a (as a part, `owl:someValuesFrom`) *MHC protein complex* (recall that the major histocompatibility complex takes fragments of antigens and presents them on the cell surface as a part of the immune response). Note that this definition uses combinations of nested nested OWL constructs to construct an elegant and precise definition of the term.

Classes can be defined by stating what they are not. For instance, one of the classes in OBI describes culturing cells that do not express the CD8 receptor on their surface. In a way similar to the above example, one can restrict the definition to cells that are the complement of cells that express the CD8 receptor using the keyword `owl:complementOf`:

```
<owl:intersectionOf rdf:parseType="Collection">
  <owl:Class rdf:about="CL:cell"/>
  <owl:Class>
    <owl:complementOf>
      <owl:Restriction>
        <owl:someValuesFrom>
          <owl:Class rdf:about="#CD8 receptor"/>
        </owl:someValuesFrom>
        <owl:onProperty>
          <owl:TransitiveProperty rdf:about="#has_part"/>
        </owl:onProperty>
      </owl:Restriction>
    </owl:complementOf>
  </owl:Class>
</owl:intersectionOf>
```

Unexpected results can be obtained from the naive use of `owl:complementOf`. For instance, although one might be tempted to think that the complement of *Man* is *Woman*, the meaning of `owl:complementOf` *Man* is actually everything that is not a *Man*, including for instance, the kitchen sink, Neptune, and the Pacific Ocean. Thus, `owl:complementOf` is often combined with other restrictions to express the intended meaning (for instance, the intersection of *Human Being* and `owl:complementOf` *Man* is *Woman*).

OWL also provides a union operator that is analogous to the set union operator. For instance, we can define the class Human to be the union of all men and women:

```
<owl:Class rdf:ID="Human">
  <owl:unionOf rdf:parseType="Collection">
    <owl:Class rdf:about="#Man" />
    <owl:Class rdf:about="#Woman" />
  </owl:unionOf>
</owl:Class>
```

Contrast the `owl:unionOf` construct to the following (incorrect) attempt to define *Human* as a subclass of *Man* and of *Woman*.

```
<owl:Class rdf:ID="Human">
  <rdfs:subClassOf rdf:resource="#Man" />
  <rdfs:subClassOf rdf:resource="#Woman" />
</owl:Class>
```

Since no one individual instance of *Human* can be an instance of both of the classes *Man* and *Woman*, this class is unsatisfiable. An *unsatisfiable class* is one that cannot be true of any possible individual; unsatisfiable classes are thus equivalent to the empty set. For example $A = \{C \cap \neg C\}$ is unsatisfiable because it implies a direct contradiction.

Individuals in OWL

We have seen how to describe individuals in RDF using the `rdf:type` keyword. In OWL, `rdf:type` is the property that connects an individual to a class of which it is a member. For instance, if we have a class of cities:

```
<owl:Class rdf:ID="City"/>
```

then we can declare individual cities to be members of this class as follows:

```
<City rdf:ID="Berlin"/>
```

This declares that *Berlin* is an individual of the class *City*. An alternative syntactical form with the identical meaning is

```
<owl:Thing rdf:ID="Berlin"/>
```

```
<owl:Thing rdf:about="#Berlin">
  <owl:type rdf:resource="#City"/>
</owl:Thing>
```

Cardinality in OWL

OWL defines constraints on cardinality using `owl:minCardinality` and `owl:maxCardinality` to specify the minimum and maximum cardinality of a set, and `owl:cardinality` to specify that a set has some exact number of elements. These constructs are relatively straightforward. In the following slightly simplified example from the WC3's wine ontology, wine is defined as a potable liquid that has exactly one maker, which must come from the class of Winery, and must consist of at least one type of grape.

```
<owl:Class rdf:ID="Wine">
  <rdfs:subClassOf rdf:resource="&food;PotableLiquid" />
  <rdfs:subClassOf>
```

Classes and Individuals in OWL

Note that different ontologies have different levels of representation of classes, subclasses, and instances. An OWL class is a collection of individuals, but this can be defined in different ways. A statement such as *John is an instance of the class HumanBeing* is easily understood. On the other hand, in our Indian food ontology, we will make *MattarPaneer* an instance of the class *VegetarianDish* (Mattar Paneer is a delicious dish with peas, cubes of the Indian fresh cheese called "Paneer," and a sweet and spicy sauce). The instance *MattarPaneer* is not a particular serving of this wonderful dish the way John is a particular human being. Rather, it is a particular recipe. This is a modeling decision; the class *VegetarianDish* is a collection of all kinds of Indian recipes, and therefore, *MattarPaneer* or *MasalaDosa* are instances rather than subclasses of *VegetarianDish*. Any subclass of *VegetarianDish* should denote a subset of these dishes, for instance the subset of vegetarian dishes that are soups, say *VegetarianSoupDish*. Users of an ontology need to be aware of the strategies used by the ontology to model the domain in question.

```
    <owl:Restriction>
      <owl:onProperty rdf:resource="#hasMaker" />
      <owl:cardinality>1</owl:cardinality>
    </owl:Restriction>
  </rdfs:subClassOf>
  <rdfs:subClassOf>
    <owl:Restriction>
      <owl:onProperty rdf:resource="#hasMaker" />
      <owl:allValuesFrom rdf:resource="#Winery" />
    </owl:Restriction>
  </rdfs:subClassOf>
  <rdfs:subClassOf>
    <owl:Restriction>
      <owl:onProperty rdf:resource="#madeFromGrape" />
      <owl:minCardinality>1</owl:minCardinality>
    </owl:Restriction>
  </rdfs:subClassOf>
</owl:Class>
```

While many low-quality wines are produced from more than one grape variety, White Burgundy wines are superb wines made in the Burgundy region in Eastern France made from Chardonnay grapes. Therefore, the following class definitions state that *WhiteBurgundy* is the intersection of the classes *WhiteWine* (itself a subclass of *Wine*) and *Burgundy* and what is more, White Burgundy is made from at most one grape variety, *viz.*, *ChardonnayGrape*.

```
<owl:Class rdf:ID="WhiteBurgundy">
  <owl:intersectionOf rdf:parseType="Collection">
    <owl:Class rdf:about="#Burgundy" />
    <owl:Class rdf:about="#WhiteWine" />
  </owl:intersectionOf>
</owl:Class>

<owl:Class rdf:about="#WhiteBurgundy">
  <rdfs:subClassOf>
    <owl:Restriction>
      <owl:onProperty rdf:resource="#madeFromGrape" />
      <owl:hasValue rdf:resource="#ChardonnayGrape" />
    </owl:Restriction>
  </rdfs:subClassOf>
  <rdfs:subClassOf>
    <owl:Restriction>
      <owl:onProperty rdf:resource="#madeFromGrape" />
      <owl:maxCardinality>1</owl:maxCardinality>
    </owl:Restriction>
  </rdfs:subClassOf>
</owl:Class>
```

Because the constraint on `madeFromGrape` of minimum cardinality 1 inherited from the class *Wine* and the maximum cardinality of 1 declared for the class *WhiteBurgundy*, this type of wine can be made from exactly one type of grape.

Property Characteristics in OWL

OWL has a number of mechanisms for specifying characteristics of properties. OWL does this by defining inference rules that pertain to subjects and objects connected by the property.

`owl:TransitiveProperty` specifies a binary relation P between two elements such that if $A \xrightarrow{P} B$ and $B \xrightarrow{P} C$ then we can infer that $A \xrightarrow{P} C$. For instance, the `part_of` relation is generally defined as transitive, which makes sense because for instance, if a mitochondrion is part of the cytoplasm and the cytoplasm is part of the cell, then it is clear that the mitochondrion is part of the cell. We might declare this in RDF/XML syntax as follows:

```
<owl:ObjectProperty rdf:ID="partOf">
  <rdf:type rdf:resource="&owl;TransitiveProperty" />
</owl:ObjectProperty>

<owl:Thing rdf:ID="Cell"/>

<owl:Thing rdf:ID="Cytoplasm">
```

```
  <partOf rdf:resource="#Cell" />
</owl:Thing>

<owl:Thing rdf:ID="Mitochondrion">
  <partOf rdf:resource="#Cytoplasm" />
</owl:Thing>
```

Thus, we have stated that *Cell, Cytoplasm,* and *Mitochondrion* are subclasses of owl:Thing, and that *Cytoplasm* is part of *Cell,* and *Mitochondrion* is part of *Cytoplasm.* From these assertions, OWL's rule about transitive properties allows us to infer the following:

```
<owl:Thing rdf:ID="Mitochondrion">
  <partOf rdf:resource="#Cell" />
</owl:Thing>
```

owl:SymmetricProperty specifies a binary relation between subject and object such that the relation holds between subject and object if and only if it holds between object and subject. Thus, if $A \xrightarrow{P} B$, then we can infer that $B \xrightarrow{P} A$. Properties such as hasSameColorAs are clearly symmetric. The inference rule is straightforward. For instance, if we know that *adjacent to* is a symmetric property and it is asserted that the supraoptic nucleus is *adjacent to* the optic chiasm, then we can infer that the optic chiasm is *adjacent to* to the supraoptic nucleus.

owl:inverseOf defines one property to be the inverse of another. If a property P is declared to be owl:inverseOf a property Q, then for any two objects $A \xrightarrow{P} B$ iff $B \xrightarrow{Q} A$. For instance, if we declare has_part to be the inverse of part_of, then the statement A part of B would imply that B has part A. The following OWL code makes such a declaration:

```
<owl:ObjectProperty rdf:about="&ro;has_part">
  <rdf:type rdf:resource="&owl;TransitiveProperty"/>
  <rdfs:label>has_part</rdfs:label>
  <owl:inverseOf rdf:resource="&ro;part_of"/>
  <rdfs:subPropertyOf rdf:resource="&owl;topObjectProperty"/>
</owl:ObjectProperty>
```

Declaring a property owl:inverseOf another property may lead to unintended inferences. For instance, while nucleus partOf cell is always true (a nucleus cannot exist outside of a cell), the inverse statement cell hasPart nucleus is only true for certain cells. Bacterial cells, among others, do not have a nucleus.

A function f is defined on a set of values called the domain of f and a rule for specifying the value $f(x)$ for each x in the domain of f. If $f(x) = y$ and $f(x) = z$ then it must be the case that $y = z$. owl:functionalProperty is defined by an analogous inference rule that states that if $A \xrightarrow{P} B$ and $A \xrightarrow{P} C$ for a property P that is an `owl:functionalProperty`, then we can infer that $B = C$. The following snippet of RDF/XML code shows how one might declare a property `encodesAminoAcid` to be functional in order to assert that a specific codon always codes for the same amino acid.

```
<owl:ObjectProperty rdf:ID="encodesAminoAcid">
  <rdf:type rdf:resource="&owl;FunctionalProperty" />
  <rdfs:domain rdf:resource="#Codon" />
  <rdfs:range  rdf:resource="#AminoAcid" />
</owl:ObjectProperty>
```

If a property P is declared to be an `owl:inverseFunctionalProperty`, then for all A, B, and C, $B \xrightarrow{P} A$ and $C \xrightarrow{P} A$ implies that $B = C$. Note that `encodesAminoAcid` is *not* an `owl:inverseFunctionalProperty` because, for instance, TAT and TAC both code for the amino acid tyrosine.

Disjoint Classes in OWL

The operator `owl:disjointWith` can be used to express the disjointness of two or more classes. If two classes are disjoint, we can infer that an individual cannot simultaneously be an instance of both classes.

```
<owl:Class rdf:ID="DNA">
  <rdfs:subClassOf rdf:resource="#Biopolymer"/>
  <owl:disjointWith rdf:resource="#RNA"/>
  <owl:disjointWith rdf:resource="#Protein"/>
</owl:Class>
```

The above definition of the class *DNA* states that *DNA* is disjoint from *RNA* and *Protein*. It does not, however, assert that *RNA* and *Protein* are disjoint. Such a statement would require a separate `owl:disjointWith` assertion.

The keyword `owl:differentFrom` declares two individuals to be distinct. A related keyword defines a set of mutually distinct individuals. For instance, if we decide to model DNA, RNA, and Protein not as classes, but as distinct individuals, we can state

```
<owl:AllDifferent>
  <owl:distinctMembers rdf:parseType="Collection">
    <:Biopolymer rdf:about="#DNA" />
    <:Biopolymer rdf:about="#RNA" />
    <:Biopolymer rdf:about="#Protein" />
  </owl:distinctMembers>
</owl:AllDifferent>
```

On the other hand, if we want to state that two individuals are identical, we can use the keyword `owl:sameAs`. Because of the open-world assumption of OWL, we cannot infer without further information that an individual named `Sam` is distinct from an individual named `Mary`. This construct can be used to equate individuals from two different documents and thereby integrate two different ontologies. For instance, the food ontology of the W3C equates `food:Red` with `vin:Red`:

```
<owl:Thing rdf:ID="Rose">
  <owl:sameAs rdf:resource="&vin;Rose" />
</owl:Thing>
```

In the wine ontology, `Rose` (i.e., Rosé wine) is declared to be an individual of type `WineColor`:

```
<owl:Class rdf:ID="WineColor">
  <rdfs:subClassOf rdf:resource="#WineDescriptor" />
  <owl:oneOf rdf:parseType="Collection">
    <owl:Item rdf:resource="#White" />
    <owl:Item rdf:resource="#Rose" />
    <owl:Item rdf:resource="#Red" />
  </owl:oneOf>
</owl:Class>

<WineColor rdf:ID="Rose" />
```

4.2.4   OBO, RDF, RDFS, and OWL

In this chapter, we have seen the major ontology languages used in bio-ontologies. We will conclude this chapter with an informal comparision of these languages.

OBO vs. OWL

Thanks in no small part to the enormous success of the Gene Ontology (GO), which has developed the OBO format as a language for the GO, OBO format ontologies have become predominant in the field of bio-ontologies, with over 50 OBO format ontologies currently available at the OBO Foundry home-page (see Chapter 7). The OBO format was initially designed with an eye towards lowering the threshold for acceptance by biologists by being compact, easily readable by humans, and easy to parse. GO, and OBO, were designed specifically for the biomedical domain and with the specific goal of annotating biomedical data. OWL was designed to describe any domain and with a focus on automated reasoning using inference.

At the risk of oversimplification, one can state that OBO ontologies have been developed by or in close collaboration with biomedical domain experts, and several of them, including especially the GO, have become standard in their field. OBO ontologies have tended to be taxonomies, classifications of a domain in biology according to subclass (`is_a`), parthood (`part_of`) relations, and algorithmic work with OBO ontologies has tended to emphasize graph-based algorithms that relate the hierarchical structures of the ontologies to annotations of biomedical concepts with the terms of the ontologies, as we will see in Chapters 8, 9, 10, and 11. On the other hand, OWL ontologies are most often developed by computer scientists, and much of the work on OWL has had a focus on developing frameworks, algorithms and software for inference as well as on developing the Semantic Web, which will use RDF and OWL ontologies to make a "smart" Web (as described in Chapter 1). With some prominent exceptions, including some prominent ontologies affiliated with the OBO Foundry effort, the content of many biomedical ontologies developed in OWL is still immature compared to ontologies such as GO [233].

One very important characteristic of OBO ontologies is that they are designed to support science. Classes (terms) of OBO ontologies generally represent universal concepts in science, such as *Nucleus*. OWL is designed for everything and nothing in particular. An OWL class can represent a universal concept just as well as it can represent the set C of empty coffee mugs on our desk.[8]

The OBO community has shown increasing concern with developing logically sound foundations for OBO syntax and ontological structure (see Chapter 6), and a number of additional structures have been added to the OBO language since its introduction, including notably the intersection syntax that was described in this chapter. The OBO language now comprises a substantial subset of OWL-DL, and efforts are being made to develop software that will be able to interconvert between OBO and OWL ontologies, which will allow tools developed for OWL to be used by the OBO community and improve interoperability between OBO and OWL ontologies [93, 251]. Chapter 15 will explain how inference is being used in OBO ontologies, and Chapters 14 and 16 will provide an in-depth introduction to inference in OWL ontologies and demonstrate how this is used in the framework of SPARQL queries to enable intelligent searching in OWL bio-ontologies.

RDF vs. RDFS vs. OWL

The W3C has developed and continues to develop a number of different standards in support of the Semantic Web. It will be helpful at this point to review the major high-level differences between them.

The Resource Description Framework (RDF) forms the basis of all of the standards. RDF is a graphical formalism that describes resources in terms of triples, that is, subject-predicate-object statements. A collection of such statements forms a labeled directed graph in which the subject and object are nodes and the predicate is an edge between them. RDF is a simple data model designed especially to allow data from various sources on the Web to be merged for data integration.

The RDF schema language RDFS extends RDF with additional specifica-

---

[8]$C \neq \emptyset$.

tions and keywords that allow some forms of inference to be performed. RDFS provides mechanisms to define classes and subclasses that provide the basis of a vocabulary to define the meaning of things in RDF graphs and to enable consistent use of vocabulary. RDFS is too weak to describe resources in sufficient detail for many forms of inference. For instance, existential restraints are lacking with which one could express the fact that a biomedical publication must have an author.

OWL further extends RDFS and existing Web standards to add more expressive power, by providing syntactic constructs that allow modelers to define classes that describe a view of how the world fits together and allow computer reasoning and inference over classes and individuals. OWL became a W3C recommendation in 2004, and OWL 2.0 became a W3C recommendation in 2009. The original W3C specifications defined three sublanguages of OWL. OWL-lite represents a "light-weight" subset of OWL-DL that was intended mainly for classification hierarchies with simple constraints on classes. OWL-lite does not have any of the `owl:hasValue` restrictions, owl:oneOf constructs, and has limited cardinality restrictions. Both OWL-DL and OWL-Full use the complete OWL vocabulary. Each statement in OWL-DL can be converted to a statement in the formal logic system *Description Logic*, and the inferences in OWL-DL are based on inference in Description logic. Roughly speaking, this means that it is possible to determine which classes are equivalent to one another (or not), and which individuals are members of which classes by means of an algorithm known as the Tableaux algorithm, which will be explained in Chapter 15. The difference between OWL-DL and OWL-Full relates to some restrictions on OWL-DL. In OWL-DL, a class cannot also be an individual, which implies that restrictions cannot be applied to the language elements of OWL itself [258, 263]. In OWL-Full, all language constructs can be used in any combination. Roughly speaking, the distinction between OWL-DL and OWL-Full is that the restrictions on OWL-DL ensure decidability (computational completeness). In some cases, the inferences that can be drawn are different if an ontology is considered as OWL-DL or OWL-Full. The full details of these distinctions go beyond the scope of this book, and interested readers are referred to the documentation of the W3C. Unless otherwise stated, all references to OWL in this book will refer to OWL-DL. Table 4.3 gives a bird's eye view of the differences between RDF, RDFS, and the dialects of the initial version of OWL.

OWL 2

OWL 2, an extension and revision of the first version of OWL published in 2004, became a recommendation of the W3C in 2009. The material presented in this chapter is valid for OWL 1 and OWL 2, and we will not attempt to comprehensively describe the differences between OWL 1 and OWL 2. We give interested readers some references in the Further Reading section of Appendix D for this.

Language	Main Purpose
RDF	Annotation
RDFS	Integration
OWL	Inference
OWL 1	
OWL-lite	Classification hierarchy and simple constraints
OWL-DL	Maximum expressiveness, computational completeness
OWL-Full	Maximum expressiveness, syntactic freedom of RDF

TABLE 4.3: RDF, RDFS, and OWL: An oversimplified comparison. See Table 4.4 for an overview of OWL 2.

Ontologies that satisfy some conditions placed on the ontology structures and are interpreted using the Direct Semantics[9] [172] are called *OWL 2 DL ontologies*. *OWL 2 Full* refers to RDF graphs considered as OWL 2 ontologies and interpreted using the RDF-Based Semantics.

OWL-lite was never widely adopted and has been considered by some to be a failure. Instead of an "OWL 2 lite," the W3C has introduced three OWL 2 profiles that are syntactic subsets of OWL 2 DL that are designed to offer advantages for certain types of applications. Each profile represents a different syntactic *restriction* of OWL 2 DL that trades some amount of expressivity for improvements in the computational complexity of inference algorithms (Table 4.4).

Language	Main Purpose
OWL 2 EL	Useful for ontologies with very large numbers of properties or classes. Implements a subset of OWL 2 for which inference can be performed in time that is polynomial with respect to the size of the ontology.
OWL 2 QL	Useful for applications with large volumes of instance data in which query answering is the most important task.
OWL 2 RL	This profile is aimed at applications that require scalable reasoning without loss of too much expressive power. OWL 2 RL can optionally be implemented using rule-based inference algorithms.

TABLE 4.4: OWL 2 profiles. The documentation of the W3C provides detailed specifications of the three profiles [170].

None of the three profiles are strict subsets of another. Any OWL 2 EL, QL or RL ontology is also an OWL 2 ontology and can be interpreted using either

---

[9]See Chapter 13.

the Direct or RDF-Based Semantics. In general, OWL 2 is very similar to OWL 1, and there is essentially complete backwards compatibility. All OWL 1 Ontologies remain valid OWL 2 Ontologies, with identical inferences in all practical cases [170].

OWL 2 does provide a number of new capabilities, including keys, property chains, enhanced annotations, and more ways of specifying the behavior of properties [94]. Many of the novelties of OWL 2 represent advanced features that go beyond the scope of this book. Almost all of the OWL features used in current bio-ontologies are already present in OWL 1 and are used identically in OWL 2.

4.3 Exercises and Further Reading

There are a number of excellent books on the Semantic Web and OWL. *The Semantic Web for the Working Ontologist* [6] provides a practical introduction to modeling domains with RDF and OWL and is highly recommended. *Foundations of Semantic Web Technologies* by Hitzler, Krötzsch and Rudolph provides a more formal explanation of the languages of the Semantic Web [117]. The best place to consult the technical specifications of RDF, RDFS, and OWL is the Web site of the W3C, which has published technical specifications and summaries of RDF, RDFS, and OWL [39, 160, 258]. Cuenca Grau and colleagues provide an overview of the shortcomings in OWL 1 that were re-engineered for OWL2 [61].

The Web site of the GO consortium[10] provides documentation about the OBO language. RDF, RDFS, OWL, and OBO are not the only ontology languages, but they are by far the most important ones for bio-ontologies. Readers will find information about frame-based ontologies in the relevant Protégé documentation on this subject [187].

Exercises

The exercises for this chapter are designed to help build intuition and understanding about OWL ontologies. Readers will be led through the process of constructing a small ontology based upon one of the example ontologies made by the W3C, the food ontology. `food.owl` contains classes and individuals that are related to one another with restrictions as described in this chapter. Readers will first be asked to examine parts of the food ontology and to compare XML/RDF and N3 syntax versions of this ontology. Then, readers will be asked to use the program Protégé to create a small ontology based on the food ontology but in which Indian food is used to illustrate the concepts.

---

[10]http://www.geneontology.org.

The exercises involving the Indian food ontology will guide the reader through a number of commonly made mistakes that can lead to imprecise or simply wrong inferences being made. Readers should take time to work through these exercises step by step and try to answer the questions themselves before going on.

4.1 There are many ways of serializing RDF and OWL documents, that is of writing them down as computer and human readable files. Probably the most widely used is the RDF/XML syntax. Notation 3, or N3, is a simplified syntax that can express essentially the same concepts as RDF/XML but is a good deal easier to understand when getting started. N3 was explained in this chapter, and additional information about N3 is available from the W3C (e.g., http://www.w3.org/2000/10/swap/Primer.html). For this exercise, download the food ontology from the Web site of the W3C (http://www.w3.org/TR/owl-guide/food.rdf). Now download and install[11] cwm (http://www.w3.org/2000/10/swap/doc/cwm.html), which is a W3C tool described as a data processor for the Semantic Web. Among other things, cwm can convert RDF/XML documents to N3, which is the use to which we will put cwm here. Assuming you have installed the program and it is available in your path, the command

```
cwm --rdf food.rdf --n3 > food.n3
```

will convert the RDF/XML version of the food ontology (food.rdf) into an equivalent N3 version. Now examine equivalent classes in the RDF/XML and N3 versions of the food ontology. For instance, both of the following definitions state that *PastaWithHeavyCreamSauce* is an instance of owl:Class and is a subclass of *PastaWithWhiteSauce*.

RDF/XML:

```
<owl:Class rdf:ID="PastaWithHeavyCreamSauce">
  <rdfs:subClassOf rdf:resource="#PastaWithWhiteSauce" />
</owl:Class>
```

N3:

```
food:PastaWithHeavyCreamSauce a :Class;
    rdfs:subClassOf food:PastaWithWhiteSauce .
```

Note that it may be necessary to consult other parts of the RDF/XML

---

[11]The Web site provides installation instructions for linux, Windows, and Mac. cwm is written using the programming language Python, which is required for the installation. Under Debian linux, cwm can be easily installed using the command aptitude install swap-cwm. If you are not currently using Debian or a related linux distribution such as Kubuntu, this exercise provides a good opportunity to upgrade your system.

Protégé

In the following exercises, we will use the program Protégé to build a small version of the food ontology for Indian food. Protégé was developed by Stanford Center for Biomedical Informatics Research at the Stanford University School of Medicine, and is one of the most widely used programs for ontology development [142, 143]. Protégé is free and open-source ontology editor, and versions for all major operating systems can be downloaded from http://protege.stanford.edu. The following exercises were developed using Protégé version 4.1.

Part of the material from these exercises was adapted from the tutorial provided by the University of Manchester team that has been involved in developing Protégé version. Their Web site: http://owl.cs.manchester.ac.uk/ contains lots of valuable information about OWL and a number of useful papers and tutorials. Readers who are interested in learning more about OWL are encouraged to work through the full tutorial available at the University of Manchester OWL Web site [121].

---

and N3 documents to fully understand the meanings of individual statements and why they are equivalent. For instance, the default namespace of food.n3 was defined using the following prefix:

```
@prefix : <http://www.w3.org/2002/07/owl#> .
```

This means that :Class is equivalent to owl:Class.

Likewise, the "#" syntax in RDF/XML refers to a class that was defined within the same document by an XML node of the form <owl:Class rdf:ID="X"/>, where "X" is the name of the class. In the above example, the *PastaWithWhiteSauce* class had been defined as follows:

```
<owl:Class rdf:ID="PastaWithWhiteSauce">
  <rdfs:subClassOf rdf:resource="#Pasta" />
  <owl:disjointWith rdf:resource="#PastaWithRedSauce" />
</owl:Class>
```

For this exercise, examine a number of classes in both RDF/XML and N3 syntax. Use the online documentation of the W3C to clarify the meaning of syntax that is unfamiliar to you. Start off with the definitions for *PastaWithSpicyRedSauceCourse* and *FraDiavolo*.

4.2 After you have installed Protégé as above, start the program. You will see a window that allows you to open an existing OWL ontology or create a new one. Before continuing with the exercise, it will be useful to open the food ontology (`food.rdf`) and explore some of the capabilities of the program. For this exercise, however, click on `Create new OWL ontology`. You will be asked to enter an IRI (Internationalized Resource Identifier), which is a generalization of a URI. You can enter an arbitrary URL for this exercise (suggestion: `http://www.example.org/indian-food.owl`), but note if you were creating an ontology for publication you would enter the URL that would point to the ontology file on your Web server. Protégé will then ask you to specify the path and file name of the ontology file on your computer. Choose an appropriate path and name the file `indian-food.owl` and create the file using the format RDF/XML. You will now see the Active Ontology tab of Protégé.

A good way of learning about XML/RDF syntax is to examine the changes in this file that are made by Protégé over the course of these exercises. Take a look at the file `indian-food.owl` in a text editor. Now go back to Protégé, and click on the "+" sign after the word annotations. Choose the annotations *creator* and *date*. Enter your name and today's date as the values. Now, return to the file `indian-food.owl`. You will see that Protégé has modified the line:

```
<owl:Ontology
    rdf:about="http://www.example.org/indian-food.owl"/>
```

to add information about the new annotations:

```
<owl:Ontology
    rdf:about="http://www.example.org/indian-food.owl">
  <dc:creator>Peter N. Robinson</dc:creator>
  <dc:date>17-08-2010</dc:date>
</owl:Ontology>
```

Additionally, Protégé has added two lines, to indicate that `dc:creator` and `dc:date` are annotation properties from the Dublin Core namespace. In OWL, annotation properties are descriptions of classes that are not intended to be property axioms for reasoning but rather to add metadata about the annotated classes or individuals:

```
<owl:AnnotationProperty rdf:about="&dc;creator"/>
<owl:AnnotationProperty rdf:about="&dc;date"/>
```

If you like, add a few more annotations to complete this exercise.

4.3 We will now create the class structure for our ontology. If necessary, open the Indian food ontology in Protégé. Switch to the *Classes* tab. You will see a single class called "Thing" (owl:Thing) which is defined in OWL as the set containing all individuals. Thus, every class in OWL is a subclass of owl:Thing. First, add a Class called *Food* as a subclass of *Thing* (Figure 4.4).

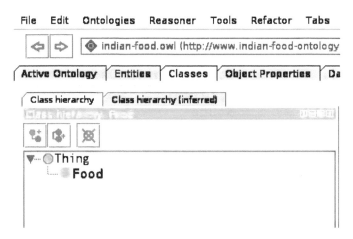

FIGURE 4.4: Protégé Tutorial (1). View of the *Classes* tab of Protégé. To create a new class, mark on the parent class and click the leftmost button ("Add subclass") on the class hierarchy panel. The other two buttons allow users to add a sibling class or to delete a class.

Now make *Dish* and *Ingredient* subclasses of Food. Make *VegetarianDish* and *NonVegeterianDish* subclasses of *Dish*. Continue in this manner to create the classes shown in Figure 4.5.

By default, OWL classes are assumed to overlap. Although obvious to a human, there is no way for a computer to "know" that a particular recipe cannot be both a *VegetarianDish* and a *NonVegetarianDish*. To specify this, we need to assert this. Click on the class *VegetarianDish* in the class hierarchy panel, and look for the button next to the phrase "Disjoint classes" in the Annotations panel on the right-hand side of the Protégé window. Click on this button and select the class *NonVegetarianDish* to be disjoint. Now save the ontology, and examine the corresponding section of the RDF/XML file. If everything went well, it should be as follows (the URIs have been shortened for clarity):

```
<owl:Class rdf:about="#NonVegetarianDish">
  <rdfs:subClassOf rdf:resource="#Dish"/>
  <owl:disjointWith rdf:resource="#VegetarianDish"/>
</owl:Class>
```

To finish this exercise, examine the RDF/XML definitions of the classes

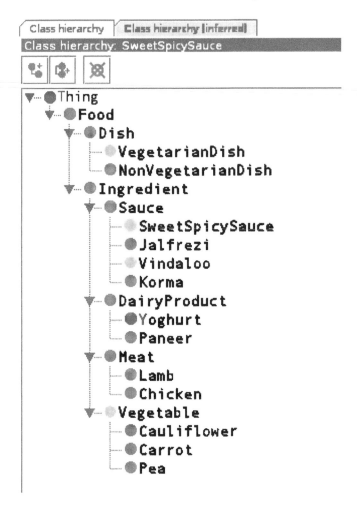

FIGURE 4.5: **Protégé Tutorial (2)**. View of the *Classes* to be created for the Indian food ontology.

you have created with particular attention to the use of the keyword `rdfs:subClassOf` to indicate subclass relations.

4.4 In this exercise, we will use object properties to help define our classes. First, create an object property called `hasIngredient` by switching to the *Object Properties* tab in Protégé, marking the node `topObjectProperty` and clicking on the button *add sub property* (top left in the Object property panel). Use the Description panel to add domain and range restrictions. Restrict the domain of this property to *Dish* by clicking on the "+" sign next to the node *Domains (intersection)* and using the class hierarchy navigator to select the class *Dish*. Restrict the range of the property to the class *Ingredient* in similar fashion. Now save the ontology and examine the RDF/XML file. If everything has gone well, you will see that Protégé has added the following lines (the URIs have been shortened for clarity):

```
<owl:ObjectProperty rdf:about="#hasIngredient">
  <rdfs:domain rdf:resource="#Dish"/>
  <rdfs:range rdf:resource="#Ingredient"/>
  <rdfs:subPropertyOf rdf:resource="&owl;topObjectProperty"/>
</owl:ObjectProperty>
```

We will now add some property restrictions. Up to now, the class *Non-VegetarianDish* has been defined as a subclass of the class *Dish*, but we have not specified whether the class has anything to do with the class *Meat*, which itself is a subclass of *Ingredient*. In order to specify that any *NonVegetarianDish* must have one *Meat* ingredient, open the Classes tab in Protégé, mark the class *NonVegetarianDish*, and click on the "+" sign next to the Superclasses node in the Description panel. Open the tab for the Class Expression Editor, and enter the words `hasIngredient some Meat`, and close the dialog. Now save the ontology and again examine the RDF/XML file. The class definition should now read as follows:

```
<owl:Class rdf:about="#NonVegetarianDish">
  <rdfs:subClassOf rdf:resource="#Dish"/>
  <rdfs:subClassOf>
    <owl:Restriction>
      <owl:onProperty rdf:resource="#hasIngredient"/>
      <owl:someValuesFrom rdf:resource="#Meat"/>
    </owl:Restriction>
  </rdfs:subClassOf>
  <owl:disjointWith rdf:resource="#VegetarianDish"/>
</owl:Class>
```

To finish this exercise, express the above OWL statement in English.

4.5 In this exercise, we will demonstrate the inference capabilities of Protégé. Consider the class structure of our ontology as it now is. Note that although it seems obvious to us, there is no way for the computer to know that vegetarian Indian dishes typically have vegetables, dairy products such as Paneer cheese, and a sauce. Therefore, if we define a new class representing the dish Mattar Paneer and define it to have peas, paneer, and sweet spicy sauce, the computer will not recognize that it is a vegetarian dish. In this exercise, we will first create such a class, and then we will modify the definition of the class *VegetarianDish* to allow this kind of inference. First go to the Classes tab in

Protégé and create a subclass of *Food* called *MattarPaneer* as above. Also as described above, go to the Description panel for the newly created class and add some restrictions by clicking on the "+" sign next to the Superclasses node, opening the class expression editor, and entering the restrictions `hasIngredient some Pea`, `hasIngredient some Paneer`, and `hasIngredient some SweetSpicySauce`. If we now go to the Protégé menu item *Reasoner*, choose one of the reasoners,[12] and click on *classify*, we can see what can now be inferred by examining the panel *Class hierarchy (inferred)*. We see that while *MattarPaneer* is a subclass of *Food* in the asserted hierarchy, it is a subclass of *Dish* in the inferred hierarchy. Recall our definitions of the range and domain of the property `hasIngredient` and consider why that might be before reading the footnote[13] (Figure 4.6).

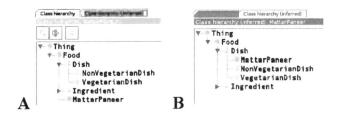

FIGURE 4.6: **Protégé Tutorial (3)**. View of asserted A) and inferred B) classes in Protégé. Note that the class *MattarPaneer* was asserted to be a subclass of *Food* but inferred to be a subclass of *Dish*.

We have thus achieved something, but we would like the reasoner to be able to infer not only that *MattarPaneer* is a subclass of *Dish*, but rather that it is a subclass of *VegetarianDish*. To do so, we will need to improve the definition of the class *VegetarianDish*. Our current definition of the class states that *VegetarianDish* is a subclass of *Dish* which in turn is a subclass of *Food*. Thus, we can infer that any instance of *VegetarianDish* is also an instance of *Dish* and *Food*. Examine the RDF/XML document before continuing. You should see the following class definitions (shortened here for clarity):

```
<owl:Class rdf:about="#Dish">
  <rdfs:subClassOf rdf:resource="#Food"/>
</owl:Class>
```

---

[12]The reasoner *hermiT* should be installed by default, and other reasoners can be installed as plugins. See the Protégé homepage for more information.

[13]Recall that range and domain restrictions do not behave as constraints but rather are used to allow inference about the subject and object of assertions using the property. Review the corresponding explanations in this chapter if this does not make sense to you.

```
<owl:Class rdf:about="#VegetarianDish">
  <rdfs:subClassOf rdf:resource="#Dish"/>
</owl:Class>
```

To convert this definition into a one that will allow the inference that *MattarPaneer* is a subclass of *VegetarianDish*, we intuitively need to assert that vegetarian dishes are equivalent to the class of things that are dishes that contain vegetables, dairy products, and sauces.[14] To do so, we need to define *VegetarianDish* using an equivalent class.[15] Mark *VegetarianDish* in the Classes tab of Protégé, and click on the "+" sign next to *Equivalent classes* in the Description panel. Open the Class expression editor, and enter the following assertion:

```
hasIngredient some (Vegetable or DairyProduct or Sauce)
```

This means that *VegetarianDish* is equivalent to the class of things that have at least one ingredient from the classes *Vegetable*, *DairyProduct*, or *Sauce*.

Now execute the reasoner as described above. Now, *MattarPaneer* is correctly inferred to be a subclass of *VegetarianDish*. Now take a look at the RDF/XML file. The class definition for *VegetarianDish* should be as follows:

```
<owl:Class rdf:about="#VegetarianDish">
  <owl:equivalentClass>
    <owl:Restriction>
      <owl:onProperty rdf:resource="#hasIngredient"/>
      <owl:someValuesFrom>
        <owl:Class>
          <owl:unionOf rdf:parseType="Collection">
            <rdf:Description rdf:about="#DairyProduct"/>
            <rdf:Description rdf:about="#Sauce"/>
            <rdf:Description rdf:about="#Vegetable"/>
          </owl:unionOf>
        </owl:Class>
      </owl:someValuesFrom>
    </owl:Restriction>
  </owl:equivalentClass>
  <rdfs:subClassOf rdf:resource="#Dish"/>
</owl:Class>
```

---

[14]Technically speaking, this is a lacto-ovo vegetarian dish.

[15]Make sure you state this as an equivalent class, which in essence represents a necessary and sufficient condition for inference. If one entered the same restrictions using the *Superclass* option in the class definition panel, we would make *VegetarianDish* a subclass of an anonymous class defined by the restriction; this would represent a necessary, but not sufficient condition.

We have made progress, but the ontology is still far from being a logically sound and comprehensive description of Indian food. To see some of the problems, create a class called *ChickenVindaloo* and define it using restrictions on the `hasIngredient` property such that the class has some `Chicken` and some `Vindaloo`. Now run the reasoner again. You will find that *ChickenVindaloo* has been inferred to be a subclass of *VegetarianDish* rather than *NonVegetarianDish*. As defined, *ChickenVindaloo* satisfies our definition of *VegetarianDish*, because it has some *Vindaloo*, which is a subclass of *Sauce*. Even though *ChickenVindaloo* also contains a meat ingredient, our definition of *NonVegetarianDish* did not state that this is a sufficient condition to be a *NonVegetarianDish*, so we cannot infer that *ChickenVindaloo* is a subclass of *NonVegetarianDish*. We will see how to deal with this in the next exercise.

4.6 One might assume that we now only need to make *NonVegetarianDish* equivalent to the class of things that have at least one ingredient from the class *Meat* to infer that *ChickenVindaloo* is a subclass of *NonVegetarianDish*. Try this as described above by adding an Equivalent class `hasIngredient some Meat` to the definition of *NonVegetarianDish*. If we now again perform reasoning as described above, we see that *ChickenVindaloo* has been inferred to be a subclass of *Nothing*[16] (Figure 4.7). Before going on, consider some reasons for this result. Which definitions are responsible for this unsatisfiable class?

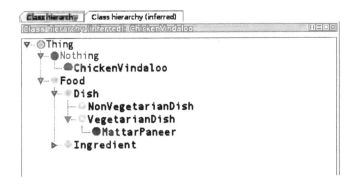

FIGURE 4.7: Protégé Tutorial (4). *ChickenVindaloo* has been inferred to be a subclass of *Nothing* because it is unsatisfiable.

4.7 This exercise will first explain the reason why our first attempt at defining the class *ChickenVindaloo* failed and will then begin to work on solutions. The definition of the class *ChickenVindaloo* contains a con-

---

[16]`owl:Nothing` is a predefined class that corresponds to the empty set. It is used here to indicate an error. It is impossible to fulfill the conditions on `ChickenVindaloo`, so there is no individual that could be an instance of the class except for the empty set.

tradiction because the assertions that have been made cannot all be true. In this case, the definition of *ChickenVindaloo* allows the inference that it is a subclass of *NonVegetarianDish* but also of *VegetarianDish*. Since we have declared these two classes to be disjoint, this is a logical impossibility. Let us consider how to fix this by amending the class definitions.

We would like to modify the class definition of *VegetarianDish* to state that an instance of this class must have at least one ingredient, and that none of its ingredients can be meat. Since a *NonVegetarianDish* also requires at least one ingredient, we can declare this requirement in their common parent class *Dish*. Enter the line `hasIngredient min 1` in the class expression editor as a superclass of *Dish*. Examine save the ontology and examine the owl file. The restriction on the number of ingredients an instance of *Dish* must have is expressed using a construct with the keyword `owl:minCardinality`:

```
<owl:Class rdf:about="#Dish">
  <rdfs:subClassOf rdf:resource="#Food"/>
  <rdfs:subClassOf>
    <owl:Restriction>
      <owl:onProperty rdf:resource="#hasIngredient"/>
      <owl:minCardinality>1</owl:minCardinality>
    </owl:Restriction>
  </rdfs:subClassOf>
</owl:Class>
```

Make sure you understand this syntax (Consult Appendix C for more information on the RDF/XML syntax and the other cardinality constraints `owl:maxCardinality` and `owl:cardinality`). Now we will alter the class definition of *VegetarianDish* to reflect the fact that it can *only* have one of the three kinds of vegetarian ingredients. Enter the following restriction as an equivalent class: `hasIngredient only (Vegetable or DairyProduct or Sauce)`.

Now run the reasoner again. You will see that *ChickenVindaloo* is now correctly inferred to be a subclass of *NonVegetarianDish* (why is this inference now possible?). However, *MattarPaneer* is no longer inferred to be a subclass of *VegetarianDish* but only of *Dish*. What happened?

The answer to this question is surprising to OWL novices, and has to do with the *open world assumption* of OWL reasoning: We cannot assume something doesn't exist until it is explicitly stated that it does not exist.

To finish this exercise, consider whether you have made a closed world assumption about any of the classes we have created for the Indian Food ontology. Consider whether an unjustified closed world assumption could

---

It's an Open World

The closed world assumption means that anything we do not know is assumed to be false. For example, if there is no mention of an allergy in a patient's chart, we can assume he or she does not have an allergy and administer Penicillin. The open world assumption is the opposite. If we do not know whether something is true or false, we cannot assume that it is false. In traditional database environments, the closed world assumption means that any information not contained in a database is false. Related to this is the *unique name assumption*, that any two different names refer to different entities, and the *domain-closure assumption*, that there are no entities in the domain being modeled other than those already represented in the database. The Semantic Web makes the opposite assumptions. After all, in the World Wide Web anything can be said by anybody, and any given Web site has no way of knowing what other servers might come online in the future with new domain knowledge.

---

be responsible for the unexpected inference. In the next exercise, we will answer this question and complete our Indian food ontology.

4.8 Under the open world assumption, it is false to conclude that all kinds of *Ingredient* can be classified as one of *Sauce, Vegetable, DairyProduct*, and *Meat*. For instance, another kind of ingredient could be *Fruit*. Because we defined the range of the hasIngredient to be the class *Dish* and the domain of the operator hasIngredient to be the class *Ingredient*, the open world assumption implies that any instance of the class *Dish* can have an *Ingredient* that we do not have listed in our ontology. Therefore, for all we know, *MattarPaneer* could have an additional *Ingredient* such as, say *CorianderLeaf*, that is not listed in the class definition of *VegetarianDish*.

Make sure you understand the explanation. Now consider a solution. We have to state that any *Ingredient* that *MattarPaneer* has must be from among our vegetarian ingredients. In Protégé, enter the restriction hasIngredient only (Paneer or Pea or SweetSpicySauce). Perform inference again in Protégé. Save the ontology, and examine the changes in the RDF/XML code for the classes made in the last several exercises. The OWL file for the complete Indian food ontology is available at the Web site of the book.

4.9 If you enjoy Indian food, add several of your favorite dishes. Add ingredients and other classes if needed. Define your classes such that the reasoner can correctly classify them as *VegetarianDish* or *NonVegetarianDish* by inference. If you run into trouble, congratulations, this is a good opportunity to learn. It may be helpful to learn how to use one of the

OWL ontology debuggers; methods for debugging ontologies to determine the reason for unsatisfiability of one or more classes have been developed [137]. The Semantic Web Ontology Editor `swoop` is a useful tool for debugging that is available at `http://code.google.com/p/swoop/`.

4.10 Convert the Indian food ontology into N3 format using the methods described in Exercise 4.1. Identify the N3 constructs corresponding to the classes developed over the course of Exercises 4.2–4.8.

4.11 Use Protégé to develop an ontology with about 30 terms for some domain of your interest. Now think of an additional class and an inference problem similar to the ones in Exercises 4.2–4.8. If inference doesn't work as expected the first time, the problem could quite well be related to an unjustified closed world assumption that you have made. Try to find the mistake on the basis of the considerations presented in the last several exercises.

Part II

Bio-Ontologies

Chapter 5

The Gene Ontology

5.1 A Tool for the Unification of Biology

"One Ring to rule them all, One Ring to find them,
One Ring to bring them all and in the darkness bind them"
-J.R.R. Tolkien, The Lord of the Rings

The Gene Ontology (GO) is being developed with the goal of providing a set of structured vocabularies for the annotation of genes and their products. Since the publication of the original paper in *Nature Genetics* in 2000 [14], GO has become one of the most widely used and mature bio-ontologies, although it is still very much a work in progress. GO was designed to enable data integration, and allows genes to be classified according to their functional properties. A primary goal is to provide a single, explicit definition of each biological concept, which means that the terms can be applied consistently by all biologists and also that computational representations of biology have a consistent interpretation.

One of the initial motivations for GO was the observation that a large fraction of genes coding for proteins involved in core biological functions are shared by all eukaryotes, a fact that has become increasingly obvious as ever more genome sequences were published beginning in the 1990s (Figure 5.1). The conservation of biological sequences suggested also that there is conservation of biological functions. That is, many biological phenomena such as DNA replication and glycolysis are common to all eukaryotic and many non-eukaryotic organisms.

In contrast to previous hierarchical systems for describing the functions of genes and gene products, such as the Enzyme Commission catalog, initially the GO was structured as a directed acyclic graph (DAG), in which a term can have multiple parents. A DAG allows a more rich and flexible representation of biological reality because specific terms can be children of multiple broader terms, which lets different biological aspects of the term to be captured in a way that would not be possible in a simple hierarchy in which a term can have at most one parent term. With the recent introduction of a **has_part** relation, the restriction on acyclic graphs has been removed as well.

By the end of the 1990s, a large number of tools for comparing the DNA and protein sequences of different organisms had come into common use. The

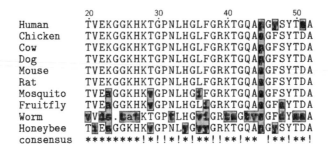

FIGURE 5.1: Multiple Sequence Alignment of Cytochrome C. A multiple sequence alignment showing part of the protein sequence of cytochrome C, a mitochondrial protein that has been highly conserved throughout the course of evolution. The last common ancestor of humans, cows, and dogs lived approximately 100 million years ago, and the last common ancestor of mammals and insects lived at least 500 million years ago. Nonetheless, the protein sequence of cytochrome C, which is involved in the oxidative phosphorylation system, is remarkably similar between humans and honeybees and a wide range of other organisms, indicating a common evolutionary origin.

Basic Local Alignment Search Tool BLAST and the DNA and protein sequence databases that can be searched with BLAST at the NCBI Web site are among the most important resources in computational biology, and were designed to be used to infer functional and evolutionary relationships between sequences as well as to help identify members of gene families [8]. Before GO, it was essentially impossible to compare functional profiles of sets of genes from different organism because genes of different organisms were annotated using incompatible vocabularies (if at all). Therefore, in 1998 the GO Consortium was founded by the model organism databases for the fruit-fly *Drosophila melanogaster* (FlyBase), the baker's yeast *Saccharomyces cerevisiae* (Saccharomyces Genome Database), and the house mouse *mus musculus* (Mouse Genome Informatics) with the goal of creating a cross-species biological ontology that would be used by multiple model organism databases to annotate genes and gene products in a consistent way. Subsequently, essentially all major genome databases have provided GO annotations, including annotations for humans, chickens, cows, parasites, and bacteria. It soon became apparent that GO and annotations to it would be useful for many things besides cross-species comparisons, and molecular genetics and bioinformatics groups began to employ GO for gene-expression profiling experiments, automatic annotation of ESTs and genomes, comparative genomics, network modeling, analysis of semantic similarity, and many other applications. GO analysis has thus become an essential part of many different areas of bioinformatics.

This chapter will provide an introduction to the goals, scope, and organization of GO. Chapter 8 and 9 will explain how GO is used to analyze the results of microarray hybridizations and other high-throughput experiments. Chapter 10 will show how semantic similarity measures can be derived from annotations to GO and other ontologies. The exercises for the present chapter will show how to use GO browsers or the book R package to search for gene annotations, and will present several other exercises designed to give readers familiarity with the most important uses of the GO for computational biology.

5.2   Three Subontologies

The GO is divided into three main branches that correspond to the three major ways in which molecular biologists talk about gene and protein function:

1. the basic biochemical activity of the encoded protein;

2. the biological objective to which it contributes; and

3. the location within or outside of the cell at which it is located.

It therefore makes sense to model these properties in separate ontologies. As they are all part of the Gene Ontology effort, they are usually referred to as the subontologies of the Gene Ontology.

5.2.1   Molecular Function

The molecular function subontology describes the biochemical activity of a gene product. This can refer to either an enzymatic reaction, or other activities such as the specific binding to a ligand or macromolecular structures. The definition applies to a potential of the gene product to perform an activity, that would occur in the correct context [115]. For instance, the protein neurofibromin in humans has been annotated with the GO term *Ras GTPase activator activity*, but it can only exert this activity in a context in which GTP-bound Ras is also present. Table 5.1 shows three example terms from the molecular function subontology. It is important to note that the molecular function subontology describes only what is done at a molecular level of granularity, but does not specify where or when the event actually occurs, or what purpose it serves the organism.

5.2.2   Biological Process

The biological process subontology describes the biological objective to which a gene or gene product contributes. A process is achieved by one or

Term	Term ID	Definition
mannosyltransferase activity	GO:0000030	Catalysis of the transfer of a mannosyl group to an acceptor molecule, typically another carbohydrate or a lipid.
zinc binding	GO:0008270	Interacting selectively and non-covalently with zinc (Zn) ions.
extracellular matrix structural constituent	GO:0005201	The action of a molecule that contributes to the structural integrity of the extracellular matrix.

TABLE 5.1: Molecular function. Three example terms from the molecular function subontology. The term *mannosyltransferase activity* describes a type of catalytic activity, *zinc binding* describes a type of binding activity, and *extracellular matrix structural constituent* describes a type of structural molecule activity.

more ordered assemblies of molecular functions and involves a collection of molecular events with a defined beginning and end. Processes generally involve a chemical or physical transformation at the level of granularity of the cell or organism. Table 5.2 shows three example terms from the biological process subontology.

The molecular function and biological process subontologies describe different aspects of biological functions. For instance, the gene *PIGV* encodes a protein called phosphatidylinositol glycan anchor biosynthesis, class V. This protein has the molecular function of *mannosyltransferase activity*, but this alone does not tell us what biological purpose is accomplished by the activity. Glycosylphosphatidylinositol (GPI) anchoring of cell surface proteins is a complex posttranslational modification of proteins such as alkaline phosphatase. The GPI anchor is synthesized by over 20 proteins, including PIGV, in the endoplasmic reticulum. Therefore, PIGV is annotated to the biological process term *GPI anchor biosynthetic process*.

5.2.3 Cellular Component

The cellular component subontology refers to the location inside or outside of the cell where a gene product is active, and comprises terms for the cytoplasm, for cellular organelles such as the nucleus or mitochondrion, for protein complexes such as the proteasome, and for the extracellular region. Table 5.3 shows three example terms from the cellular component subontology.

The GO is being developed to describe gene products in terms of their associated biological processes, cellular components and molecular functions in a species-independent manner. This means that not all terms apply to all

Term	Term ID	Definition
ossification	GO:0001503	The formation of bone or of a bony substance, or the conversion of fibrous tissue or of cartilage into bone or a bony substance.
regulation of glial cell proliferation	GO:0060251	Any process that modulates the frequency, rate or extent of glial cell proliferation.
B cell selection	GO:0002339	The process dependent upon B cell antigen receptor signaling in response to self or foreign antigen through which B cells are selected for survival.

TABLE 5.2: Biological process. Three example terms from the cellular component subontology. The term *ossification* describes a type of developmental process, *regulation of glial cell proliferation* describes a type of biological regulation, and *B cell selection* describes a type of immune system process.

Term	Term ID	Definition
Golgi apparatus	GO:0005794	A compound membranous cytoplasmic organelle of eukaryotic cells, consisting of flattened, ribosome-free vesicles arranged in a more or less regular stack.
presynaptic active zone membrane	GO:0048787	The membrane portion of the presynaptic active zone; it is the site where docking and fusion of synaptic vesicles occurs for the release of neurotransmitters.
viral capsid	GO:0019028	The protein coat that surrounds the infective nucleic acid in some virus particles.

TABLE 5.3: Cellular component. Three example terms from the biological process subontology. The term *Golgi apparatus* describes an intracellular organelle, *presynaptic active zone membrane* describes a part of the synapse, and *viral capsid* describes a part of the virion, or complete virus particle.

organisms. For instance, the term *viral capsid* clearly applies only to viruses, and the term *ossification* only applies to organisms that have bones.

5.3   Relations in GO

Ontologies such as GO can be expressed using description logics but also conceived of as graphs in which the terms are represented by the nodes or vertices of the graph that are linked together by one or more types of semantic relations encoded by the edges of the graph. A number of different relations are used by different bio-ontologies [238]. The two most important relations for GO are is_a and part_of.

The terms of bio-ontologies such as GO refer to general classes or types, rather than specific instances of the class. The edges between the terms of GO represent relations that obtain between the classes in reality, independently of the experimental methods used to measure them. A relation between two classes expresses a general statement about the corresponding biological classes [238].

The is_a relation represents a subclass relation. If we say, A is_a B, we mean that any instance of A is also an instance of B. For instance, the biological process term *phosphate transport* is defined as, "The directed movement of phosphate into, out of, within or between cells by means of some external agent such as a transporter or pore." The term *phosphate transport* is connected to the term *inorganic anion transport* by an is_a relation. The latter term is defined as "The directed movement of inorganic anions into, out of, within or between cells by means of some external agent such as a transporter or pore." Since phosphate is a kind of inorganic anion, it is clear that every instance of a biological process that involves the transport of phosphate also can be said to involve the transport of inorganic anions.

It is important to differentiate between the is_a relation and the instance_of relation, which relates an instance to a class. If we say that *James Tiberius Kirk* instance_of *Captain*, we mean that Kirk is one particular instantiation of the general class of *Captain*. In contrast, *Captain*, *Commander*, *Lieutenant*, and *Ensign* represent subclasses of a class *Star Fleet Officer* that are instantiated by Kirk, Spock, Sulu, and Chekov in the original Star Trek series, and by Picard, Riker, Tasha Yar, and Wesley Crusher in Star Trek: The Next Generation.

The part_of needs to invoke instance-level relations to define the class-level relation. It is used to express part-whole relations. By saying that *nucleus* part_of *cell*, we mean that any instance of *nucleus* (i.e., an actual nucleus) must occur as part of an instance of *cell*. That is, whenever a nucleus exists, it exists as part of a cell. In description logics, an equivalent definition of the concept *nucleus* would be that *nucleus* ⊑ ∃part_of.*cell*. Thus the statement does not imply that every instance of *cell* must contain an instance of *nucleus*.

Field	Content	Explanation
DB Object Symbol	FPR4	Gene name for Peptidyl-prolyl *cis-trans* isomerase
GO ID	GO:0000415	Negative regulation of histone H3-K36 methylation
DB: Reference	PMID:16959570	Reference to publication with PubMed ID
Evidence Code	IMP	Inferred from mutant phenotype

TABLE 5.4: The regulates Relation in GO. FPR4 negatively regulates histone H3 methylation at the lysine at position 36 (K36).

In addition to is_a and part_of, GO uses the relation regulates to denote that one process directly affects the manifestation of another process. Thus, the relation can only be placed between two terms of the biological process subontology. For instance, histones are proteins that bind to DNA and order it into structural units called *nucleosomes*. Methylation of specific amino acid residues of histones affects the conformation of the nucleosomes and regulates processes such as transcription and DNA repair. In yeast the proline isomerase Fpr4 was shown to catalyze the isomerization of proline residues in histone H3 in such a way that inhibits the methylation of the lysine (K) residue at position 36 of histone H3 (Table 5.4).

A relatively new addition to the set of relations within GO is the has_part relation. It is used to express whole-part relations between terms. If we say that *nucleus* has_part *chromosome* we mean that all instances of *nucleus* have one or more *chromosomes*. In description logics, an equivalent definition of the concept *nucleus* would be that *nucleus* ⊑ ∃has_part.*chromosme*. The has_part relation doesn't imply that all *chromosomes* are necessarily part of a *nucleus*. Hence, the has_part relation does not simply express the inverse of part_of. Note that has_part relations directed in the opposite orientation as part_of or is_a. The resulting ontology graph therefore is no longer acyclic.

5.4 GO Annotations

The development of bio-ontologies has enabled and facilitated the analysis of very large data-sets. This utility comes not only from the ontologies per se, but from the use to which they are put during the curation process that results in *annotations*. The terms of GO do not directly refer to specific genes or proteins but rather to their attributes. A GO annotation is a statement that a functional attribute represented by a GO term inheres in a particular gene product.

The Gene Ontology Annotation (GOA) database and several other groups provide annotations for over 50 species [47]. GO annotations are generated by biocurators, who are biologists involved in the organization and representation of biological information who attempt to make biological information from the biomedical literature accessible to both humans and computers. PubMed is a major database provided by the United States National Library of Medicine that offers citations, abstracts and some full-length articles from the biomedical literature. PubMed currently contains over 20 million citations, but this data is only useful if researchers are able to find relevant citations and integrate the knowledge into their own work. The goal of biocuration is therefore to extract information from published literature, develop ontologies and other structured vocabularies to tag data, and to provide databases that will allow researchers quick, comprehensive, and accurate access to the information contained in the literature. Biocurators typically read the full text of articles in their area of expertise, and transfer the information into a database using a structured vocabulary such as GO [129].

A Work in Progress

Following the publication of the Gene Ontology in 2000, when it comprised about 4500 terms, GO has been under continual development and now contains over 27,000 terms. GO is now used by over 20 model organism consortia that have joined to provide GO annotations of genes and proteins. Curators of model organism databases and genome annotation centers have annotated gene products using the terms of GO in order to capture information about the biological characteristics of gene products as reported in the scientific literature [115], and there are now nearly 900,000 manually curated annotations and 43 million computationally derived annotations [86]. Work on GO and GO annotations is expected to continue at this pace into the foreseeable future.

Model organism databases attempt to provide annotations for all genes in the organism's genome. For instance, at the time of this writing, the SGD provides 89,978 GO annotations for 6382 genes using a total of 4304 distinct GO terms. Each annotation is given as a single tab-separated line with 17 fields (Table 5.5).

The individual fields are important for different kinds of analysis, but four items are particularly important: the gene identifier, the GO term it is being associated with, the type of evidence used to support the annotation, and the reference for the evidence.

As we will see in Chapter 8, one of the most important uses of GO is for the analysis of genome-scale experiments such as microarray analysis or

	Content	Description	Example
1	DB	The contributing database	SGD
2	DB Object ID	A unique identifier in the DB	S000005342
3	DB Object Symbol	Gene symbol or other unique valid symbol	MNT3
4	Qualifier	Flag that modifies the interpretation of an annotation	NOT
5	GO ID	GO identifier	GO:0016021
6	DB:Reference	An identifier for the source of the annotation	PMID:10521541
7	Evidence Code	One of the GO evidence codes	IDA
8	With or From	An additional identifier required for annotations using certain evidence codes	GO:0000346
9	Aspect	One of M, F, B	F
10	DB Object Name	Name of gene or gene product	Notch
11	DB Object Synonym	Gene symbol or other text	x|y|z| …
12	DB Object Type	What type of entity is being annotated (gene,protein)	gene
13	Taxon	NCBI taxon (species) ID	taxon:7227
14	Date	Date on which annotation was made (YYYYMMDD)	20060803
15	Assigned By	The database which made the annotation	FlyBase
16	Annotation Extension	Cross-references to other ontologies to qualify or enhance the annotation	part_of(CL:0000576)
17	Gene Product Form ID	A specific form (variant) of the gene or gene product being annotated	UniProtKB:Q9NUD9

TABLE 5.5: GO annotation file format. Contributing model organism database groups provide annotations as tab-delimited files with the fields shown here. Fields 4, 8, 10, 11, 16, and 17 are optional. Some fields can accept multiple items. For instance, the synonym field accepts a list of synonyms separated by pipes, e.g., x|y|z| …. Further information is available at the GO homepage http://www.geneontology.org.

Field	Content	Explanation
DB Object Symbol	Mybbp1a	One of several identifiers for MGI:106181
GO ID	GO:0005515	Protein binding
DB: Reference	PMID:14744933	PubMed ID
Evidence Code	IPI	Inferred from physical interaction
With or From	UniProtKB:O70343	This field indicates that Mybbp1a binds **With** UniProtKB:O70343, i.e., Ppargc1a

TABLE 5.6: Inferred from physical interaction. This annotation is based on the publication by Fan and coworkers, *Suppression of mitochondrial respiration through recruitment of p160 myb binding protein to PGC-1α: modulation by p38 MAPK*, which was published in 2004 in *Blood* **18**:278–289. In this paper, which received the PubMed ID 14744933, a protein binding assay was used to show that Mybbp1a physically interacts with Ppargc1a.

next-generation sequencing analysis of gene expression. Users query the GO database or similar programs with a list of differentially expressed genes in order to retrieve a list of biological functions that are common among the genes. Unfortunately, there is a profusion of naming conventions for genes, and of course the software must be able to recognize the gene names. The fields DB Object ID, DB Object Name, DB Object Synonym, and Gene Product Form ID each contain different synonyms for the same gene, and most GO software packages use the information to attempt to identify the genes in the list uploaded by the user.

5.4.1 Evidence for Gene Functions

The evidence code indicates the level of confidence for an annotation. As in many areas of science, there is a tradeoff between coverage and accuracy. High-quality annotations have been produced for many model organisms based on biocuration of experimental evidence in published literature. However, the vast majority of the nearly 65 million GO annotations currently available were generated automatically. Depending on the analysis to be performed, it may be desirable to use all GO annotations, or only the highest quality ones.

The highest-quality annotations correspond to experimental evidence. There are six experimental annotation codes, each of which indicates a different type of "wetlab" experiment used to support the association of a GO term to the gene. Table 5.6 shows a detailed example of an experimental annotation.

A second class of annotations is represented by a group of computational analysis codes, which indicate that the annotation is based on an *in silico*

Field	Content	Explanation
DB Object Symbol	Myc	One of several identifiers for MGI:97250
GO ID	GO:0005737	Cytoplasm
DB: Reference	MGI:MGI:4417868	Mouse Genome Informatics ID
Evidence Code	ISO	Inferred from sequence orthology
With or From	UniProtKB:P09416	This field indicates that the inference is based on UniProtKB:P09416, the rat homolog of myc.

TABLE 5.7: **Inferred from sequence orthology**. Sequence orthology means that genes in two different species have a common evolutionary origin. The entry in the With field indicates that mouse myc is orthologous with rat myc. In this case, the curators of MGI have based the claim that mouse myc can be localized in the cytoplasm on the analogous claim of the Rat Genome Database (RGD) that rat myc is a cytoplasmic protein.

analysis. These evidence codes do imply that a curator has evaluated the computational prediction. Table 5.7 gives a detailed example of an annotation with a computational analysis evidence code.

Author statement evidence codes are used when a biocurator bases an annotation on an article in which the author does not directly present the scientific evidence, be it experimental or computational, for the claim from which an annotation is derived. An example of this would be a description of a function in a review article that references several other papers that present original research. Table 5.8 gives a detailed example of an annotation with an author statement evidence code.

Curatorial evidence statement codes are used where an annotation is not supported by any direct evidence, but can be reasonably inferred by a curator from other GO annotations, which are supported by direct evidence (Table 5.9).

Finally, there are still many genes whose function remains unknown. It can be useful to record the fact that a biocurator searched for but could not find information that would allow a GO annotation at a certain date. This is indicated by the evidence code ND: No biological Data available. ND annotations can be made separately for each of the three subontologies molecular function, biological process, and cellular component. For instance, at the time of this writing, the SGD had provided a total of 89,951 annotations for 6378 yeast genes including 4105 ND annotations for 2317 genes. Thus, no functional information is available for roughly one third of all yeast genes. As research in molecular genetics progresses, the proportion of ND annotations will decline.

Field	Content	Explanation
DB Object Symbol	Myf5	One of several identifiers for MGI:97252
GO ID	GO:0003705	RNA polymerase II transcription factor activity, enhancer binding
DB: Reference	PMID:1846704	PubMed ID
Evidence Code	TAS	Traceable author statement

TABLE 5.8: Traceable author statement. In this case the PubMed ID refers to the review article by Weintraub and coworkers, *The myoD gene family: nodal point during specification of the muscle cell lineage* published in 1991 in *Science* **251**:761–6. Among other things, this article describes how MyoD binds cooperatively to muscle-specific enhancers and activates transcription, and cites articles from the primary literature to support this claim.

Field	Content	Explanation
DB Object Symbol	Ncor2	One of several identifiers for nuclear receptor co-repressor 2
GO ID	GO:0005634	Nucleus
DB: Reference	PMID:15681609	PubMed ID
Evidence Code	IC	Inferred by curator
With or From	GO:0016564	This field indicates that the inference is based on the annotation of Ncor2 to the GO term *transcription repressor activity* (GO:0016564)

TABLE 5.9: Inferred by curator. In this case the curator inferred that the protein Ncor2 can be localized in the nucleus because it has been annotated to the term *transcription repressor activity*, which is an activity that takes place in the nucleus. The latter annotation was based on an article with PubMed ID 15681609 (in this paper, Ncor2 is designated with the alternate name SMRT).

Field	Content	Explanation
DB Object Symbol	Lst1	Gene name for MGI:1096324
GO ID	GO:0016020	Integral to membrane
DB: Reference	MGI:1354194	Reference to keyword mapping at MGI
Evidence Code	IEA	Electronic annotation
With or From	SP_KW:KW-0472	Accession number of the SwissProt keyword *membrane*

TABLE 5.10: Inferred by electronic annotation: keyword mapping.
In this case, the protein Leukocyte-specific transcript 1 protein was annotated to the keyword *membrane* by UniProt biocurators. The MGI database automatically transferred this annotation to the corresponding gene *Lst1* and the GO term *integral to membrane* on the basis of a mapping between UniProt keywords and GO terms. Since this annotation was not reviewed individually by a biocurator at MGI, it has received the evidence code IEA.

5.4.2 Inferred from Electronic Annotation

Except for the best-annotated genomes such as that of the Baker's yeast, by far most annotations have the evidence code *Inferred from Electronic Annotation* (IEA). These are annotations that have been generated computationally but have not been checked by a biocurator for accuracy. There are two main classes of IEA annotations. The first involves mapping functional data from other databases that use vocabularies other than GO to describe biological functions. One of the most important databases in molecular biology is UniProt (previously known as SwissProt), which provides comprehensive, high-quality information on protein sequence and function [254]. Each UniProt knowledge base (UniProtKB) entry can be tagged with keywords from ten different categories including biological process, cellular component, molecular function, developmental stage, coding sequence diversity, disease, domain, ligand, post-translation modification, and technical terms. The goals of UniProt are different from those of GO, and even though there is a good amount of overlap in the terms in the categories biological process, cellular component, and molecular function with the terms in the respective subontologies of GO, they are not identical. The Mouse Genome Informatics (MGI) database has created a mapping from UniProt keywords to GO terms, and used this to automatically generate GO annotations for some mouse proteins based on the UniProt keywords for the proteins. Since these annotations were generated by automatic parsing of the UniProt data and have not been individually checked by a biocurator, they have received the annotation code IEA. Table 5.10 offers an example of such an annotation.

There are relatively few model organism databases compared to the number of species for which gene and protein sequence information is available. In the UniProtKB, there are currently over a half a million protein sequence entries from over 12 thousand different species. The Gene Ontology Annotation project provides annotations for UniProtKB entries using GO terms, currently containing over 32 million annotations [22]. Many of these are based on the transfer of GO annotations from well-studied organisms such as the mouse to other, less-studied organisms. One-to-one orthology as predicted by the Ensembl project [130] is used to transfer experimentally verified manual GO annotation data to orthologs.[1] The UniProtKB protein accession number of the protein that was investigated experimentally is indicated in the "With" columns of the annotation file. The GO reference code `GO_REF:0000019` is used in the DB:Reference field of the annotation file to reference this type of IEA annotation.

In all, there are currently 17 different evidence codes for GO annotations. Table 5.11 gives an overview of the codes and shows their distribution among three genomes.

5.4.3 The True Path Rule and Propagation of Annotations

The GO consortium has defined rules for reasoning over GO annotations.[2] Among other things, these rules specify the correct behavior of programs that perform GO term overrepresentation analysis, which is will be covered in detail in Chapter 8.

Subsumption

To **subsume** is to incorporate something under a more general category. For instance, RNA and DNA are subsumed under the category nucleic acid. In ontologies, if some class C is-a D, we say that C is a subclass of D, which is equivalent to saying that D subsumes C.

If a gene g is annotated to a GO term X and X is_a Y, then g is also annotated to Y. This inference rule is also to be applied transitively, so that g is annotated not only to X and Y but also to all of the is_a ancestors of X in the GO DAG. Annotations also propagate over part_of relations. If g is annotated to X and X is part_of Y, then g is annotated to Y. This means that a GO annotation to some term t implies annotation to all ancestors of t along any path. These inferred annotations are not explicitly stated in gene association files (this would lead to a substantial increase in size of these files), but rather it is expected that software for GO analysis will perform the above-described inference steps automatically. Since occasionally there may

---

[1]The assumption is that orthologous genes, i.e., genes in different species that derive from a common ancestor gene, are likely to have the same functions.

[2]See the page about gene association file (GAF) inference in the GO wiki at http://wiki.geneontology.org.

Code	Description	Human	Fly	Thale cress
Experimental Evidence				
EXP	Inferred from experiment	6464	0	0
IDA	Inferred from direct assay	24013	5725	18981
IPI	Inferred from physical interaction	21962	1143	2449
IMP	Inferred from mutant phenotype	4315	11449	7143
IGI	Inferred from genetic interaction	293	1290	1927
IEP	Inferred from expression pattern	482	450	3215
Computational Evidence				
ISS	Inferred from sequence or structural similarity	6007	15292	25923
ISO	Inferred from sequence orthology	0	1	0
ISA	Inferred from sequence alignment	0	141	0
ISM	Inferred from sequence model	0	40	0
IGC	Inferred from genomic context	0	0	0
RCA	Inferred from reviewed computational analysis	35	19	46
Computational Evidence (automatically assigned)				
IEA	Inferred from electronic annotation	100584	14923	39056
Author Statement Evidence				
TAS	Traceable author statement	14359	5126	6067
NAS	Non-traceable author statement	8227	11045	850
Curator Statement Evidence				
IC	Inferred by curator	831	208	213

TABLE 5.11: GO evidence codes. The table shows 16 evidence codes as well as the number of annotations to each code for humans, the fruit-fly *Drosophila melanogaster*, and *Arabidopsis thaliana* (Thale cress), which is a small flowering plant that is currently the most studied model organism among plants. The seventeenth evidence code, ND (no biological data available), used for annotations when information about the molecular function, biological process, or cellular component of the gene or gene product being annotated is available.

be more than one path from a given GO term to an ancestor term, software must be written to ensure that redundant inferred annotations are avoided. In formal description logics, these rules can be expressed using regular role inclusions axioms. For DL, this feature is commonly abbreviated by $\mathcal{R}$ (see subsection 2.5.3).

Annotation Propagation Rule

Genes or gene products are annotated to the most specific GO terms possible according to the current state of knowledge. The annotations are implicitly propagated to all ancestor terms. For instance, if a gene is explicitly annotated to the GO term *beta-catenin binding*, it is expected that computer programs using GO annotations will implicitly add the gene to the list of genes added to all the ancestors of *beta-catenin binding*, including for instance its parent term *protein binding* (Figure 5.2). Annotations are also propagated along part_of relations according to the GO inference rules that will be explained in Chapter 12.

The *true path rule*[3] is a constraint on the structure of the links in GO that ensures logical consistency of the ontology.[85] Genes are annotated with the most specific GO terms possible to reflect current knowledge on the gene's function. Because of the annotation propagation rule, genes are implicitly understood to be annotated to all ancestors of the term up to the root, and therefore, the pathway from a child term through all ancestors back to the root must be true (biologically accurate). That is, child terms inherit the meaning of all of their parent terms, which implies that every annotation to a child term should be true for every parent of that child (Figure 5.2).

It is important that the relations in GO are defined such that the true path rule is not violated in order to ensure correct and comprehensive search results and to allow useful inferences to be made about the functions and activities of gene products. Figure 5.3 shows an example of how poorly defined terms and relations can violate the true path rule and how restructuring of the GO structure was performed to correct this.

As we will see in the course of this book, ontology development is a creative task that demands a profound knowledge of the domain being modeled and a number of decisions on how to model the entities and relations of the domain [186]. The GO is intended to be useful for all species, and the modeling error shown in Figure 5.3 presumably resulted from a *Drosophila*-centric modeling decision to make *chitin metabolism* a child of *cuticle synthesis* (chitin is used only for cuticle synthesis in the fly). GO terms should not be species-

---

[3]We note that the phrase "annotation propagation rule" is not currently used in the literature on GO. Instead, the phrase "true path rule" is occasionally used both in the sense of the "annotation propagation rule" given here as well as with the sense given by the documentation on the GO Web site that is also presented here.

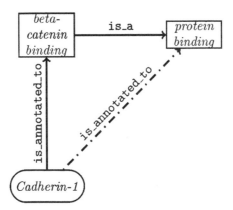

FIGURE 5.2: Annotation Propagation along is_a. The protein *Cadherin-1* is_annotated_to the GO term *beta-catenin binding*. Since GO defines *beta-catenin binding* to be an is_a child of the term *protein binding*, the protein Cadherin-1 is implicitly annotated to *protein binding* as shown with a dashed line. As we will see in later chapters, this is a form of computational *inference*.

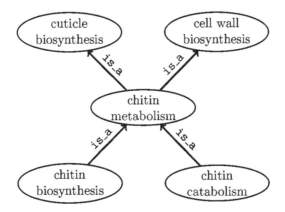

FIGURE 5.3: A Violation of the True Path Rule. In GO, every path from a node back to the root must be biologically accurate. In an early version of the GO, the structure of terms that were used to describe chitin metabolism violated the true path rule. Chitin metabolism is a part of cuticle (i.e., exoskeleton) synthesis in the fly and is also part of cell wall organization in yeast. A yeast chitin synthase gene might be annotated to the term *chitin biosynthesis*, which would implicitly imply the annotation of that yeast gene to *cuticle biosynthesis*, which is nonsense because yeast does not have cuticles.

specific, and after this problem was recognized [114], the structure of the ontology for the terms involved in the problem was modified (Figure 5.4).

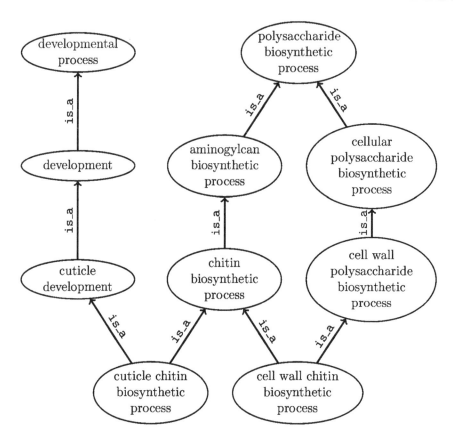

FIGURE 5.4: A Correction of a Violation of the True Path Rule.
The recognition of the problem shown in Figure 5.4 led to the addition of new
GO terms and restructuring of the ontology. A yeast gene involved in chitin
biosynthesis can now be annotated to *cell wall chitin biosynthetic process*. All
of the implicit annotations to the ancestors of that term are now correct, and
the true path rule is fulfilled.

5.5 GO Slims

GO slims are reduced versions of the entire GO that consist of the most important general terms of the original ontology. The main use of GO slims is to provide a broad functional overview of the annotations of all the genes of a genome, or of other comprehensive gene sets. This provides biologists with a bird's eye view of a given genome and allows them to compare the general functional classes that are most common in two or more different organisms. Figure 5.5 was generated by mapping all of the GO annotations to genes in the zebrafish genome to only 127 GO terms in the generic GO slim ontology. It shows biological process terms with at least 250 annotations in the zebrafish genome. In the exercises, we explain how to perform this kind of analysis using program code and data from the GO Web site and ask the reader to use GO slim analysis to compare the functional categories of two different genomes.

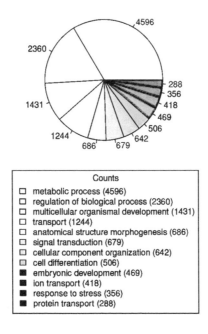

FIGURE 5.5: GO Slim Annotations for the Zebrafish. GO Slim analysis displays the broad general categories of the *Danio rerio* (zebrafish) genome.

5.6   Exercises and Further Reading

The original publication on GO by the GO Consortium [14] remains a good introduction to the goals and scope of GO. David Hill and coauthors provide an excellent description of GO annotations and how they are produced [115]. Barry Smith and coworkers provide a highly readable treatment of relations in biomedical ontologies which has become standard for GO and other OBO ontologies [238]. The GO Web site at http://www.geneontology.org is the best source of information about GO, and also offers downloads of the ontology file, annotation files, and links to GO browsers and other software. It is quite worthwhile to explore the GO using the GO consortium browser AmiGO. The European Bioinformatics Institute has developed another highly recommended browser for GO called QuickGO. A tutorial for using QuickGO for data mining was published by Rachael Huntley and colleagues [131].

Exercises

5.1 Browsing GO terms and annotations can be a useful way to learn about biology and to get a feeling for the scope and intent of GO. A complete list of GO browsers is available from the Tools section of the GO Web site at

http://www.geneontology.org.

One of the best browsers was developed by the GO Consortium and is available directly at the GO Web site. QuickGO is another very useful browser that is available at

http://www.ebi.ac.uk/QuickGO/

The QuickGO Web site offers a number of written and video tutorials that are highly recommended for learning more about GO. For this exercise, enter the word "microfibril" into the main search window of QuickGO. This will return a number of items including the GO term *microfibril* (GO:0001527). Clicking on this will take you to a Web page with information on this term. Click on the tab for Protein Annotation to explore lists of proteins annotated to this term. Limit the results by clicking on the blue boxes under the column names: DB: UniProtKB and Taxon:9606 (human). Now, click on the Refresh button to get a list of annotated proteins. Then, spend some time to explore the online tutorials (as can be seen, this exercise is non-directive).

5.2 Use the QuickGO or AmiGO browser to examine the implicit (indirect) annotations of genes to all ancestors of any term a gene is annotated to. To do so, first look at all genes annotated to a given GO term (it will be easier to complete this exercise if you limit annotations to some species (taxon) and perhaps only one evidence code - use the online help systems of QuickGO or AmiGO to learn how to do this). Confirm that if a gene is annotated to a GO term, it is always also annotated to all of the ancestors of that term.

5.3 GO Slim (1). In this exercise you will learn how to create GO slim annotations for any annotated organism and to display the distribution of functional categories using R. The **go-perl** package was written by Chris Mungall for performing a number of different kinds of analysis with GO and OBO style ontologies. Among other things, it provides a simple way to create GO slim annotations. The simplest way to install go-perl from CPAN (a freely available archive of Perl modules) is directly from the Bash shell:

```
$ perl -MCPAN -e shell
$ install GO::Parser
```

If this does not work or if you would prefer not to install from CPAN, download the go-perl package and follow installation instructions appropriate for your system:

> http://search.cpan.org/~cmungall/go-perl/

We will need the perl script map2slim from this distribution. The CPAN installer will generally install it in a location such as /usr/local/bin/map2slim, so that it is available from the shell. Now go to the GO Web site

> http://www.geneontology.org

and download the generic GO slim ontology file goslim_generic.obo and the standard ontology file **gene_ontology.1_2.obo**. Also download the gene association file for the zebrafish *Danio rerio* and run the map2slim script as follows (as a single command line):

```
$map2slim -t-c goslim_generic.obo gene_ontology.obo
   gene_association.zfin -o zebrafish_counts.txt
```

The -c flag causes map2slim to output the counts of annotations to the GO slim terms. The -t flag causes it to indent the lines of output according to the depth of the tree hierarchy in the GO slim file. The -o flag indicates the name of the output file.

Examine the output file **zebrafish_counts.txt**. What are the most common terms in the molecular function and cellular component ontologies?

5.4 GO Slim (2). In this exercise you will learn how to use R code to create a simple graphic representing the distribution of functional categories in GO slim analysis. Although pie charts are not always a good way to convey information and are not intended to display exact proportions, they are a visually simple way to display large amounts of data and are often used for GO slim analysis (the bar plot is another commonly used option).

We first examine the output of the map2slim script. In this example, we will recreate the GO slim analysis shown in Figure 5.5.

We first note the counts of all biological process terms with at least some threshold value of annotations (here > 250), and assign the values to a vector we will call `ct`. We also set the names of the elements of `ct` to the counts, because the `pie` function of R that we will use to make the piechart uses the "names" attribute of the elements to create the labels in the piechart.

```
ct<-c(4596,2360,1431,1244,686,679,642,506,469,418,356,288)
names(ct)<-ct
```

We now define a vector of gray-scale colors for the various sections of the piechart, and expand the bottom of the clipping rectangle to make room for the legend using the par function.

```
colors<- c("white","grey95","grey90","grey85","grey80",
    "grey75","grey70","grey65","grey60","grey55","grey50",
    "grey45")
par(xpd=T, mar=c(17,0,2,0))
```

Finally, we create the piechart and add a legend. The fill argument to the `legend` command adds small boxes next to each item with the same color as in the pie.

```
pie(ct,col=colors)
legend(-1.25,-1.2,c("metabolic process (4596)",
    "regulation of biological process (2360)",
    "multicellular organismal development (1431)",
    "transport (1244)",
    "anatomical structure morphogenesis (686)",
    "signal transduction (679)",
    "cellular component organization (642)",
    "cell differentiation (506)",
    "embryonic development (469)",
    "ion transport (418)",
    "response to stress (356)",
    "protein transport (288)"),
```

```
    fill=colors,
    title="Counts")
par(mar=c(5, 4, 4, 2) + 0.1) # restore margin to default
```

For this exercise, execute this R code and generate the graphic seen in Figure 5.5. Use R's help functions to learn about unfamiliar commands. For instance, to learn about legend, type the following at the R command line:

>?legend

It is reasonably easy to write a Perl script that will generate the above R code on the basis of the output of the map2slim script. This is left for extra credit.

5.5 GO Slim (3). Use map2slim and the R code from the previous exercises to compare the distribution of top-level functional categories from two different organisms. Use the generic GO Slim ontology mentioned in exercise 5.3. Go to the GO Web site and download annotation files for two distantly related organisms. For instance, you could consider comparing the worm *Caenorhabditis elegans* with the human genome. What are the categories which differ most between the two organisms? Can you think of biological explanations for your observations? It may be useful to perform the analysis separately for each of the three GO ontologies: biological process, molecular function, and cellular component.

5.6 Review Figures 5.3 and 5.4 and imagine there is a yeast gene that is annotated to the GO term *chitin biosynthesis*. Explain why this leads to error according to the ontology structure shown in Figure 5.3, but is correct according to the ontology structure shown in Figure 5.4. Use the annotation propagation rule and the true path rule to formulate your explanations.

Chapter 6

Upper-Level Ontologies

"Reality is like cheese: it can be cut in many ways."
-Pierre Grenon, Barry Smith, and Louis Goldberg [96]

Formal ontology deals with the interconnections of things, with objects and properties, parts and wholes, relations and collectives [234]. Computational ontological analysis depends implicitly or explicitly on the way the ontologies being used have implemented formal ontological principles. In this chapter we discuss two upper level ontologies that have been by far the most influential in the field of bio-ontologies, and then demonstrate how they are critical to understanding the structure of GO and of GO annotations.

6.1   Basic Formal Ontology

The Basic Formal Ontology (BFO[1]) is designed as a top-level ontology for biomedical ontologies in order to provide a common framework for defining and structuring the concepts of specific domain ontologies. One can think of an upper-level ontology as a guideline for how to carve up reality into coherent pieces. Upper-level ontologies are therefore theories of reality, and represent an aspect of ontologies that has close connections to philosophy, reaching back at least to the time of Aristotle. On a more practical level, upper-level ontologies offer a sort of a how-to manual for constructing a good domain ontology.

---

[1]Some pronounce it 'bufo' ("boofoe"), after *bufo*, a large genus of about 150 species of true toads in the amphibian family *Bufonidae*. This pronunciation of BFO also echoes the fact that much of the work on BFO has been done at the University of Buffalo.

Aristotle's Categories

The Greek philosopher Aristotle (384 BC–322 BC) can be considered one of the first ontology researchers, and in his many works he provides insights into ontological topics such as the relations between Universals and Particulars, as well as classifications of various domains (e.g., of animals) that are the earliest surviving comprehensive ontologies. In the work entitled *Categories*, Aristotle proposes ten high-level classes such as *substance*, *quantity*, and *quality*, into which he claimed all entities could be classified. *Categories* can thus be regarded as the first upper-level ontology.

The strategy the BFO employs to carve up reality has turned out to be very useful for biomedical ontologies, and many important bio-ontologies such as GO, the Protein Ontology, and the Cell Ontology have used the BFO to help structure their domains.[2] An understanding of the concepts surrounding BFO is important especially for those who will be participating in the construction of new ontologies, annotating data with ontology terms, or using ontologies for inference.

6.2   The Big Divide: Continuants and Occurrents

The BFO divides all entities into objects and processes, which it calls continuants and occurrents. Although newcomers might wonder why the BFO chose to use unusual (and perhaps "difficult") words to name its classes, one soon realizes there is a big advantage to it. Familiar words such as "object" and "process" have very many different meanings (for the noun forms alone, there are 10 and 13 meanings in *The Oxford English Dictionary*), and so it is quite likely that different readers would understand these words differently.

Intuitively, the big divide in the BFO lies between entities in three-dimensional space (continuants) and entities in four-dimensional space, i.e., in space and time (occurrents). In biomedicine, this is like the difference between a three-dimensional anatomical object such as the heart (a continuant), and the physiological functioning of the heart to pump blood (an occurrent). In a sense, continuants and occurrents represent two different ways of viewing the same objects. An ontology for three-dimensional objects is like a snapshot of the world (e.g., a photograph of a heart), while an ontology for four-dimensional objects is like a videoscopic view of reality (e.g., a video of a beating heart). BFO aims to be a complete and adequate ontology of reality which is divided into these two orthogonal and complementary categories [95].

---

[2]A full list of ontologies using BFO is available at http://www.ifomis.org/bfo/users.

6.2.1 Continuants

Continuants are entities that continue to exist over time and preserve their identity despite change. For instance, you, dear reader, are a continuant, and have preserved your identity despite the fact that you have changed in many ways small and large since you were born.[3]

The BFO defines three main classes of continuant. An *Independent continuant* is a continuant that is a bearer of quality and realizable entities, in which other entities inhere and which itself cannot inhere in anything. Examples of independent continuants are an organism, a heart, a leg, a person, a symphony orchestra, a glucose molecule, a leukocyte, and a fly's eye. Let us examine the meaning of the components of the definition of an independent continuant:

- *Bearer of quality*: Independent continuants can have characteristics (qualities) that are essential attributes of them (that "inhere" in them). For instance, a fly's eye can bear the quality "red." Note that **inheres in** is the inverse of **bears**. If a fly's eye *bears* the quality red, then the quality red *inheres in* the fly's eye.

- *Bearer of a realizable entity*: Independent continuants can have the ability to perform a specific functioning or process that is realized under certain conditions. For instance, the independent continuant blood bears the realizable entity "ability to coagulate" that is realized only if a blood clot is formed.

- *Other entities can inhere in an independent continuant, but an independent continuant cannot inhere in other entities*: We have seen how independent continuants can bear quality entities or realizable entities. An independent continuant cannot itself inhere in (be an essential attribute of) another entity. Note that an independent continuant can be a physical part of another independent continuant (e.g., a heart is a part of a human being).

A *dependent continuant* is continuant that is either dependent on one or other independent continuant or inheres in or is borne by another entity. As we have seen above, a *quality* and a *realizable entity* are dependent continuants that can be borne by an independent continuant. Qualities such as the redness of wine are dependent in the sense that the redness cannot exist without the wine.

There are three main classes of realizable entity. A *function* is a realizable entity which inheres in a continuant entity and which is manifested in a specific context to achieve a specific end. A biological function inheres in an independent continuant that is part of an organism (e.g., a cell). The realizations of the function form part of the life plan of the organism [13]. For

---

[3]At least since you were born [237].

instance, the protein PTPN11 is the bearer of the function *phosphoprotein phosphatase activity*, which is realized in certain cellular contexts (PTPN11 can remove the phosphate group from phosphorylated proteins in the cytosol). The function is only realized if a PTPN11 protein comes into contact with a phosphorylated target protein; otherwise, the function is not realized.

A *role* is a realizable entity the manifestation of which brings about some result or end that is not essential to a continuant. Rather, the continuant can participate in the *role* in some kinds of natural, social or institutional contexts. For instance, a restriction enzyme may play a role of helping to identify fragment length polymorphisms in a molecular biology experiment, which is quite different from the *function* restriction enzymes have evolved to fulfill in nature. Similarly, a snake's venom only plays the role of a toxin if the snake bites you.

A *disposition* is a realizable entity that can cause a specific process or transformation in the object in which it inheres, under specific circumstances and in conjunction with the laws of nature. A general formula for dispositions is: X (object) has the disposition to initiate a process P under conditions C. Examples are the disposition of blood (X) to coagulate (P) if there is a wound (C) or the disposition of a patient with a weakened immune system (X) to contract disease (P) if he is exposed to micro-organisms (C). Thus a disposition is a potentiality, grounded always in some actual physical basis.

6.2.2   Occurrents

An *occurrent* is an entity that happens, unfolds or develops through time. Occurrents have a beginning, middle, and end and include processes that change continuant entities. Occurrents are the events or happenings in which continuants participate [13]. Occurrents may also be called events, processes or activities. Examples include the life of an organism and the process of cellular meiosis.

The current (1.1) version of the BFO contains a total of 39 terms. Figure 6.1 shows a simplified view of the top levels of the BFO.

6.3   Universals and Particulars

Philosophy distinguishes between *Universals* (sometimes called types, species, classes, or kinds) and *Particulars* (sometimes called instances or individuals). An example of a Universal is "the liver," and an example of a Particular is the liver located in your abdomen.

Science studies instances in order to discover things about universals. That is, we do experiments on individual molecules, cells, or organs, but we generalize the results of the experiments by induction. That is, if we show by

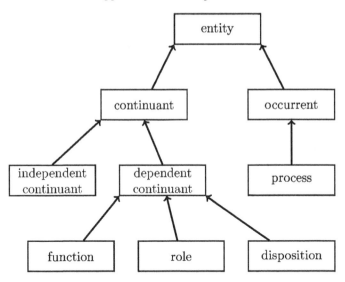

FIGURE 6.1: A Simplified View of the BFO. The BFO divides all entities into continuants and occurrents, which very roughly are equivalent to a photographic snapshot (continuant) or a video film (occurrent) of entities that may depend on one another as the process of pumping (occurrent) depends on the anatomic structure of the heart (continuant). All of the edges represent subclass (is_a) relations.

an experiment that Cyclin D is a protein involved in the cell cycle in yeast, we have only investigated a certain number of yeast cells (instances), but we conclude that Cyclin D is involved in the cell cycle in all yeast cells. Scientific ontologies such as the GO represent universals. Whether or not we want to believe that there is a "universal" yeast cell, the only things that we can observe are actual yeast cells thus universals exist in reality through their instances.

6.4 Relation Ontology

As we have seen, ontologies have two main components: terms (concepts) and semantic relations between the terms. The Relation Ontology (RO) is to the relations what the BFO is to terms, that is an attempt to provide a logically sound common framework for the interconnections between terms in domain ontologies and to ensure consistency among all ontologies using relations based on RO. We have seen that the GO uses two main relations, is_a and part_of, but many more relations are used by current bio-ontologies.

It is essential for logical inference and interoperability between ontologies that relations are used in a consistent and well-defined manner. There are multiple reasonable interpretations of the English words "is a." For instance, "Mary is a human being" means that Mary is an instance of the universal called human being. The sentence, "A woman is a human being" refers to two universals, and the English words is_a mean "subclass of" rather than "instance of." Likewise, the English words "part of" might be taken to mean possible parthood or necessary parthood, among other meanings: Does "A *part of* B" mean that all instances of A only occur as a part of instances of B? Do instances of B necessarily have an instance of A? To address these problems, the RO defines a theory of Universals and Particulars and uses a small set of basic, axiomatic relations to provide unambiguous and computable definitions of relations. RO has been adopted by the GO and many other bio-ontologies that will be presented in the next chapter. In the following text, we will present the RO's definitions of the is_a and part_of relations. Readers are referred to the original publication [238] for explanations of other relations such as located_in, derives_from, and adjacent_to.

Terms in scientific ontologies such as the GO refer to Universals. In some cases, however, relations pertain between Particulars. We denote a directed binary relation from A to B as $A \xrightarrow{R} B$.

We can now define three types of binary relations, using the abbreviation U for Universal (class) and P for Particular (instance):

1. $U \xrightarrow{R} U$ denotes a relation between two Universals, for instance *cofactor transporter activity* is_a *transporter activity*.

2. $P \xrightarrow{R} U$ denotes a relation between a Particular and a Universal, for instance *Mary* instance_of *human being*.

3. $P \xrightarrow{R} P$ denotes a relation between two Particulars, for instance, one particular nucleus part_of one particular cell.

We can now define primitive relations which will serve as axioms with which all other relations can be defined. For our purposes, we need only define two.

1. x instance_of X – This signifies that x is an instance (Particular) of the Universal X. If X is a continuant it may be necessary to specify the time t (e.g., *Abba* instance_of *Top-ten band* may have been true in the 1970s, but, thankfully, time goes by).

2. x part_of y – This is a primitive relation between two instances meaning that x is a part of y at a specified time t.

We can now use these axioms to provide a definition of the *is a* relation.

Definition 6.1 *A is_a B* $\overset{\text{def}}{=}$ *For any instance x and time t, if x* **instance_of** *the class A at t, then x* **instance_of** *the class B at t.*

This definition implies that if we know that Snoopy is an instance of *Beagle* and that *Beagle* is a subclass of *Dog*, then we can infer that Snoopy is an instance of *Dog* (Figure 6.2).

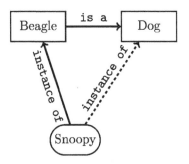

FIGURE 6.2: **Definition of is_a in the Relation Ontology**. If *A* is a subclass of *B* and something is an instance of *A* at time *t*, then it is also an instance of *B* at time *t*. The instance_of edge drawn with a solid line corresponds to an asserted fact, and the edge drawn with a dashed line to an inferred fact.

The definition for processes does not require the time to be specified. This is because a continuant, but not a process, can instantiate different classes in the course of its existence while still preserving its identity. For instance, a continuant called John might instantiate the classes "Baby," "Child," "Teenager," and "Adult" at different times of John's existence. Processes, on the other hand, have a beginning, middle, and end, and thus of necessity occur at some particular time *t*. The definition of the relation is_a for processes in the Relation Ontology is thus:

Definition 6.2 *P is_a Q* $\overset{\text{def}}{=}$ *For any instance p, if p* **instance_of** *the process P, then x* **instance_of** *the process Q.*

For instance, the GO term *protein metabolic process* is a subclass of the GO term *macrometabolic metabolic process*. Any process that is an instance of *protein metabolic process* is thus also an instance of *macrometabolic metabolic process* from the beginning to the end of the process. Because of the way the BFO defines processes (occurrents), there is no way of the process instance ceasing to be a *protein metabolic process* while still retaining its identity in the way that continuants can. Therefore, there is no need to specify the time during which the is_a relation is valid; it is valid throughout the lifespan of the entire process.

In addition to is_a, the part_of relation is so commonly found in ontologies that it is called a foundational relation in the RO. The axiomatic part_of relation described above is a relation between instances, e.g., Joe's appendix is part_of Joe at time t (one never knows if the appendix might need to be removed from Joe at time $t + 1$). It is a little less obvious what a parthood relation should mean for universals. The RO defines parthood as a relation between classes by means of how parthood must be defined for instances of those classes. If we say that the class *Nucleus* is part of the class *Cell*, then we mean that each and every instance of *Nucleus* must be part of some particular instance of *Cell*, that no instance of *Nucleus* exists that is not part of some instance of *Cell*. The formal definition for continuants is

Definition 6.3 *A part_of B at t $\stackrel{\text{def}}{=}$ For any instance x and time t, if x instance_of the class A at t, then there exists some instance y, such that y instance_of of B at time t and x part_of y at t.*

Note the asymmetry in the definition. **All** instances of *Nucleus* are part_of **some** instance of instance of *Cell*.[4] Figure 6.3 illustrates the parthood definition.

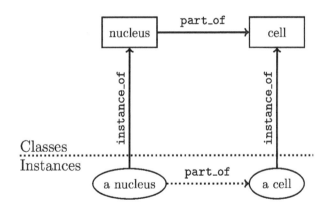

FIGURE 6.3: Definition of part_of in the Relation Ontology. The class *Nucleus* is *part of* the class *Cell* if for every instance n of *Nucleus* there is an instance c of the class *Cell* c such that n is part of c on the instance level (shown as a dashed line). Note that this relation is valid for all times t such that n is an instance of *Nucleus*.

The initial version of RO comprised 10 relations. Table 6.1 shows examples from the biomedical domain for each of these relations. The original publication [238] provides definitions for all relations in terms of a small set of instance relations in a way that allows logically clear inference rules to be derived.

---

[4]Although this may seem curious at first, note that it would make no sense to state that **all** instances of *Nucleus* are part_of **all** instances of the class *Cell* because any one nucleus is part of one particular cell. Likewise, to say that **some** instance of *Nucleus* is part_of **some** instance of *Cell* would not make a general statement about the class of *Nucleus*, only about some particular nucleus.

Foundational relations	
Relation	**Example**
is_a	DNA helicase activity is_a helicase activity
part_of	cardiac chamber development part_of heart development
Spatial relations	
Relation	**Example**
located_in	intron located_in gene
contained_in	liver contained_in abdominal cavity
adjacent_to	intron adjacent_to exon
Temporal relations	
Relation	**Example**
transformation_of	platelet transformation_of megalokaryocyte
derives_from	osteoblast derives_from mesenchymal stem cell
preceded_by	cell division preceded_by DNA replication
Participation relations	
Relation	**Example**
has_participant	translation has_participant tRNA
has_agent	transcription has_agent RNA polymerase

TABLE 6.1: Relations in the RO. Examples of the 10 relations of the Relation Ontology [238] are shown. Further relations have been proposed for future versions of the RO.

6.5 Revisiting Gene Ontology

It should now be apparent that the cellular component ontology of the GO consists of independent continuants – objects such as a flagellum or a chromosome [240]. The molecular function ontology consists of dependent continuants. If a molecular has a function, this is a realizable entity that persists over time and does not unfold itself over time as an occurrent does. Thus, a molecular function is regarded as an enduring potential to carry out an activity in an appropriate biological context. This potential continues to exist even if it is not currently being exercised (e.g., an enzyme continues to have the potential to catalyze a reaction, even if its substrate is not currently present within the cell).

If on the other hand, the function is realized, the result is a *process* which does have a beginning, middle, and end. The realization of a function in this way is often referred to as a *functioning*. In general, biological processes, which are accomplished by realizations of one or more molecular functions to some biological end, are occurrents [240].

While terms of the cellular component ontology represent independent entities, the molecular function and biological process entities depend on other entities to allow them to occur. The molecular functions inhere in molecules, and the biological processes are dependent on molecules, aggregate, cells, organs or organisms for their realizations.

6.6   Revisiting GO Annotations

We have seen in chapter 5 that one of the main applications of GO has been to enable computer-readable annotations about genes and genes products. Nearly a million annotations have been created by biocurators at model organism databases, genome annotation centers, and other databases for molecular biology, and many more millions of annotations have been generated computationally. We will see in Part III of this book that annotations are the basis for a number of extremely useful ontological algorithms for biomedical data. Therefore, it seems important to define exactly what is meant by a GO annotation using the formal ontological framework of the BFO and RO.

Definition 6.4 *An annotation is the statement of a connection between a type of gene product and the types designated by terms of GO. Biocurators create annotations on the basis of observations about the gene product and inferences drawn from the observations.*

Note that the annotation is about universals (types). As noted above, scientists are able to perform actual experiments only on instances of the types, but generalize the results to universals.

In order to create an annotation about some gene or gene product, biocurators first identify the relevant literature and information about whatever experiments have been performed about the gene, or in some cases other information such as experiments performed about an orthologous gene in another organism. The second major step is to identify the best GO term or terms for the annotation and to decide upon the appropriate evidence codes (see Chapter 5). The following guidelines were developed by Hill and coworkers [115] to promote consistent annotation practices among different groups involved in biocuration. They also form a good basis for bioinformaticians who may use this information for inference.

Annotations are about a type (universal) of a gene product. An entry in a database such as UniProt is taken to represent this universal class. For instance, the UniProt entry CAH8_HUMAN (P35219) is taken to represent the universal class *Carbonic anhydrase-related protein CA8 in humans* rather than specific instances of this protein in a specific test tube.

Molecular Function Annotations

An annotation of a protein P to a GO molecular function term means that a specific instance p of P has a disposition (propensity) to realize this function in certain contexts. For instance, the enzyme called *catalase* catalyzes the decomposition of hydrogen peroxide into water and oxygen:

$$2H_2O_2 \xrightarrow{\text{catalase}} 2H_2O + O_2$$

Thus, a single protein of type *Catalase* has as its function an instance of the Molecular Function type *Catalase activity*. By annotating the protein *Catalase* with a term called *Catalase activity*, we mean that in the appropriate context, a protein of type *Catalase* will realize the function *Catalase activity*.[5] Note that a particular catalase protein may not always be exerting catalase activity, for instance if there is no hydrogen peroxide in the cytosol, a particular instance of *Catalase* might not exert catalase activity even once. Therefore, the annotation actually refers to the *potential* to realize the activity, rather than to the activity itself. In BFO, this kind of potential is referred to as a *disposition*.

The kinds of experiments that are generally cited to document molecular function GO terms are usually assays that test whether a collection of instances of some protein realizes some molecular function, say by measuring the concentration of the substrates or products of a reaction after adding a certain amount of the protein to a solution.

Biological Process Annotations

A molecular function instance is the potential of an instance of a gene product to realize a certain function. In contrast, a biological process instance is the actual execution of such a molecular function instance or of several such instances in order to accomplish some biological objective. Biological process instances act on a higher level of granularity (cell or organism) than do molecular function instances (molecule). Similar to molecular function annotations, biological process annotations denote relations between universals: a link between a gene product type and process types at the level of the cell or organism. If we annotate a protein class P with a biological process term B, we mean that instances of P are associated with known or unknown molecular functions whose realization contributes to the occurrence of an instance of the biological process B. For instance, the protein called forkhead box C1 (Foxc1) is annotated to the biological process term *eye development* in the mouse. An instance of the term *eye development* would be the development of an eye in a particular mouse, and individual Foxc1 proteins contribute to this.

---

[5]Note that commonly the name of many proteins, and especially enzymes, is similar or identical to their molecular function. For example, the word "catalase" refers both to a protein with a certain amino acid sequence and structure, as well as to the function that can be realized by that protein of catalyzing the reaction shown above.

Molecular function annotations are typically made on the basis of experiments that measure one specific kind of molecular reaction, which can be conducted completely *in vitro* (e.g., in a test tube). This is often not possible for biological process terms, which may comprise numerous known and unknown molecular functions. For instance, the *eye development* is an intricate process that depends on hundreds of genes and hundreds of different molecular functions, not all of which are known. Therefore, biological process annotations are often based on the results of genetic perturbations. Genetic manipulations in the mouse including knocking out genes (making them inactive by targeted mutation) have become a mainstay in molecular biology research aimed at understanding the functions of genes. Biological process annotations are often based on the descriptions of mice with knockouts of the gene in question: What doesn't work if gene x is inactivated? As we have seen in Chapter 5, these annotations receive the evidence code IMP (inferred from mutant phenotype).

For instance, the mouse protein Foxc1 received the annotation *eye development* on the basis of experiments in a mouse model in which the *Foxc1* gene had been knocked out. The affected mice die at birth with hydrocephalus, eye defects, and multiple skeletal abnormalities [146]. Because the knockout mice had defective eye development, it was inferred that Foxc1 must in some way contribute to the normal process of eye development.

Cellular Component Annotations

Cellular component annotations are generally made on the basis of an observation of instances of the cellular component using a microscope in which instances of the protein in question have been made visible by an appropriate technique such as visualization with fluorescently labeled antibodies. For instance, one might observe a fluorescent signal for some protein in the nucleus in a number of images taken in the course of some experiment, and infer that the class of the protein (a universal) localizes in the corresponding cellular component.

6.7   Exercises and Further Reading

Barry Smith is a philosopher who has been highly influential in the field of bio-ontologies by working with the Gene Ontology community to promote the development of logically rigorous frameworks for OBO ontologies on the basis of BFO and RO. Professor Smith runs the National Center for Ontological Research at the University of Buffalo (http://ncor.us/), and has put a large number of resources for research and training in ontology, including several highly recommended streaming video courses, at the Web site of the Buffalo

Ontology Site (http://ontology.buffalo.edu/smith/). Much of the material in this chapter was based on articles on the BFO and RO available at that Web site, which is highly recommended for readers who would like to learn more about the philosophical and logical aspects of bio-ontologies.

The aim of this chapter was to illustrate what upper-level ontologies are and how they are used in the field of bio-ontologies. We have by no means provided a comprehensive treatment of the field, and readers should be aware that there are several different upper-level ontologies with different ways of viewing reality. One particularly interesting top-level ("foundational") ontology for biomedical applications is the General Formal Ontology (GFO). The GFO integrates objects and processes by unifying a three-dimensional and four-dimensional ontology into a single coherent framework based on the philosophy of integrative realism. GFO contains a number of other unique features, among them an ontology of categories, including categories of different types, as universals, concepts, and symbols, but also categories of higher order. GFO includes levels of reality, and contains ontological modules for functions, for roles, and for processes [113, 112]. There are many other upper-level ontologies [184, 83, 161]. Each upper-level ontology has different ways of "cutting up reality," with distinct consequences for domain ontologies based on them.

There are a number of useful articles describing how bio-ontologies are used to handle biological data [34, 36, 52, 152, 216, 233], desirable characteristics for bio-ontologies [46, 243, 154], experiences with building bio-ontologies [85, 90, 86], and naming conventions for ontology terms [222].

Exercises

6.1 Choose an article from a domain you know well, and make a list of the important concepts (universals) in the article as well as the relations between them. Try to determine the occurrents and the independent and dependent continuants. Can you identify subclass (is_a) relations? What about part_of relations?

6.2 Choose an article from the molecular genetics literature that has been used for GO annotation (it is easy to find such articles by looking through GO annotations at the GO Web site; you may have to search a bit to find an article that is freely available for download. If you have any trouble, search for PubMed ID 12117821 at the PubMed Web site). Try to understand the process of GO annotation for the gene products described in the article based on the explanations in this chapter.

Chapter 7

A Selective Survey of Bio-Ontologies

"If a little knowledge is dangerous, where is the man who has so much as to be out
of danger?"

-Thomas Henry Huxley

The Gene Ontology (GO) is the most widely used bio-ontology, as measured by the number of users, number of annotations, or number of citations in scientific articles. The success of GO has been one of the reasons for the proliferation of bio-ontologies since the initial publication of GO in 2000 [14]. At the time of this writing, several hundred bio-ontologies are available. Since biologists and bioinformaticians often need to integrate data from a variety of sources, it is therefore desirable that bio-ontologies are constructed to be interoperable. For this reason, Suzanna Lewis and Michael Ashburner, two of the initiators of the GO project, began to develop the Open Biological Ontologies (OBO) Foundry in the years following the publication of GO. This chapter introduces a number of ontologies that are of particular interest to the authors. There are so many ontologies covering so many areas of biomedicine that a complete survey would require a book of its own. We therefore begin the chapter with a description of the OBO Foundry and of the National Center for Biomedical Ontology (NCBO), which are currently the two main comprehensive resources for biomedically relevant ontologies.

7.1   OBO Foundry

Biomedical ontologies serve to integrate clinical and experimental data and to facilitate structuring and analysis. The value of any one ontology grows with the amount of data annotated using it, because it becomes possible to compare and integrate ever larger amounts of data if the data has been annotated using the same standards. The OBO Foundry has become the most successful group of ontology developers internationally. The single most successful ontology of the OBO Foundry is without doubt the Gene Ontology (see Chapter 5), but many of the other ontologies that are members or aspiring members of the OBO Foundry are also widely used [236].

OBO Foundry does not intend to be an official standards body for bio-ontologies (this does not exist), but rather is a consortium of ontology developers who are developing their ontologies according to a set of shared principles in a way that will allow ontologies to be interoperable and logically well formed. Interoperability of ontologies is the prerequisite for the cross-product definitions that define GO or HPO terms using terms from other OBO ontologies (see Chapter 12). Many essential biomedical research questions involve the comparison of data from two different but related domains. For instance, model organisms are used to study human disease. Therefore, it is essential that the ontologies used to annotate data from domains that will be compared with one another are also logically compatible with one another. This is done by agreement on a number of basic principles.

OBO Foundry has developed a number of principles that are designed to help participating ontologies achieve the goals of interoperability and logical soundness. Ontologies should use a common shared syntax (either the OBO or OWL formats presented in Chapter 4), provide textual definitions for terms and be open to collaborative development by other Foundry members. Furthermore, ontologies should strive to develop formal definitions (see chapters 4 and 12 for information on cross-product definitions), and textual definitions should be made according to the genus-differentia pattern. The OBO Foundry additionally provides guidelines and advice about naming conventions for ontology terms [222].

Another important tenet of the OBO Foundry is that ontologies are designed to incorporate accurate representations of biological reality. There have been criticisms of the OBO Foundry's definition of realism [154], but in essence this principle reflects the needs of biologists who use the ontologies. Instances of the terms of an OBO ontology should correspond to instances in reality. Any ontology is a simplified model of reality at a certain level of abstraction depending on the uses for which it is intended. Ontologies in the OBO Foundry are intended to be useful for scientific research by providing a common language for communication, data annotation, and computational analysis.

There are many quite useful ontologies and classification systems that were designed to fulfill other needs. For instance, in the International Classification of Diseases (ICD), there are many categories that are useful for medical statistics or billing purposes but that do not refer to instances in reality. Here is a list of shoulder lesions in the ICD-10.

```
(M75.) Shoulder lesions
    * (M75.0) Adhesive capsulitis of shoulder
    * (M75.1) Rotator cuff syndrome
    * (M75.2) Bicipital tendinitis
    * (M75.3) Calcific tendinitis of shoulder
    * (M75.4) Impingement syndrome of shoulder
    * (M75.5) Bursitis of shoulder
    * (M75.8) Other shoulder lesions
    * (M75.9) Shoulder lesion, unspecified
```

While M75.0 to M75.5 refer to specific kinds of shoulder lesion, M75.8 and M75.9 do not. Rather, these terms reflect the clinical reality that it is not always possible to diagnose the precise cause of a disease, and symptomatic treatment may work even if the physician does not know the exact diagnosis. It can be important for billing purposes to know that a patient had a shoulder lesion (M75.) that was not one of M75.0–M75.5. Nonetheless, there is no such thing as an "other shoulder lesion" (M75.8) in medical reality, this is merely a convenience classification. This kind of term is not allowed in OBO Foundry ontologies.

7.2 The National Center for Biomedical Ontology

The National Center for Biomedical Ontology (NCBO)[1] is dedicated to providing software and support for the application of ontologies in the biomedical domain [215]. One of the most important tools of the NCBO is the BioPortal, an open repository of biomedical ontologies that provides access via Web services and Web browsers to ontologies developed in OWL, RDF, OBO format and Protégé frames. BioPortal functionality includes the ability to browse, search and visualize over 200 biomedical ontologies [189]. The authors have found BioPortal to be particularly useful for searching for ontology terms for specific concepts amongst all or subsets of the over 200 bio-ontologies.

A number of useful tools have been developed at the NCBO, including the biomedical ontology recommender, which uses textual metadata or a set of keywords describing a domain of interest and suggests appropriate ontologies for annotating or representing the data [136] and the NCBO Annotator, which processes text submitted by users, recognizes relevant biomedical ontology terms in the text and returns the annotations to the user [230].

7.3 Bio-Ontologies

This section presents selected ontologies of the OBO Foundry, which illustrate one or another aspect of successful ontology design or application. Many of the ontologies described in this chapter are also useful for ontology-driven research on human phenotypes and disease.

---

[1] http://www.bioontology.org/.

7.3.1   Ontologies for Anatomy: The FMA and Model Organisms

The Foundational Model of Anatomy Ontology (FMA) provides a unifying framework for human anatomy that describes the various entities that make up the body together with the relations between them. The FMA is designated as *foundational* because many of its classes generalize to vertebrates other than humans. Additionally, anatomy is the basis for understanding physiology and medicine, and ontologies such as the Human Phenotype Ontology make reference to the FMA to define terms that describe the signs and symptoms of disease.

The FMA distinguishes entities in the anatomical and related domains according to the Basic Formal Ontology (BFO; see Chapter 6). The FMA describes *universals* (types) in anatomy. That is, the FMA does not describe the anatomy of some particular individual, who may have had his appendix removed or may display some minor or major deviation from "normality." Rather, *canonical anatomy* refers to universals which are idealizations of the human body and its component parts.

A continuant is an entity that continues to exist over time and preserves its identity despite change. Instances of the type "liver" as well as the universal they instantiate are independent continuants. On the other hand, bile acid secretion by liver cells is an occurent (process) that unfolds from a beginning through successive temporal phases to its ending. The BFO distinguishes between independent and dependent continuants. The lumen (inner cavity) of the gallbladder cannot exist without some gallbladder also existing, and is thus a dependent continuant.

Dictionaries list their content in alphabetical order, regardless of the meaning of their terms. The standard term list for anatomy is the *Terminologia Anatomica* [77]. The FMA arranges its classes in an inheritance hierarchy or taxonomy in a strictly *structural* context. This means that the definitions of the FMA are to be understood within the context of the ontology structure. The FMA formulates its definitions using the genus-differentia pattern using a strictly structural framework.[2] The conventional dictionary definition of the heart usually refers to its function to pump blood. A typical definition might be along the lines of the statement, *The chambered muscular organ that pumps blood received from the veins into the arteries.* This definition mixes structural ("chambered muscular organ") and functional ("pumps blood") aspects, and would present a number of difficulties for an ontology taxonomy based on genus-differentia definitions. Classifying anatomical structures according to function is problematic, not least because many organs have multiple functions. For instance, the kidney functions to remove metabolic waste products from the body, to maintain balance (homeostasis) of electrolytes in blood, to release hormones, to regulate blood pressure, to regulate the production

---

[2]Recall from that a genus-differentia definition has the form "X is a G that D," meaning that X is a subclass of G with the differentia D.

of red blood cells, and to produce vitamin D, which plays a major role in bone metabolism. Should the kidney be classified as an excretory organ or an endocrine organ or something else?

The FMA defines *Heart* based purely on structural considerations: *Organ with cavitated organ parts, which is continuous with the systemic and pulmonary arterial and venous trees.* The genus is *Organ with cavitated organ parts*, and the differentia is purely structural: *continuous with the systemic and pulmonary arterial and venous trees.* No organ other than the heart satisfies this definition.

In addition to the `is_a` hierarchy, the FMA has a rich set of structural anatomical relations that are used to describe the relative positions of different anatomical structures to one another. For instance, the heart is defined to be a `constitutional_part_of` the mediastinum (a group of structures in the thorax, including the heart, the esophagus, and the trachea, that are surrounded by loose connective tissue). Additionally, the heart `attaches_to` the pericardial sac (a sac that contains the heart and the roots of the great vessels).

In its native form, the FMA is a frame-based ontology, whereby each frame stores all the information in the ontology about a named type. The syntax of frame-based ontologies is not covered in this book, but interested readers can consult the homepage of the Protégé editor (which works with frame-based or OWL ontologies) for more information [187]. OWL and OBO versions of the FMA have been developed to help integration with other OBO Foundry ontologies. A Web browser, the FMA Explorer, provides access to a streamlined version of the full FMA. The FMA is one of the largest current biomedical ontologies, and is exemplary in its consistent use of sound ontological methods [56, 164, 188, 239].

Anatomics: Anatomy Ontologies for Model Organisms

Jonathan Bard coined the term *anatomics* to describe the formalization of anatomy in ontologies and the analysis of the complete set of tissues and organs of an organism (the so-called *anatome*) with computers. The relationships of tissues to one another are spatial, developmental, functional and partitive, may be dependent on the developmental stage, and may differ from individual to individual [20].

To date, nearly two million species of plants and animals have been named, and estimates of the total number of species on earth range from 5 million to well over 20 million. Scientific investigations have been concentrated on a much smaller number of model organisms, many of which have served as models of human disease. For instance, the laboratory mouse has been pivotal in making progress in human genetics because of the relative ease with which genes can be manipulated in this organism, the sheep has been extremely useful as a model for studying fractures because of the biomechanical similarities between the human and sheep skeletal systems, and the rat has been used as a model

Ontology	Scope	Reference
Foundational Model of Anatomy (FMA)	Human anatomy	[55, 214]
Mouse Anatomy (MA)	Mouse adult anatomy	[44, 107]
Xenopus (XAO)	Xenopus (frog) anatomy and development	[228]
Zebrafish (FA)	Zebrafish anatomy and development	[62]
C. elegans gross anatomy (WBbt)	C. elegans (worm) anatomy	[105]
Drosophila gross anatomy	Drosophila melanogaster (fruit fly)	[101]
Teleost Anatomy Ontology	A multi-species ontology for ostariophysan fishes	[62]
Cross-Species Anatomy Ontologies		
Common Anatomy Reference Ontology (CARO)	Facilitation of interoperability between existing anatomy ontologies for different species and a template for building new anatomy ontologies	[103]
UBERON	A multi-species anatomy ontology created to facilitate comparison of phenotypes across multiple species	[102]

TABLE 7.1: **Anatomy ontologies**. These are some of the ontologies available to describe macroscopic anatomy of humans and many model organisms.

for many acquired diseases such as hypertension and diabetes, because of the relatively large size of its organs. Thus, much of the biomedical research of the last decades has been concentrated on a few organisms, for many of which extensive genetic, anatomical, and physiological data has been collected. In as much as these organisms serve as models for human disease, it is extremely interesting to use ontologies to serve as a bridge between the genes, anatomy, and physiology of the model organisms and of humans.

Species-centric ontologies are available for the fruit fly *Drosophila melanogaster* [101], the zebrafish *Danio rerio* [62], the frogs *Xenopus laevis* and *Xenopus (Silurana) tropicalis* [228], the worm *Caenorhabditis elegans* [149], and the mouse [68, 107, 207], among others (Table 7.1).

In order to compare the phenotypes of two organisms for medical research, several things are needed. Anatomy ontologies are needed for each of the organisms, because many of the phenotypic abnormalities of disease are abnormalities of anatomical parts. We will see below that there are phenotype ontologies for humans and for the mouse, which is the preeminent model organism for studying human disease. The phenotype ontologies describe abnormalities of anatomy, physiology, cells, small molecules, and other things. With the exception of physiology, for which at present no mature bio-ontology is available, all of these aspects can be described by ontologies. We will see in the coming sections how this information can be put together to compare the phenotypes of humans and model organisms. Perhaps the most important component of such a comparison involves a method for comparing the anatomical structures of humans and model organisms.

Homologous anatomical structures are structures that are similar in different species because of common evolutionary descent. For instance, the last common ancestor of mice and men lived about 70 million years ago, and our hands have evolved from the same structures from which the paws of modern mice have developed. The structures of the internal organs of the mouse are remarkably similar to the structure of our own organs, which is one reason why the laboratory mouse is such a valuable model organism for the study of human disease.

There are two ontologies within the OBO Foundry that are of high relevance for interspecies anatomical comparisons. The Common Anatomy Reference Ontology (CARO) is an upper level ontology for anatomy that consists of abstract structural classes which can be extended by classes in individual anatomical ontologies in any taxon [103]. Cross-species standardization of anatomy ontologies is essential for comparison of phenotypic, gene-expression, and other data between humans and model organisms. Uberon is an uber[3]-ontology for multi-species metazoan anatomy that was created primarily to support translational research by allowing comparison of phenotypes across species and to provide logical cross-product definitions for GO biological process terms [102]. We will see in Section 7.3.10 how Uberon is being used together with PATO to provide a bridge between multiple species anatomies to compare phenotypes of different organisms.

7.3.2   Cell Ontology

The Cell Ontology[4] (CL) was designed as a structured controlled vocabulary for cell types. CL is not organism specific; rather, it describes cell types from the prokaryotic, fungal, animal and plant worlds. CL includes the major cell types from the major model organisms including human, mouse, Drosophila, Caenorhabditis, zebrafish, *Dictyostelium discoideum*, Arabidopsis, fungi, and prokaryotes.

The parent and child terms are connected to each other by is_a and develops_from relationships. The former is a subsumption relationship, in which the child term is a more restrictive concept than its parent and the

---

[3] Actually, über...

[4] The Cell Ontology has occasionally been referred to as the "Cell Type Ontology," but the preferred name is "Cell Ontology."

latter is used to code developmental lineage relationships between concepts (Figure 7.1).

CL aims to include cell types from all the major model organisms within a common framework. CL is providing a common referential framework for databases such as the Array Express database of functional genomics and gene expression profiles [196]. The GO consortium is using CL to provide logical cross-product definitions of GO terms that involve cell types [177], and the HPO is using CL to provide logical definitions of phenotypic abnormalities of cells (see Section 7.3.8 later in this chapter). The CL team is in turn currently developing logical definitions for CL terms based on cross-products formed with multiple ontologies, including the PRO, GO, ChEBI, and UBERON.

7.3.3 Chemical Entities of Biological Interest

The Chemical Entities of Biological Interest (ChEBI) ontology [66, 67] is an ontology of molecular entities such as molecules, groups, rings and atoms, apart from those that have been directly encoded by the genome such as proteins. Each entry is manually annotated with multiple synonyms, and where feasible a molecular graph is provided, accompanied by the chemical structural representations InChI, InChIKey, and SMILES. Additional chemical data such as formula, mass and charge are provided, and entries are then extensively cross-referenced. The ontology incorporates a structural classification according to IUPAC guidelines beneath a root *molecular structure*, and a classification according to bioactivity or chemical reactivity beneath a root *role*, with chemical entities being linked to particular roles by means of the has_role relationship. All data in ChEBI is open and freely available in a number of download formats. ChEBI may be accessed online at http://www.ebi.ac.uk/chebi.

The following slightly shortened stanza describes one of the favorite small molecules of the authors by naming the superclass (trimethylxanthine) and describing three biological roles of the molecule. The synonyms provide the molecular formula and the SMILES representation of the molecule (Figure 7.2).

```
[Term]
id: CHEBI:27732
name: caffeine
synonym: "C8H10N4O2" RELATED FORMULA [KEGG COMPOUND:]
synonym: "Cn1cnc2n(C)c(=O)n(C)c(=O)c12" RELATED SMILES [ChEBI:]
is_a: CHEBI:27134 ! trimethylxanthine
relationship: has_role CHEBI:25435 ! mutagen
relationship: has_role CHEBI:35337 ! central nervous system stimulant
relationship: has_role CHEBI:38809 ! ryanodine receptor modulator
```

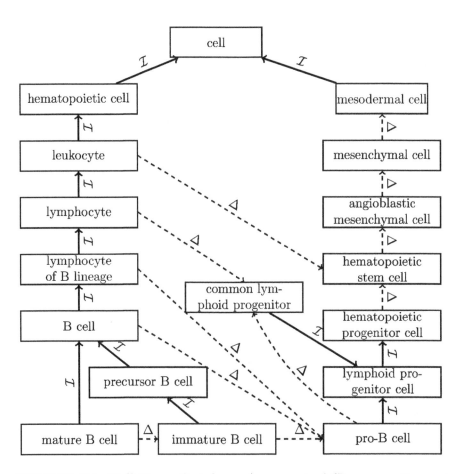

FIGURE 7.1: Cell Type Ontology. An excerpt of CL representing part of the B cell lineage is shown. is_a links are shown as edges labeled with the symbol $\mathcal{I}$, and develops_from links are shown as dashed edges labeled with the symbol Δ. The structure of the CL graph shown here is slightly simplified. In the original, all CL terms are connected to a subsuming parent term by an is_a link, and some are connected to other terms by develops_from relations.

7.3.4   OBI

Although standards are being developed for the description of experiments such as microarray hybridizations [37], the descriptions of the vast majority of experiments reported in scientific journals are written in free text, which means that automated computerized analysis and comparison of experiments from different sources is essentially impossible. The Ontology for Biomedical Investigations (OBI) is being developed as a global, cross-community effort to provide a resource that will offer an ontological framework for capturing data about biological and clinical investigations [41].

OBI is written in the OWL ontology language using many of the methods for constructing classes that we have described in Chapter 4 and will see in Chapter 14 (see also Appendix D). We have already shown some classes from OBI in Chapter 4, and a close perusal of OBI is highly recommended to readers who would like to learn more about OWL as a language for bio-ontologies.

7.3.5   The Protein Ontology

The Protein Ontology (PR) is designed as a formal ontology for proteins that classifies proteins on the basis of evolutionary relationships among one another and additionally classifies the different forms of a protein that can be derived from a single gene [12, 180, 181]. The genes and proteins of organisms living today can be grouped into families derived from common ancestors. Members of a set of orthologous genes are recognized on the basis of similarity in protein sequences (see Figure 5.1). In addition, related proteins may be grouped in evolutionarily related families. PR's classification scheme is based on that of the Protein Information Resource (PIR), which developed the PIR SuperFamily (PIRSF) classification system, as well as on the PANTHER (Protein ANalysis THrough Evolutionary Relationships) Classification System [250]. Protein family members are homologous (sharing com-

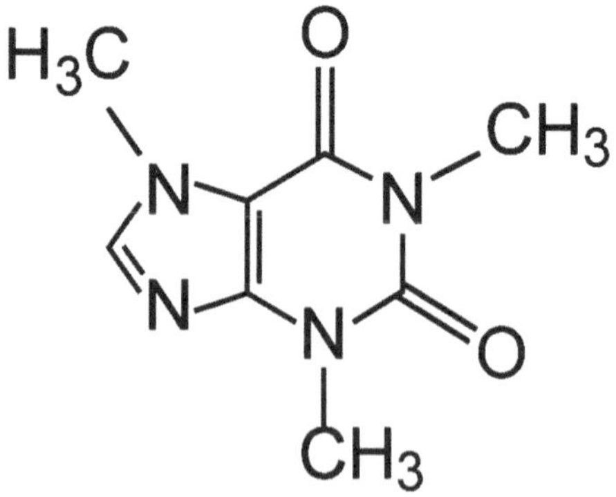

FIGURE 7.2: CHEBI:27134 Cn1cnc2n(C)c(=O)n(C)c(=O)c12.

mon ancestry) and also homeomorphic (sharing full-length sequence similarity with common domain architecture). Parent superfamily nodes connect more distantly related families based on common domains [265]. This hierarchical scheme allows annotation of family-specific biological functions and sequence features (Figure 7.3).

Any given gene can give rise to multiple proteins with different biological activities. Different protein forms can arise both from alternative splicing with inclusion or exclusion of protein coding sequences represented by alternatively spliced exons as well as from posttranslational modifications such as phosphorylation of specific amino acid residues. At present, GO annotations for the most part are made to a single entry standing for a protein without differentiating which protein form actually has the function. For instance, the protein SMAD2 is cytoplasmic and inactive if unphosphorylated, but phosphorylated SMAD2 forms a nuclear complex that leads to transcriptional activation of target genes. PR also has an annotation file (PAF) that gives annotations for proteins that are types of a given class. For instance, only the phosphorylated forms of SMAD2 are annotated with the GO terms *SMAD protein heteromerization* and *signal transduction*.

PR represents membership in protein superfamilies with the standard is_a subclass relationship from the Relation Ontology (see Chapter 6). PR uses an interesting ontology design pattern to represent protein forms that takes advantage of OBO's intersection syntax (see Chapter 4).

For instance, the protein *TGF-beta 1 isoform 1, signal peptide removed form* is defined as the intersection of all proteins that are *TGF-beta 1 isoform 1* and those that lack a signal peptide:

```
[Term]
id: PR:000025454
name: TGF-beta 1 isoform 1, signal peptide removed form
def: "A TGF-beta 1 isoform 1 that has had the signal
     peptide removed." [PR:DAN]
comment: Category=modification.
intersection_of: PR:000000397 ! TGF-beta 1 isoform 1
intersection_of: lacks_part SO:0000418 ! signal_peptide
```

Similarly, the unmodified form is defined as the intersection of all proteins that are TGF-beta 1 isoform 1 *TGF-beta 1 isoform 1* and those that have an immature protein part (i.e., the unmodified protein chain) and those that lack a modified amino acid. Further, the unmodified form is declared to be disjoint from the class of proteolytic cleavage products (for instance, the form that has had its signal peptide removed is a proteolytic cleavage product).

```
[Term]
id: PR:000002555
name: TGF-beta 1 isoform 1 unmodified form
def: "A TGF-beta 1 isoform 1 that has not been subjected
```

to any co- or post-translational residue modification
or peptide bond cleavage." [PR:DNx]

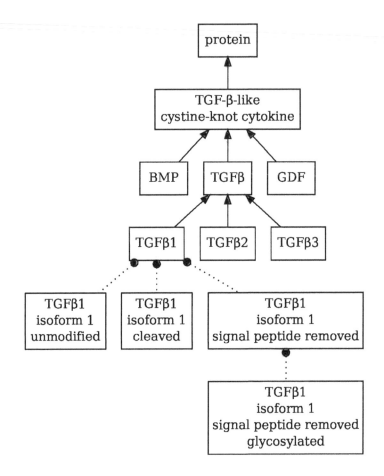

FIGURE 7.3: Representation of TGFβ in PR The figure shows a small portion of the representation of the TGFβ superfamily in PR. *TGF-beta-like cystine-knot cytokine* is defined in PR as "A protein with a core domain composition consisting of a signal peptide, a variable propeptide region and a transforming growth factor beta like domain," and is a synonym for the TGFβ superfamily, which comprises signaling molecules such as TGFβ, the bone morphogenetic proteins (BMP), the growth/differentiation factor (GDF) family, and others. The edges depicted as solid lines with arrowheads represent is_a relations in the sense that *TGFβ1* is a subclass of *TGFβ* which in turn is a subclass of *TGF-β-like cystine-knot cytokine*. The relations depicted by dotted lines with a circle represent protein forms. For instance *TGFβ1, isoform 1, signal peptide removed* is a specific form of the protein *TGFβ1*. See text for details on how this is represented in the ontology.

```
comment: Category=modification.
intersection_of: PR:000000397 ! TGF-beta 1 isoform 1
intersection_of: has_part PR:000021935 ! immature protein part
intersection_of: lacks_part PR:000025513 ! modified amino
                                        acid chain residue
disjoint_from: PR:000018264 ! proteolytic cleavage product
```

PR is freely available from the Protein Information Resource Web site[5] as well as from the OBO Foundry.

7.3.6 The Sequence Ontology

The Sequence Ontology (SO) is a structured controlled vocabulary for the parts of a genomic annotation. Traditional biological terminology for biosequences is just as ambiguous as it is for other areas of research. The SO is primarily intended for genome annotation. Like the GO, the SO originated as a collaborative project between several model organism databases who desired to increase the interoperability of their data. SO provides a standardized set of terms and relationships with which to describe genomic annotations and provide the structure necessary for automated reasoning over their contents, thereby facilitating data exchange and comparative analyses of annotations.

The SO originally comprised three relations, `kind_of`, `part_of` and `derives_from`. The SO has undergone a significant amount of development since its initial publication in 2005. A recent article [178] describes how the SO, which predated the BFO, was made compliant to the BFO in order to improve interoperability with other OBO ontologies, and extended to a larger number of relations to increase the expressivity of the ontology (37 at the time of this writing), and to have logical cross-product definitions (see Section 4.1.2). This allows for computable definitions of biological sequences. A simple example of this would be the term for mitochondrial gene,[6] which is defined as equivalent to the intersection of genes and mitochondrial sequences. Note the comment in the `is_a` line, which means that this relation was inferred computationally based on the logical definition of the term (see Chapter 12 for more information on the use of logical cross-product definitions for ontology maintenance).

```
[Term]
id: SO:0000088
name: mt_gene
def: "A gene located in mitochondrial sequence." [SO:xp]
synonym: "mitochondrial gene" EXACT []
```

---

[5]http://pir.georgetown.edu/pro/pro.shtml.

[6]Recall that the mitochondrion contains a small circular genome; in humans, the mitochondrial genome comprises 16,569 base pairs with 13 genes encoding proteins involved in oxidative phosphorylation and 22 genes encoding tRNAs.

```
is_a: SO:0000704 ! implied link automatically realized ! gene
intersection_of: SO:0000704 ! gene
intersection_of: has_origin SO:0000737 ! mitochondrial_sequence
```

SO also has a comprehensive vocabulary for describing sequence variants, which is quickly increasing in importance with the advent of next-generation sequencing technologies with which it is possible to determine the sequence of (nearly) all exons of all genes (the so-called "exome") or even of the entire genome in a matter of days. Since the number of variants to reference sequences range from tens of thousands for the exome to up to a few millions for a genome, it is important to have a standardized scheme for recording and sharing this information. SO has about 100 terms that describe sequence variants, such as non_synonymous_codon, which is defined as *A sequence variant whereby at least one base of a codon is changed resulting in a codon that encodes for a different amino acid*, and is part of a hierarchy of terms for describing transcript variants. SO is being used by the Genome Variation Format (GVF), and extension of the Generic Feature Format (GFF) for describing genome variation data for genome sequences or DNA microarray-based variant calls [202].

7.3.7   Mammalian Phenotype Ontology (MPO)

The mouse is probably the most important model organism for the study of human monogenic disease. The International Mouse Phenotyping Consortium (IMPC) intends to identify the function of every gene in the mouse genome, and in 2010 a major funding initiative was announced to set aside over 100 million dollars of the estimated 900 million dollars that will be required to create viable strains in which each of the more than 20,000 genes of the mouse genome is knocked out. The resulting knockout strains will be put through systematic phenotypic screens to check for physical and behavioral differences [1, 167]. Phenotype ontologies will play a major role in making the information that comes out of this project useful for research into human disease by providing a standardized vocabulary that can be related to the human phenotype in a well-defined fashion (see below).

The Mammalian Phenotype Ontology (MPO) was designed to classify and organize phenotypic information related to the mouse and other mammalian species [241]. The MPO is used by the main centers for research on mouse genetics including the Mouse Genome Informatics (MGI) resource [45] and EuroPhenome, which is a repository for high-throughput mouse phenotyping data of the European Mouse Disease Clinic (EUMODIC) [167]. In addition to mouse data, the MPO has been used to classify rat phenotype data from the Rat Genome Database [253] as well as data from other species [150].

7.3.8   Human Phenotype Ontology (HPO)

Many of the medical terminologies developed over the last two centuries aim to describe disease entities, such as the International Classification of Diseases (ICD). More recent efforts to use computational algorithms to analyze human phenotypic features have additionally used vocabularies that define the signs, symptoms, and other manifestations of the diseases. A disease is characterized by one or more phenotypic abnormality. For instance, the common cold (a disease) is characterized by the phenotypic abnormalities *fever, cough,* and *runny nose.* Any given phenotypic abnormality can occur with multiple diseases; for instance, *runny nose* can occur with a cold or with hay fever (two different diseases). In mathematical terms, this is an m-to-n relationship, and together with the fact that a person with some disease may not have all possible phenotypic abnormalities associated with the disease, or indeed may have additional, unrelated abnormalities, is the main reason why finding the correct diagnosis is often quite difficult. The process of considering the combination of the medical history and the physical and laboratory findings to identify the correct diagnosis out of a longer differential diagnosis is arguably the single most important task of the physician.

Phenotypic analysis plays a fundamental role in human genetics diagnostics and research. Phenotype descriptions in peer-reviewed publications describing new disease genes or genotype-phenotype correlations in known syndromes have generally relied on free text descriptions of phenotypic features. In recent years, computational analysis of the spectrum of phenotypic features found in human disease, the so-called "phenome" [192], has been taken up by a number of groups for a number of goals including prioritizing candidate disease genes [144, 148], investigating modularity of the genetic disease-phenotype network, and comparing phenotypic data across species [260]. However, the terms that clinicians use to describe phenotypic manifestations have evolved in a haphazard and uncoordinated manner [5, 33], and comprehensive databases of phenotypic information about humans with hereditary disease collected in a standardized fashion do not exist [82].

At the time of this writing, there is no database for human genetics that contains comprehensive data on gene mutations, diseases, and phenotypic features, which presumably could be used to improve clinical diagnostics and biomedical research. Calls for a human phenome project have emphasized the need for standards in reporting and international integration [82, 225], but at present the data – if recorded at all in databases – is dispersed into databases that concentrated primarily on mutations with minimal phenotype data such as the Human Gene Mutation Database [245], into locus specific databases that cover one or a few genes [89], and into some general knowledge resources and databases such as the Online Mendelian Inheritance in Man (OMIM) database, which was developed over decades by Professor Victor McKusick and many colleagues at Johns Hopkins University [9]. Several databases record genotype and phenotype information from genetic association studies including the HGVbaseG2P [81] and NCBI's dbGAP [158]. Although some of these databases make some use of controlled vocabularies, for the most part phenotypic data is stored in an unstructured and mutually incompatible way, a situation which has been called a "terminology Babel" in human clinical databases [225].

Computational analysis of phenotypic data has in the past been difficult owing to the lack of a controlled vocabulary using consistent annotations with well-defined relationships to one another. For instance, the descriptions "generalized amyotrophy," "generalized muscle atrophy," "muscular atrophy, generalized," and "muscle atrophy, generalized" all refer to the same medical finding, but it would be difficult to recognize this based on a computational (text mining) analysis of different medical articles using these phrases. Therefore, one of the goals of the HPO is to offer a standardized vocabulary for describing human phenotypic abnormalities, and to associate synonyms with the preferred terms.

The HPO currently comprises over 10,000 terms referring to different signs and symptoms of human disease [211]. The terms have been used to annotate over 5,000 Mendelian diseases.[7] Each term in the HPO describes a phenotypic abnormality, such as *atrial septal defect*. These terms are related to parent terms by "is a" relationships, meaning that they represent a subclass of a more general parent term. In contrast to strict hierarchies, the data structures used to represent ontologies (e.g., DAGs) allow a term to have multiple parent terms. In the HPO, multiple parentage allows the different aspects of phenotypic abnormalities to be represented. The phenotypic feature *atrial septal defect* for instance has the parent terms *abnormality of the cardiac septa* and *abnormality of the cardiac atria*, both describing a *cardiac abnormality* (Figure 7.4).

The HPO can be used for a number of purposes including as a framework for aiding clinical diagnostics. The differential diagnosis can be particularly challenging in the field of medical genetics, because many hereditary diseases can be characterized by numerous, highly variable clinical abnormalities. The HPO was originally constructed using data from OMIM, whereby synonyms were merged and semantic links were created between the terms to create the ontological structure, and nearly all of the annotations were derived from OMIM. The HPO is now being refined, corrected, and expended by a process of manual curation in which term definitions are being created, terms for concepts not originally found in the OMIM data are being made (Figure 7.5).

In chapters 10 and 11, we will see how search algorithms based on semantic similarity metrics for ontologies have been adapted to create applications intended to help physicians with the differential diagnostic procedure. Later in this chapter, we will see how logical entity-quality definitions are being used

---

[7]These are diseases that are caused by mutation in single genes and that follow an autosomal dominant, autosomal recessive, or X-linked mode of inheritance. Currently, many thousands of Mendelian diseases have been described, and the genetic basis has been discovered for about 3,000 diseases.

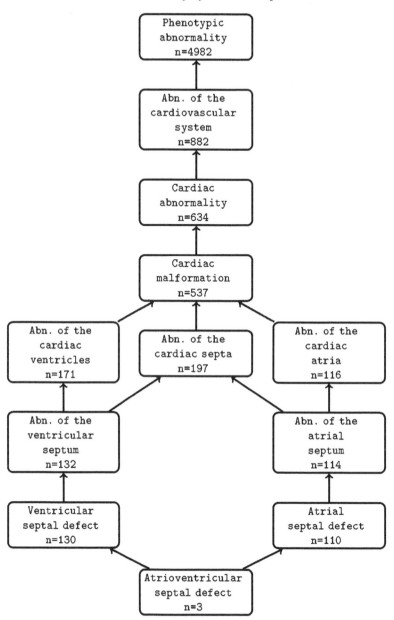

FIGURE 7.4: Representation of Some Cardiovascular Phenotypic Abnormalities in the HPO. The numbers under the term names show the number of diseases annotated to the term at the time of this writing. *Abn.*: abnormality.

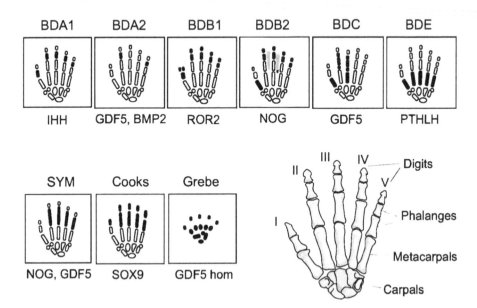

FIGURE 7.5: **Brachydactyly: Precomposed Terms.** The figure shows the patterns of phenotypic abnormalities affecting the bones of the hand in nine different hereditary forms of brachydactyly, which is a Greek word that refers to shortening of the hands or feet owing to missing, deformed, or shortened bones [176]. Hereditary brachydactylies are divided into groups A-D including several subgroups. The brachydactylies are annotated to distinct but partially overlapping subsets of terms in the HPO. For instance, brachydactyly A1 (BDA1) is annotated to the term *hypoplasia of the proximal phalanx of the thumb*, brachydactyly A2 (BDA2) is annotated to the term *aplasia/hypoplasia of the middle phalanx of the second finger*, and brachydactyly type E (BDE) is annotated to *hypoplasia of the metacarpal bones*. Each of the diseases is annotated with 5–20 HPO terms each representing a phenotypic abnormality. A comprehensive and detailed description of the phenotypic abnormalities is required to differentiate these disorders computationally. Figure kindly provided by Stefan Mundlos, Institute for Medical Genetics and Human Genetics, Charité Universitätsmedizin Berlin.

to compare phenotypic abnormalities in the mouse and zebrafish to abnormalities seen in humans as a tool for medical research.

7.3.9 MPATH

The field of pathology comprises the clinical and scientific study of disease processes at a cellular and tissue level. Medical pathologists typically examine biopsies or specimens of bodily fluids in order to come to a diagnosis of disease processes or etiologies. Pathological examinations often involve the macroscopic or microscopic examination of tissues affected by disease, and findings of pathological examinations often describe changes affecting individual cell types, tissues, or organs.

Many of the difficulties arising from the lack of a controlled vocabulary apply to pathology as well. The MPATH ontology was originally developed to serve as a controlled vocabulary for describing mouse pathological lesions within Pathbase, which is a database of histopathology photomicrographs and macroscopic images derived from mutant or genetically manipulated mice. MPATH is a reference terminology for pathologists of mutant mice. As mentioned above for the HPO, the MPATH, term definitions and synonyms help clarify the often disparate set of names used to describe the same type of lesion by different pathologists [224, 223]. MPATH currently has 643 terms arranged as a DAG using subclass (is_a) relations.

The way that MPATH is used in Pathbase nicely illustrates some issues surrounding pre-composition and post-composition in phenotypic abnormalities. Pre-composition creates all terms needed for annotation in advance. The HPO uses a pre-composition approach, meaning that all terms that might be needed to annotate human disease are explicitly created. While this arguably makes the task of annotation and searching in a phenotypic database somewhat easier, it does have the disadvantage of ontology bloat – the presence of large numbers of similar terms. For instance, the HPO contains sets of terms such as *Abnormality of the epiphysis of the distal phalanx of the 2nd toe, Abnormality of the epiphysis of the proximal phalanx of the 2nd toe, Abnormality of the epiphysis of the distal phalanx of the 3rd toe*, and so on. Detailed phenotypic descriptions are important for the differential diagnosis of certain kinds of diseases (as can be appreciated by comparing the distribution of affected hand bones in the nine hereditary forms of brachydactyly shown in Figure 7.5). In the HPO, it was felt that pre-composed terms were necessary because of the complexity of many of the terms required to describe human disease features would make a post-composition approach impossible.

With the post-composition approach, complex terms such as those mentioned above are broken down into components and combined at the time of annotation. We will see in the next section how the PATO ontology can be used for post-composition of terms involved in phenotypic abnormalities. For now, we will mention that in pathology, many tissue responses are common to multiple anatomical sites. MPATH does not have a separate term for any given response in each of the tissues of the body. Rather, the additional topological or anatomical information is expected to be provided separately. In Pathbase, each record in the database will have one MPATH term as well as additional information on the organ or tissue affected. For instance, Pathbase Image PB 3401 shows a case of *pancreatic islet cell hyperplasia*, and annotates the image with the MPATH term MPATH:134 for *hyperplasia* as well as with the mouse anatomy term MA:0000127 for *pancreatic islet* as well as other information on the mouse strain, age, and genotype status of the mouse in which the lesion was described.

7.3.10 PATO

The original motivation for the development of PATO, the *Phenotype And Trait Ontology*, was to develop a schema according to which phenotypes could be represented as qualifications of descriptive nouns [90, 91]. It soon became apparent that a major advantage of the PATO approach would be the ability to link phenotypic abnormalities in humans and model organisms to ontologies of anatomy, biochemistry, cell types and components, pathology, as well as molecular functions and processes in a way that would enable an integrative computational analysis of phenotypic abnormalities and disease.

PATO consists of a single hierarchy of qualities and currently offers 2209 terms. PATO is designed to be used in conjunction with ontologies of "quality-bearing entities," prominently including ontologies of anatomical entities such as the Foundational Model of Anatomy (FMA) ontology [214], GO [14], or the cell type ontology [19]. PATO is used in combination with one of these ontologies to create composite terms. For instance, to describe a "red-eye" phenotype in Drosophila, one can combine the PATO term *red* (Q=Quality) with the Drosophila gross anatomy (FBbt) term *eye* (E=Entity).

```
E= FBbt:eye    Q= PATO:red
```

This is called "composing" (or "coordinating") the description by using existing terms as elements of the description.[8]

Many terms in the HPO describe abnormalities of anatomic structures. For these HPO terms, therefore, terms from the FMA human anatomy ontology are used to identify the affected entity (bearer of the abnormal quality). An appropriate PATO term is then chosen to describe the abnormal quality that the anatomic structure possesses, which can be described either in qualitative or quantitative terms, and are said to inhere_in the bearer. For instance, for the HPO term *Reduced bone mineral density* (HP:0004349), we combine the FMA term for *bone* with the PATO term for *decreased density*. This combination is expressed in OBO format notation[9] as follows:

---

[8]If this is done at the time of annotation, it is referred to as "post-composition," as mentioned in the previous section.

[9]See Section 4.1.2.

```
[Term]
id: HP:0004349 ! Reduced bone mineral density
intersection_of: PATO:0001790 ! decreased density
intersection_of: inheres_in FMA:5018 ! Bone organ
```

This states that the phenotype denoted by HP:0004349 is equivalent to all instances of decreased density occurring in bones (the FMA term *bone organ* refers to the set of all bones in the body).

The annotation model can additionally include further qualifiers concerning the developmental stage at which a phenotype holds or qualifiers indicating the expressivity of the phenotype with respect to some baseline (e.g., "abnormal"). For instance, the HPO term *Osteosclerosis* refers to an abnormal increase in bone density. The intersection of the FMA term for *bone* with the PATO term *increased density* does not faithfully reflect the meaning of the HPO term *Osteosclerosis*, because not all increased density is *abnormal*. One can therefore add a further intersection with the PATO term *pathological*:

```
[Term]
id: HP:0002796 ! Osteosclerosis
intersection_of: PATO:0001788 ! increased density
intersection_of: qualifier PATO:0001869 ! pathological
intersection_of: inheres_in_part_of FMA:5018 ! bone organ
```

This definition is an intersection of the entity *bone organ* with the PATO quality *increased density*. The HPO term is defined to comprise the occurrence of osteosclerosis in any or all of these bones.

The ability to manipulate the mouse genome has rendered the mouse one of the most important model organisms for studying human disease. In the gene driven approach to model discovery targeted mutants are made in specific genes associated with disease in man. Determination of the accurate relationship of mouse phenotypes to human diseases is essential for projects that endeavor to understand human disease on the basis of experiments in the mouse models. Although many mouse models display phenotypes that are reminiscent of the phenotypes of humans with inherited mutations at the same genes, important differences between human and mouse phenotypes resulting from mutation in homologous genes are frequently observed [213, 226]. Bridging the gap between mouse phenotypes and human diseases is therefore problematical; partly because formal disease nomenclature differs between mouse and man, but more importantly because not all of the aspects of cognate diseases will be manifested in both species. One approach to discovering a disease model is therefore to break down the summative (precomposed) diagnosis into its component parts and to search for matches within the resulting pool of constituent phenotype elements across both species. PATO-based E/Q decompositions are currently being developed for mouse and human phenotype data [25, 91, 92].

Let us consider an example of how this mapping can be performed. The MPO has a term for *aortic valve stenosis*. This feature, which is also known as *aortic stenosis*, is a construction of the opening of the aortic valve. The logical definition of this term states that *aortic valve stenosis* is equivalent to the intersection of the PATO term *constricted* and all qualities that inhere in the mouse *aortic valve*, which is defined by a term in the adult mouse anatomy ontology (MA:0000087).

```
[Term]
id: MP:0006117 ! aortic valve stenosis
intersection_of: PATO:0001847 ! constricted
intersection_of: inheres_in MA:0000087 ! aortic valve
```

The corresponding term in the HPO is called *Aortic stenosis*. This term is defined as the intersection of the PATO term *constricted* and all qualities that inhere in the human *aortic valve*, which is defined by a term in the FMA (FMA:7236).

```
[Term]
id: HP:0001650 ! Aortic stenosis
intersection_of: PATO:0001847 ! constricted
intersection_of: inheres_in FMA:7236 ! Aortic valve
```

Thus, we can see that both terms are describing the identical phenotypic feature, one in the mouse and one in humans. We can now use UBERON to create a mapping between the mouse and human anatomy ontologies. The following UBERON term groups the terms for the aortic valve from a number of different ontologies (they are shown as external references with the xref key).

```
[Term]
id: UBERON:0002137
name: aortic valve
alt_id: UBERON:FMA_7236-MA_0000087
def: "Cardiac valve which has as its parts the anterior,
      right posterior and left posterior cusps, attached
      to the fibrous ring of aortic valve."
      [Wikipedia:Aortic_valve]
comment: This class was created automatically from a
      combination of ontologies
subset: uberon_slim
xref: EHDAA:4410 ! aortic valve
xref: FMA:7236 ! Aortic valve
xref: GAID:563 ! aortic valve
xref: galen:AorticValve
xref: HOG:0000815 ! aortic valve
xref: MA:0000087 ! aortic valve
xref: OpenCyc:Mx4rv5914JwpEbGdrcN5Y29ycA
is_a: UBERON:0000946 ! cardial valve
```

```
xref: ncithesaurus:Aortic_Valve ! Aortic Valve
xref: OpenCyc:Mx4rv5914JwpEbGdrcN5Y29ycA ! aortic valve
xref: EHDAA2:0000134 ! aortic valve
```

The PATO definitions and UBERON mappings then allow mapping between the phenotypes of model organisms and humans (Figure 7.6). This is one component of algorithms for comparing the manifestations of disease in model organisms with the signs and symptoms observed in human patients [179, 260].

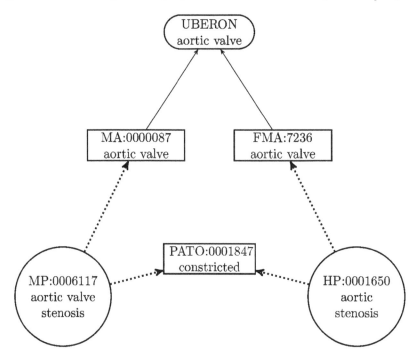

FIGURE 7.6: **Interspecies Phenotypic Mapping with PATO.** The UBERON mapping between the species-specific terms for the aortic valve and the PATO term for constricted allow a mapping between the MPO term *aortic valve stenosis* and the HPO term *Aortic stenosis.*

7.4 What Makes a Good Ontology?

We do not believe it is possible to provide a general recipe for ontology development that would be of much help to researchers who need to develop an ontology for their field of research. Nevertheless, it can be helpful to consider some of the requirements for good ontology development. A profound knowledge of the domain being modeled is probably the single most important factor. There is no one single correct way to model a domain; although it is desirable for an ontology to be useful in many contexts, inevitably the purpose the ontology designer had in mind when creating the ontology will affect the contents and structure of an ontology. For instance, a medical ontology designed primarily for electronic patient records and medical billing will inevitably be different from a medical ontology designed for research on the pathophysiology of disease.

The concepts (terms) of an ontology should correspond to physical or logical objects of the domain being modeled. These are likely to be nouns in sentences that describe the domain, and the relations are likely to be verbs in such sentences [186]. Therefore, it is useful simply to make a list of terms and statements about the domain to be modeled by the ontology. At this point, it is useful to use upper-level ontologies such as the BFO and the RO to assist in the categorization of entities and relations, which is an important starting point for the construction of a new ontology, and defining the hierarchy of classes. What are the instances and which types (universals) do they belong to? If the ontology is to be used for the annotation of biological data, what are the attributes of the data that need to be captured? Which aspects are to be modeled in the ontology, and which aspects will be recorded as part of the annotations?

It should be noted that an ontology designed to support scientific research should endeavor to represent entities as they exist in reality. That is, a term in such an ontology much correspond to a *universal* rather than to a set which is convenient for some particular purpose. Thus, the terms *human* and *mammal* represent universals, and the meaning of a statement such as *human* $\xrightarrow{is\ a}$ *mammal* is clear and can be said to convey a law between the universals that has been discovered by scientific research. On the other hand, a class such as *patient whose hospital bill is over $27,000* conveys a relation based on a particular set inclusion rule rather than denoting a universal [235].

Naming conventions should be defined for the terms and used consistently. Each term should have a clear and precise definition, which should reflect the semantics of relations involving the term.

It is highly recommended that an ontology have a complete backbone of is_a relations between all terms to which optionally additional relations (such as part_of) can be added. This kind of structure lends itself to genus-differentia or so-called Aristotelian definitions of the terms. That is, if a term U has a parent V, then the definition of U is that U is a V (genus) that has some characteristic that differentiates it from other kinds of V. The process of making Aristotelian definitions starts from the root of the ontology. For each subsuming (parent) term, the term has at least two child terms that are defined on the basis of some differentiating qualities [135]. The use of unambiguous definitions according to this scheme can help to generate clean classification hierarchies in which all children of a given parent are disjoint, so that nothing can be an instance of both a class and its sibling.

For those considering creating a bio-ontology from scratch, it is advisable to first search OBO Foundry and the NCBO to determine if there already is an available ontology that will fulfill the requirements. If not, it would be extremely helpful to first study a few successful bio-ontologies including GO and perhaps another ontology that is most relevant to your field of study. Finally, make sure you have a deep understanding of the field you intend to model ontologically, or initiate a close collaboration with a domain expert.

7.5 Exercises and Further Reading

In the last five years or so, the field of bio-ontologies has undergone an enormous expansion, and good ontologies are now available for many (but certainly not all) areas of medical practice and biomedical research. This chapter has presented a small selection of relatively mature bio-ontologies that illustrate the wide range of applications in which bio-ontologies are being put to use. Readers searching for an ontology for a specific area are advised to consult the OBO Foundry Webpage as well as the BioPortal. An article by Barry Smith and coworkers provides a good introduction to OBO Foundry and its goals [236]. The OBO Foundry currently lists eight OBO Foundry ontologies and numerous candidate ontologies. More information about the individual ontologies can usually found by following the links from the OBO Foundry Web site to the homepages of the ontologies.

Exercises

The exercises for this chapter are intended to help readers to explore the resources at the OBO Foundry and NCBO Web sites.

7.1 Go to the BioPortal Website at `http://bioportal.bioontology.org/`. Use the search engine to search all ontologies for a biomedically relevant concept (for instance, "aortic stenosis"). For many concepts you will see that there are corresponding terms in multiple ontologies. Compare the definitions for the terms from several ontologies by clicking on the symbol in the *Details* column of the search result window, and examine the location of the terms in the graph structure of several ontologies by clicking on the graph symbol in the *Visualize* column. What are some of the problems that result from the fact that multiple ontologies have terms with the same or overlapping meaning? What are some potential solutions to the problems? (You may want to consult [236] to help answer this question.)

7.2 Go to the BioPortal Web site, and use the Annotator service,[10] to find ontology terms that could be used to annotate an abstract in PubMed. You can use the abstract from the article by Reed Pyeritz on the Marfan syndrome that is available under PubMed id 10774478.[11] Copy the abstract text and paste it into the search window of the BioPortal Annotator tool. First use all available ontologies (the default setting), and then use the *Choose* button next to the *Ontologies* text-entry field to select the Human Phenotype Ontology and search again. You should see that the Annotator identifies multiple ontology terms that could be used to annotate the abstract.

7.3 Go to the Web site of the Ontology Lookup Service (OLS) at `http://www.ebi.ac.uk/ols`. The OLS provides interactive and programmatic interfaces to query, browse and navigate biomedical ontologies that are complementary to the BioPortal. For this exercise, read the article by Côte and colleagues about the OLS [59], which is freely available, and explore the OLS according to the descriptions in the article.

7.4 Go to the OBO Foundry Web site at `http://www.obofoundry.org/`. Spend some time browsing through the list of ontologies, and go to the homepage of several ontologies that are related to your own fields of interest. The OBO Foundry Web site offers automatically generated OWL versions of the OBO ontologies. Examine several terms in both the OBO and OWL versions and compare the syntax.

7.5 OBO-Edit [64] is an extremely useful application for viewing and searching in OBO ontologies (and also for developing OBO ontologies). Go to the OBO-Edit homepage at `http://oboedit.org/`, and download the version of OBO-Edit that is appropriate for your system (versions for Linux, Mac, and Windows are available). Download `gene_ontology_edit.obo` from the Gene Ontology Web site or the OBO Foundry Web site. Then, load this file into OBO-Edit and explore the structure of the ontology as well as the definitions and synonyms of the terms.

---

[10]Click on the *Annotate* button on the BioPortal Web site, or go to http://bioportal.bioontology.org/annotator#.

[11]http://www.ncbi.nlm.nih.gov/pubmed/10774478.

Part III

Graph Algorithms for Bio-Ontologies

Chapter 8

Overrepresentation Analysis

The practical importance of the Gene Ontology (GO) is linked to the fact that it has been used to annotate millions of gene products from thousands of organisms. The growth of GO parallels that of high-throughput technologies in molecular biology that allow essentially all genes in the genome to be measured experimentally. DNA microarrays consist of thousands of probes of either short oligonucleotides (roughly 25 to 60 nucleotides long) or longer cDNA probes that are complementary to the sequences to be measured. A cDNA or cRNA sample is hybridized to the probes on the microarray to detect and quantify the amount of the corresponding sequences in the sample. For gene expression profiling, the expression levels (mRNA concentrations) of thousands of genes are measured simultaneously. Often, microarray experiments are used to compare gene expression profiles under two or more biological conditions, say a comparison between healthy and diseased tissue or at different developmental stages. A typical experiment involves three or four replicate microarray experiments for each biological condition, and the statistical analysis would involve performing a t test or variant thereof on each of the genes on the microarray [7]. More recently, next-generation sequencing methodologies including especially RNA-seq [169] are further extending the range of transcription profiling experiments that can be performed. The outcome of transcription profiling experiments, whether by microarray or RNA-seq, is often a list of hundreds or even thousands of differentially expressed genes.

A typical application of GO is in the analysis of lists of differentially expressed genes coming from exploratory experiments in which the transcriptional activity of all or at least most of the genes in the genome have been measured. The basic question is whether one or more specific GO terms annotate more of the differentially expressed genes than one would expect by chance. For instance, say 221 of 6000 yeast genes (3.7%) represented on a microarray are annotated to the GO term *sporulation*. If we perform some experiment and observe 100 differentially expressed genes, then we might expect about 3 or 4 of them to be annotated to sporulation purely by chance, because this would correspond to the percentage of all genes that are annotated to this term. What if instead of 3 or 4, 35 of the 100 differentially expressed genes are annotated to *sporulation*? This seems unlikely to be pure chance, and we conclude that the GO term *sporulation* is *overrepresented* among the differentially expressed genes. This may motivate us to develop the hypothesis that

---

Hypothesis-Driven vs. Exploratory Experiments

The scientific method involves the formulation of a hypothesis, i.e., a proposed explanation for some observable phenomenon, and the design of experiments to test that hypothesis. At the risk of oversimplification, it can be said the result of a hypothesis-driven experiment is designed to be such that one can accept or reject a hypothesis such as "the enzyme hexokinase catalyzes the conversion of glucose to glucose 6-phosphate." On the other hand, much of bioinformatics involves the support of exploratory experiments in which data about the expression levels of thousands of genes or proteins is related to some experimental intervention. The goal is not to test a hypothesis, but to generate an answer to the age-old question, "what's going on?" Optimally, bioinformatic analysis will suggest a specific hypothesis that can be tested experimentally. For this reason, exploratory experiments are often referred to as *hypothesis-generating*.

---

our experiment in some way stimulated *sporulation* in the yeast, and perhaps we will consider ways of designing specific hypothesis-driven experiments to test our hypothesis. In a typical GO analysis of this sort, multiple overrepresented GO terms may be identified. Often, they are taken to represent the salient biological features of the genes in the study set.

Although the goal of this kind of analysis is simple enough, statistical dependencies in the GO ontology affect the results of the analysis, generally inflating the number of GO terms called as significantly overrepresented. While a list of 5 or 10 GO terms can be remarkably helpful in the interpretation of a set of hundreds of differentially expressed genes, a list of 50 or 100 GO terms is probably much less helpful to the average biologist who is trying to design the next experiment, because it is simply difficult to choose which of the many terms is most characteristic of the biology of the experiment. For this reason, a number of algorithms have been developed to compensate for such dependencies. These algorithms all are based upon the hypergeometric distribution and related concepts that were presented in Chapter 3. The next chapter will present a different approach to GO analysis that does not test each of the GO terms for overrepresentation, but instead searches for the optimal combination of GO terms that together best explain the observed pattern of differential expression.

8.1 Definitions

In this chapter, we refer to the set of genes which are investigated in an experiment as the *population set*. Note that we use the term *gene* only for simplicity. The items annotated to the ontology terms may also be proteins or other biological entities. We denote this set by the uppercase letter M, while the cardinality of the population set is denoted by m. If, for example, a microarray experiment is conducted, the population set will comprise all genes that the microarray chip can detect. The outcome of the experiment is referred to as the *study set*. It is denoted by N and has the cardinality n. In the microarray scenario the study set could consist of all genes that were detected to be differentially expressed.

8.2   Term-for-Term

The standard approach to identify the most interesting terms is to perform Fisher's exact test for each term separately. For this reason, we refer to this procedure as the term-for-term (TfT) approach [98]. The TfT approach has been the most commonly used method for GO analysis [139, 206].

For each GO term t, the genes in the population set M are either annotated to term t or not. This divides the population set into two classes of items, comparable to an urn with white and black marbles. We denote the set of all genes in the population that are annotated to GO term t as M_t and the cardinality of the set as m_t.

The study set is assumed to be a random sample that is obtained by drawing n genes without replacement from the population. The random variable that describes the number of genes of the study set that are annotated to t in this random sample, denoted by X_t, therefore follows the hypergeometric distribution, i.e.,

$$X_t \sim h(k|m; m_t; n) := P(X_t = k) = \frac{\binom{m_t}{k}\binom{m - m_t}{n - k}}{\binom{m}{n}}. \tag{8.1}$$

That is, $P(X_t = k)$ gives the probability of observing exactly k annotated genes in a study set of size n if the population set of size m contains m_t genes that are annotated to term t. As explained in detail in Chapter 3, the equation for $P(X_t = k)$ is equivalent to the product of the number of ways of choosing k genes from the m_t genes that are annotated to the term t and the number of ways of choosing the remaining genes, i.e., $n - k$, from genes not annotated to term t, i.e., $m - m_t$, divided by the total number of ways of choosing n genes of the study set from the m genes of the population.

The number of genes in the study set that are annotated to t represents the observation. This number is denoted by n_t. Now we want to assess whether the study set is enriched for term t, i.e., whether the observed n_t is higher than one would expect. In order to formulate a statistical test, we need to formulate a null hypothesis (H_0) and an alternative hypothesis (H_1). The null hypothesis in this case simply states that there is no positive association between genes annotated to the term t and the study set; that is, there is no overrepresentation of term t. The alternative hypothesis is that there is overrepresentation of the term.

Note that the alternative hypothesis is formulated in terms of a one-sided test: Is the representation of term t in the study set *higher* than expected by chance? It would also be possible to formulate the alternative hypothesis in terms of a two-sided test: Is the representation of term t in the study set *different* than expected by chance? However, the biological interpretation of underrepresentation of a term is often not clear. What would it mean, for instance, if we find a lower-than-expected number of genes annotated to the GO term *sporulation*? For the most part, this kind of observation does not have a clear biological interpretation (for instance, it cannot be taken to mean that the cell is actively "avoiding" sporulation). There is also a statistical disadvantage of using a two-sided test if one is only interested in *over*-representation: The critical region at the upper end of the distribution which allows us to reject the null hypothesis is only half as big (Figure 8.1). For this reason, a one-sided test is used except in certain rare situations in which *under*- or *over*-representation of a term is deemed interesting.

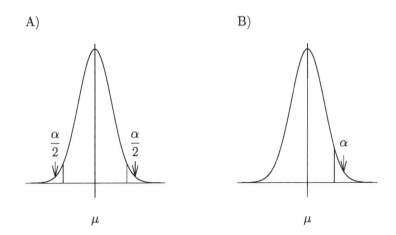

FIGURE 8.1: **One-Sided vs. Two-Sided Hypothesis Testing. A)** In a two-sided test, the alternative hypothesis is that there is either an increase *or* a decrease. The critical region has two parts. **B)** In a one-sided test, the critical region has just one part. In the figure, the critical region represents an increase. Note that the terms one-tailed and two-tailed test are synonymous with the terms one-sided and two-sided test.

Recall from Chapter 3 that the null hypothesis corresponds to the probability of observing a test statistic that is at least as extreme as the one that was observed given that the null hypothesis is true. Therefore, the null hypothesis corresponding to a one-sided test is rejected if the probability of observing n_t *or more* genes annotated to term t in the study set by chance is less than α (By convention, α is usually set to 0.05). This is given by:

$$P(X_t \geq n_t | H_0) = \sum_{k=n_t}^{\min(m_t,n)} \frac{\binom{m_t}{k}\binom{m - m_t}{n - k}}{\binom{m}{n}}. \qquad (8.2)$$

Therefore, if $P(X_t \geq n_t | H_0) < \alpha$, the null hypothesis is rejected, and we declare the term t to be *significantly overrepresented* in the study set. The biological interpretation of significant overrepresentation is usually taken to be that the term t represents an important biological characteristic of the study set. Note that the sum in Equation (8.2) is taken from the number of genes actually observed to be annotated to t in the study set (n_t) up to the total number of genes annotated to t (m_t) or the total number of genes in the study set (n), whichever is less. This is because there is zero probability of finding more than m_t genes annotated to t or of finding a number of annotated genes larger than the actual size of the study set.

Example 8.1 *Suppose that there is a population set of $m = 18$ genes, of which $m_t = 4$ genes are annotated to the term t. The outcome of an experiment conducted on all 18 genes of the population yields a set of 5 genes. This means that the study set consists of $n = 5$ genes. Moreover, we observe that a total of $n_t = 3$ genes from the genes of the study set are annotated to term t (Figure 8.2).*

We would now like to analyze whether term t is significantly overrepresented in the study set and thus can be interpreted to represent an important result of the experiment. The application of Equation (8.2) yields a p-value of term t

$$P(X_t \geq 3 | H_0) = \frac{\binom{5}{3}\binom{13}{2}}{\binom{18}{5}} + \frac{\binom{5}{4}\binom{13}{1}}{\binom{18}{5}} = 0.044.$$

Since $P(X_t \geq 3 | H_0) = 0.044 < 0.05 = \alpha$, the null-hypotheses is rejected and the term may be interpreted as being characteristic of the experiment.

8.3 Multiple Testing Problem

As we have seen in Chapter 3, conducting multiple statistical tests can lead to spurious results. This is a major issue for GO analysis, because there are currently over 32 thousand terms, and the term-for-term approach conducts a separate statistical test for each term that is represented in the population.

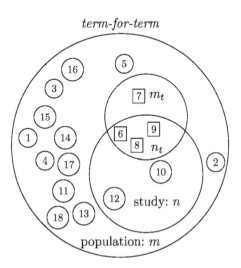

FIGURE 8.2: **Study and Population Sets in the *Term-for-Term* Approach.** In this example, the population set consists of $m = 18$ genes, whereby $m_t = 4$ genes of the population are annotated to term t. The size of the study set is $n = 5$. $n_t = 3$ genes in the study set are also annotated to term t. The model of the *term-for-term* approach is that all of the n genes of the study set are drawn from the population. The null hypothesis is that there is no association between the number of genes that are in the study set and the number of genes that are annotated to the term t, i.e., the study set is merely a random sample of the population set. We therefore would expect that it contains the same proportion of annotated terms as the population set does. The probability under the null hypothesis of observing at least n_t genes that are annotated to t can be assessed via Equation (8.2).

This means that the number of false-positive results will also be high unless appropriate correction for multiple testing is performed.

How Many Tests are Performed?

Although the GO currently has over 32 thousand terms, it is not the case that we need to perform a test for each term. For instance, it would not make any sense to perform a test for overrepresentation of the GO term *skeletal development* if we are analyzing an experiment done with yeast cells. Therefore, the analysis is performed only on GO terms that actually annotate the genes under study. Different software tools for GO analysis do this kind of filtering in different ways, but a common procedure is to perform a statistical test for overrepresentation only on GO terms which annotate at least one gene of the study set. Others recommend that terms that appear only once in the population set should not be included in the statistical analysis because overrepresentation is not possible [51].

Depending on the organism being investigated and the size of the study set, filtering for terms that appear at least once in the study set (and, optionally, those that also appear at least twice in the population set) typically reduces the number of GO terms being tested from the approximately 32 thousand of the full GO ontology down to one or a few thousand terms. This still represents a substantial multiple testing problem.

To see this, suppose that there are T tests to be performed. We assume that the null hypothesis is true for all of those tests. Before its actual determination, any p-value can be considered as a random variable as well, for which $P(p \leq \alpha | H_0) \leq \alpha$ holds [75]. This implies that it can be expected that αT tests lead to the rejection of the null hypothesis although it is true.

Example 8.2 *As explained in Chapter 3, if we conduct T tests at a significance level of α, and the null hypothesis is true for all the tests, then we can still expect αT tests to be called statistically significant. If we choose a study set of 200 genes at random from a population set of 20,000 genes, then we would not expect to find significant overrepresentation of GO terms by chance in this study set. However, if we perform an overrepresentation test for 10,000 GO terms with a significance level of $\alpha = 0.05$ and do not perform multiple testing correction, then we will expect to find $\alpha \times 10,000 = 500$ terms called significant by chance. Therefore, this represents the "background" number of terms called significant in any GO analysis unless we perform multiple testing correction.*

Therefore, the results of term enrichment analysis should always be subjected to multiple test correction. The most simple, but also the most conservative is the Bonferroni correction [2]. Here, the p-value is simply multiplied by the number of tests. The Bonferroni correction and a number of other multiple testing correction procedures were explained in detail in Chapter 3.

8.4 Term-for-Term Analysis: An Extended Example

We will now demonstrate how term-for-term overrepresentation analysis is performed using a dataset in which the gene expression profiles of the developing mouse aorta were investigated using microarray hybridization [193]. In this experiment, the thoracic aorta was harvested from 15 newborn and 15 six-week old C57BL/6 wildtype mice. Five samples each were combined to get three pooled aortic samples for each group. For gene-expression analyses, 500 ng total RNA of each RNA sample was labeled using the Agilent single-color Quick-Amp Labeling Kit and hybridized on Agilent Whole Mouse Genome Microarrays (4x44K). Following normalization, a subset of genes for data interrogation was generated that excluded probes that were absent or marginal in all of the six samples. The relative expression of each probe in aortic samples of newborn versus six-week old mice was determined. A t-test was performed followed by Benjamini and Hochberg multiple-testing correction in order to determine which genes were differentially expressed.[1]

The result of the above experiments and analysis is a list of differentially expressed genes (the study set) as well as a list of all genes that were measured by the microarray hybridization (the population set). Note that it is possible to construct a study set either from all differentially expressed genes or to construct separate study sets from differentially expressed genes that were up-regulated and those that were down-regulated. Which is "correct" will depend on the particular experiment and the questions of the researcher. These data represent the input for the GO overrepresentation analysis. The actual expression levels of the genes or the p-values of the differentially expressed genes are not needed.

There are numerous commercial as well as freely available programs for performing GO analysis. Almost all of them perform the term-for-term procedure per default. A comprehensive list of these programs is available at the GO Web site.[2] There are browser-based programs as well as downloadable applications for every major operating system. In the exercise section of this chapter, a tutorial will be presented in which readers will learn how to perform the analysis using the Ontologizer, which is a Java Web Start application developed by the authors of this book [24, 212]. In the following text we will explain the general steps that any GO analysis program needs to perform in order to do the analysis.

There are three kinds of data required to perform GO overrepresentation

---

[1] The microarray data presented here are available in raw form the ArrayExpress database (www.ebi.ac.uk/arrayexpress) under the accession number E-MEXP-2342. Files containing lists of differentially expressed genes (separate study sets for up-regulated and down-regulated genes) and all genes represented by probes with present calls (the population set) are available at the book Web site.

[2] http://www.geneontology.org.

analysis. The population and study sets are gene lists representing the outcome of some experiment and are provided as input by the user. In addition, definitions of the Gene Ontology are required with all of its more than 32 thousand terms in the ontology including is_a and other relations between the terms. Third, annotations specific to the organism whose genes are listed in the study and population set are needed. As described in Chapter 5, an annotation is statement that a gene has the property described by the GO term of the annotation. The GO definition file and the annotation file can be downloaded from the Gene Ontology Web site. Explanations of the formats and contents of these files were also given in Chapter 5.

The population set and one or more study sets are provided as input to the programs. The first task is to identify and count the genes. Many microarrays have multiple probes per gene, and in this case it is possible that the list of genes in population or study set has duplicate entries. In this case, the analysis procedure should only count each gene once. The quantities m and n for the size of the population and study set are calculated. Then, the analysis procedure loops over each of the GO terms in turn, and for each term t decides whether to perform a statistical test on the basis of whether the term is used to annotate at least one gene in the study set. If so, the number of genes annotated to the term t in the population (m_t) and in the study set (n_t) are required.

In order to calculate n_t or m_t, it is not enough merely to look up which genes are annotated to t. Instead, the outcome of the annotation propagation rule needs to be taken into account. Recall from Section 5.4.3 starting at page 128 that this rule states that genes that are annotated to subclasses of t are also implicitly annotated to t and that additionally annotations are also propagated along the part_of relation. Therefore, the analysis procedure needs to identify all is_a descendants of term t. Similar remarks apply to terms connected by part_of relations. If any of the genes in the population or study set are annotated to these descendants, n_t or m_t is adjusted to also include these genes. Note that although GO is currently being expanded to include new relations and logical definitions via cross-products, other relations between terms are not used to define study set or population set annotations in this way.

Another issue is that not all genes currently have even a single GO annotation, because their function is still completely unknown. In general, these genes are excluded from further analysis and m_t and n_t are calculated based on genes with at least one GO annotation.

The analysis procedure can now use the quantities m, n, m_t, and n_t to calculate the p-value for term t according to Equation (8.2). The p-value is calculated for each term in the same way and recorded. After all p-values have been determined, a multiple-testing correction procedure is applied, and the results are reported to the user, usually in tabular or graphical form. Algorithm 8.1 provides an overview of the most important steps of the procedure.

The primary result of this analysis is an adjusted p-value for each tested

---

Data: study set, population set, Gene Ontology $\mathcal{G}$, annotations $\mathcal{A}$
1   $m \leftarrow$ number of unique genes in population set;
2   $n \leftarrow$ number of unique genes in study set;
3   $j \leftarrow 0$;
4   **foreach** $t \in \mathcal{G}$ **do**
5   Count annotations to t using $\mathcal{A}$ and *is a/part of* relations in $\mathcal{G}$;
6       $m_t \leftarrow$ number of genes annotated to t in population set;
7       $n_t \leftarrow$ number of genes annotated to t in study set;
8   **if** $n_t < 1$ **then**
9 | jump to next t;
10   **end**
11       $p_t \leftarrow$ p-value calculated by equation 8.2;
12       $j \leftarrow j + 1$;
13   **end**
14   Perform multiple testing correction for the j nominal p-values ;
15   **return** *corrected p-values*

Algorithm 8.1: Overview of the main steps performed for term-for-term procedure analysis of a study set and population set.

GO term. Usually, the results are shown ranked according to the p-value, whereby terms with the smallest (most significant) p-values are shown at the top of the list. For the analysis of the aorta data, a study set consisting of 1494 significantly down-regulated annotated genes was compared against the 16,359 annotated genes in the population set using the term-for-term procedure followed by Bonferroni multiple testing correction. A total of 126 GO terms were found to be significantly overrepresented even after Bonferroni correction (Table 8.1). The table shows a summary of the results. For each GO term, the ID and the name of the term are shown, followed by the Bonferroni-corrected p-value and the number of genes annotated to the term in the study set (n_t) and population set (m_t). The percentage of genes in study and population set annotated to the GO term are given in parentheses.

We see that three times as many genes in the study set (genes down-regulated in the aorta of 6-week old mice) as in the population set (4.8% vs. 1.6%) are annotated to the GO term *extracellular matrix*. This reflects the fact that much of the synthesis of the extracellular matrix (which is an important component of aortic tissue) occurs early in development and is characteristically down-regulated later in development. Therefore, *extracellular matrix* can be taken to be one of the most important characteristics of the set of genes down-regulated in the adult aorta of the mouse. Similar things can be said about the other significant terms. For instance, the fifth term in the list, *developmental process*, presumably reflects that fact that genes involved in developmental processes are down-regulated in the adult aorta compared to the aorta in newborn mice because certain developmental processes in the aorta have been completed by the age of six weeks.

ID	Name	Adj. p-value	n_t	m_t
GO:0031012	extracellular matrix	3.228×10^{-13}	71 (4.8%)	269 (1.6%)
GO:0016043	cellular component organization	4.323×10^{-13}	237 (15.9%)	1548 (9.5%)
GO:0005515	protein binding	2.184×10^{-12}	581 (38.9%)	4847 (29.6%)
GO:0005578	proteinaceous extracellular matrix	6.862×10^{-12}	68 (4.6%)	265 (1.6%)
GO:0032502	developmental process	9.473×10^{-9}	311 (20.8%)	2373 (14.5%)
GO:0005634	nucleus	2.432×10^{-8}	455 (30.5%)	3791 (23.2%)
GO:0010468	regulation of gene expression	2.932×10^{-8}	273 (18.3%)	2039 (12.5%)
GO:0009653	anatomical structure morphogenesis	3.792×10^{-8}	147 (9.8%)	926 (5.7%)
GO:0019222	regulation of metabolic process	1.196×10^{-7}	310 (20.7%)	2413 (14.8%)
GO:0006996	organelle organization	1.689×10^{-7}	138 (9.2%)	869 (5.3%)
GO:0007275	multicellular organismal development	2.216×10^{-7}	280 (18.7%)	2142 (13.1%)
GO:0048856	anatomical structure development	3.178×10^{-7}	244 (16.3%)	1814 (11.1%)
GO:0060255	regulation of macromolecule metabolic process	3.637×10^{-7}	288 (19.3%)	2227 (13.6%)
GO:0045449	regulation of transcription	4.612×10^{-7}	253 (16.9%)	1904 (11.6%)
GO:0080090	regulation of primary metabolic process	9.260×10^{-7}	284 (19.0%)	2208 (13.5%)
GO:0019219	regulation of nucleobase, nucleoside, nucleotide and nucleic acid metabolic process	1.304×10^{-6}	260 (17.4%)	1989 (12.2%)
GO:0048731	system development	1.344×10^{-6}	229 (15.3%)	1702 (10.4%)
GO:0048523	negative regulation of cellular process	1.402×10^{-6}	141 (9.4%)	921 (5.6%)
GO:0043283	biopolymer metabolic process	1.557×10^{-6}	530 (35.5%)	4658 (28.5%)
GO:0051171	regulation of nitrogen compound metabolic process	1.570×10^{-6}	261 (17.5%)	2002 (12.2%)
GO:0034960	cellular biopolymer metabolic process	1.784×10^{-6}	479 (32.1%)	4141 (25.3%)
GO:0006350	transcription	1.812×10^{-6}	256 (17.1%)	1958 (12.0%)
GO:0031323	regulation of cellular metabolic process	1.918×10^{-6}	292 (19.5%)	2299 (14.1%)
GO:0031326	regulation of cellular biosynthetic process	2.002×10^{-6}	269 (18.0%)	2082 (12.7%)
GO:0009889	regulation of biosynthetic process	2.440×10^{-6}	269 (18.0%)	2086 (12.8%)

TABLE 8.1: Term-for-term analysis of the significantly downregulated genes in the aorta experiments. 16,359 genes annotated to at least one GO term were in the population set, and 1494 annotated genes were in the study set. Only the first 30 significant GO terms are shown; a total of 126 GO terms were found to be significant at a significance level of $\alpha = 0.05$. The column *Adj. p-value* shows the Bonferroni-adjusted p-values. The columns n_t and m_t additionally show the percent of all genes in the study and population set that are annotated to the GO term in question. Analysis was performed using the Ontologizer [24].

On closer inspection, we can detect a problem with the interpretation of this analysis. Many of the terms are highly similar to one another. For instance, *proteinaceous extracellular matrix* is a subclass (is_a child) of *extracellular matrix* in the GO ontology. Therefore, any gene annotated to *proteinaceous extracellular matrix* is automatically also annotated to *extracellular matrix*. Which GO term should we take as being representative of our experiment? The most significant term (*extracellular matrix*)? The most specific term (*proteinaceous extracellular matrix*)? Both terms? Similarly, multiple GO terms related to development are flagged as significantly overrepresented: (*developmental process, anatomical structure morphogenesis, multicellular organismal development, anatomical structure development, system development*). The genes annotated to each of these terms, which are close to one another in the GO graph structure, show a high degree of overlap. It is unclear whether to take one of these terms or all of them as providing the best summary of the salient biological characteristics of the dataset.

Another problem is the sheer number of GO terms that have been flagged significant by the method. While a list of 5, 10, or even 20 GO terms characterizing an experiment can be extremely helpful as a way of summarizing the main results of an experiment and suggesting areas for follow-up experiments, a list of 100 terms is more likely to be confusing. This has motivated the development of a number of algorithms whose main goal is to identify a smaller, "core" set of GO terms that provide a good concise description of the biological features of the study set. In the following section, we will explore some of the reasons behind the problem of term-for-term analysis before presenting two algorithms designed to compensate for the problem.

8.5 Inferred Annotations Lead to Statistical Dependencies in Ontology DAGs

Multiple testing correction reduces the number of false positives in a post hoc fashion by adjusting the P-values calculated by the individual statistical tests. However, as we saw in the example of the last section, standard tests for GO term overrepresentation based on the exact Fisher test can still call a high number of terms significant even after correction for multiple testing. It is not unusual for the standard test to return 50, 100, or more significant terms for one experiment even after Bonferroni correction, which is the most conservative form of multiple testing correction. While such results are correct from a purely statistical standpoint, they are not always useful for biologists who need to interpret the biological ramifications of the GO analysis.

The root cause of the problem is that if a GO term shares genes with a second term, and one of the terms is overrepresented, then it is not too

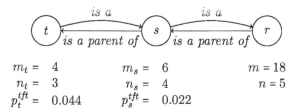

$$m_t = 4 \qquad m_s = 6 \qquad m = 18$$
$$n_t = 3 \qquad n_s = 4 \qquad n = 5$$
$$p_t^{tft} = 0.044 \qquad p_s^{tft} = 0.022$$

FIGURE 8.3: Implicit Annotations Inferred Because of the Annotation Propagation Rule. Continuing the Example 8.1 and Figure 8.2, let us now say that term t *is a* s and therefore s *parent of* t. For completeness, r, which is the root of the ontology, is also shown. It is the only parent of s. As indicated in the last row, the *term-for-term* procedure determines a p-value below 0.05 for both terms. Thus, both terms will be flagged as significantly overrepresented and would be interpreted as being related to the biology of the underlying experiment.

surprising that the other term is also detected as overrepresented. To put it in statistical language, if two GO terms share many annotated genes, then tests for overrepresentation of the terms are not statistically independent. Because of the annotation propagation rule,[3] if a gene is annotated to a given GO term, we can infer that it is also annotated to all of the ancestors of the term (see Chapter 5). For instance, if we know that 10 genes in a study set are annotated to the GO term *ATP-dependent DNA helicase activity*, then we know that at least these genes (and perhaps others) in the study set are annotated to the parent of this term, *DNA helicase activity*. We will refer to this type of statistical dependency as the *propagation problem*,[4] because GO annotations are propagated up the GO graph due to the propagation of annotations to the ancestors of an explicitly annotated GO term.

Example 8.3 (continuation of Example 8.1) *In addition to $m = 18$ and $n = 5$, and a term t with $m_t = 5$ and $n_t = 4$, there is second term s, which is the only parent of t. Term s has six annotated genes in the population ($m_s = 6$) and four in the study set ($n_s = 4$). The graphical structure of this situation is depicted in Figure 8.3. The p-value of terms t is 0.044, and that of term s is 0.022. Thus, both terms are flagged as significant for $\alpha < 0.05$ if no multiple test correction is performed. Four of the five genes annotated to t are in the study set ($n_t = 4$). Term s shares three of these genes ($n_s = 3$). Therefore, one can argue that the result that term s is identified as overrepresented is a consequence of the result that t is overrepresented.*

---

[3]Recall that the annotation propagation rule states that if a gene g is annotated to a GO term t_1 and $t_1 \xrightarrow{isa} t_2$, then g is inferred to be annotated to t_2 as well.

[4]The name *inheritance problem* was used to describe the problem in [97, 98].

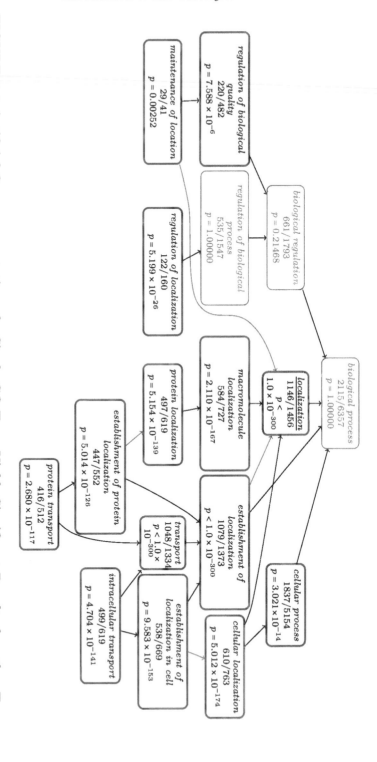

FIGURE 8.4: Artificial Overrepresentation of a Single GO Term Yields Significant Results for other Terms.
A study set that was artificially enriched for the term *localization* was analyzed using the *term-for-term* approach followed by a Bonferroni correction. In addition to the signal term, 184 other terms were found to have an adjusted p-value below the significance level of 0.05. Within the graph, terms that are among the top 10 or children of *localization* are shown. Significant ones are shaded. Obviously, the result suggests a specificity that was not put in as a signal.

In order to further demonstrate the impact of the problem, we construct a study set in which the term *localization* is artificially overrepresented. The study set is created by taking all genes of the term and then removing each gene with a probability of $\beta = 0.2$. In addition, each gene that is not annotated to the term, is added to the study set with a probability of $\alpha = 0.2$. This study set is then subjected to the *term-for-term* approach following a Bonferroni correction. Figure 8.4 shows a graph in which terms with $p < 0.05$ are colored gray. The fact that so many terms are flagged as having statistically significant overrepresentation despite that fact that only a single term was intentionally overrepresented in the simulated data provides a good illustration of the ways that statistical dependencies influence GO overrepresentation analysis.

8.6 Parent-Child Algorithms

The parent-child algorithms were developed to address the effects of the *propagation problem* that was introduced in the previous section. In the remainder of this section, let $pa(t)$ be the set of parents of term t. In order to introduce the principal ideas of the *parent-child* approaches, we first assume in the following that there is only a single parent of t, i.e., $pa(t) = \{s\}$.

The *parent-child* approaches address the *propagation problem* by conditioning the probability of the term t on properties of its parental terms. The statistical tests that are conducted are very similar to those that are applied for the *term-for-term* approach except that the different sets used to calculate the exact Fisher test. Instead of drawing the items from the full population M, we allow the items to be drawn just from the set of items that are annotated to the parents of t, which is written as $M_{pa(t)}$ and whose size is $m_{pa(t)}$. That is, the statistical test only regards genes from the study set that are annotated to the parent of t ($n_{pa(t)}$) and those in the population set that are annotated to t ($m_{pa(t)}$). Because of the propagation of gene annotations up the GO graph, genes annotated to t are also annotated to all parents of t, so $n_t \leq n_{pa(t)}$ and $m_t \leq m_{pa(t)}$.

If the number of genes annotated to t is higher than expected given that $n_{pa(t)}$ genes are annotated to the parent of t, then the term is assigned a significant p value by the parent-child approach. There are $\binom{m_t}{k}$ ways of choosing k genes annotated to GO term t from the set of genes that are annotated to the parent of t (which contains a total of m_t such genes). There are $\binom{m_{pa(t)} - m_t}{n_{pa(t)} - k}$ ways of choosing $n_{pa(t)} - k$ genes from the $m_{pa(t)} - m_t$ genes in the population set that are annotated to the parent term of t but are *not* themselves annotated to t. There are a total of $\binom{m_{pa(t)}}{n_{pa(t)}}$ ways of choosing the $n_{pa(t)}$ genes in the study set annotated to the parent of t from the $m_{pa(t)}$ genes in the population annotated to the parent of t.

This consideration yields the following equation, which plays the same role for parent-child approach as Equation (8.1) does for the term-for-term approach:

$$P(X_t = k|pa(t)) = \frac{\binom{m_t}{k}\binom{m_{pa(t)} - m_t}{n_{pa(t)} - k}}{\binom{m_{pa(t)}}{n_{pa(t)}}}. \tag{8.3}$$

Figure 8.5 summarizes the differences of the principal setting of the *term-for-term* approach with the setting of the *parent-child* approaches. Effectively, in the parent-child approaches, the population that underlies Fisher's exact test is changed from the total population set to the genes annotated by the parent of the term being analyzed (which is necessarily a superset of the genes annotating the term itself). Obviously, this also alters the involved sets for the study set. As in *term-for-term* approach, one asks for the probability of seeing the observed number of items or a more extreme event. As with Equation (8.2) in the term-for-term approach, one sums up the probabilities of observing n_t or more genes annotated to t by chance, conditioned on $n_{pa(t)}$. The sum thus goes from the observed number of terms annotated to t (n_t) up to either the total number of terms in the population annotated to t (m_t) or the total number of terms in the study set annotated to the parent of t, because it is impossible to draw more than this number of terms annotated to t in the study set.

$$P(X_t \geq n_t|H_0) = \sum_{k=n_t}^{\min(m_t, n_{pa(t)})} \frac{\binom{m_t}{k}\binom{m_{pa(t)} - m_t}{n_{pa(t)} - k}}{\binom{m_{pa(t)}}{n_{pa(t)}}}. \tag{8.4}$$

Example 8.4 (continuation of Example 8.3) *As shown in Figure 8.3, the parent of term s is the root r of the ontology, which is always annotated to all genes of the population. Therefore, its p-value for the* parent-child, p_s^{pc} *approaches is identical to the p-value of the* term-for-term approach, p_s^{tft}, *i.e.,* $p_s^{pc} = p_s^{tft} = 0.22$.

However, for term t, Equation (8.4) yields:

$$P(X_t \geq n_t|H_0) = \frac{\binom{4}{3}\binom{2}{1}}{\binom{6}{4}} + \frac{\binom{4}{4}\binom{2}{0}}{\binom{6}{4}} = 0.6.$$

Thus, the null hypothesis for term t is not rejected, which is in contrast to the result of the term-for-form *approach. Given the initial observations that the study set is already skewed to the parent s of t, the enrichment of term t is less surprising, which the* parent-child approach *correctly reflects by returning a higher (i.e., less significant) p-value.*

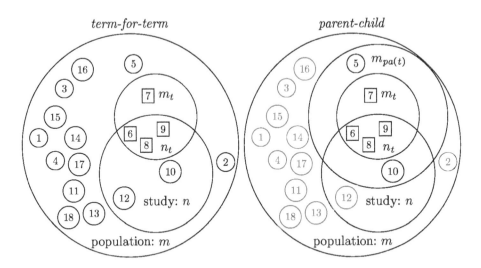

FIGURE 8.5: Differences between *Term-for-Term* and *Parent-Child* **Analysis.** In contrast to the *term-for-term* approach that is visualized in the left part of the figure, *parent-child* approaches, which are depicted in the right part, shift the focus to a smaller set of genes, for instance to the genes that are annotated to at least one of the parents of term t. Genes that are not part of this set do not contribute to the calculation. Effectively, for each term, we alter the population of the association test. In term-for-term analysis, M_t is the set of genes in the population that are annotated to term t, i.e., $\{6, 7, 8, 9\}$, with cardinality $m_t = 4$. The study set N comprises all genes of interest from the experiment (e.g., all genes found to be differentially expressed). In this example, $N = \{6, 8, 9, 10, 12\}$. Three of these genes are annotated to GO term t and thus are members of the set $N_t = \{6, 8, 9\}$, with cardinality $n_t = 3$. The significance of this observation is calculated with Equation (8.2). In parent-child analysis, m_t and n_t are defined in the same way. In addition, the set of all genes annotated to the parent of term t in the population is $M_{pa(t)} = \{5, 6, 7, 8, 9, 10\}$, with cardinality $m_{pa(t)} = 6$. $N_{pa(t)} = \{6, 8, 9, 10\}$ is the set of genes in the study set annotated to the parent of t, with the cardinality $n_{pa(t)} = 4$. The significance of this observation is calculated with Equation (8.4).

If term t has more than one parent term, it is not immediately apparent how to calculate $m_{pa}(t)$ and $n_{pa}(t)$ in Equations (8.3) and (8.4). There are two simple modifications of Equation (8.5) that lead to solutions with a similar formal and computational complexity as the single-parent solution.

For the first approach, called *parent-child-union*, one defines the sets of parents of a term t in the population and study set as the union of genes annotated the parents of t:

$$M_{pa(t)}^{\cup} = \bigcup_{u \in pa(t)} M_u, \quad N_{pa(t)}^{\cup} = N \cap M_{pa(t)}^{\cup} \qquad (8.5)$$

Therefore, $m_{pa(t)}$ and $n_{pa(t)}$ refer to the number of genes annotated to any of the parents of the respective sets.

For the second approach, called *parent-child intersection*, one defines the sets of parents of a term t as the intersection of genes that are annotated to the parents of t:

$$M_{pa(t)}^{\cap} = \bigcap_{u \in pa(t)} M_u, \quad N_{pa(t)}^{\cap} = N \cap M_{pa(t)}^{\cap} \qquad (8.6)$$

Hence, $m_{pa(t)}$ and $n_{pa(t)}$ refer to the number of genes annotated to all of the parents.

The parent-child algorithms were designed especially to avoid false-positive results related to the statistical dependencies related to the annotation propagation rule ("the propagation problem"). The parent-child approaches conceptually measure the overrepresentation of terms in a different way than the term-for-term approach, and it is important to keep this in mind when interpreting results. In almost all datasets we have analyzed, the parent-child approaches identify a smaller number of terms as significantly overrepresented, and the term-for-term approach will flag many of the descendants of these terms as being overrepresented as well. The term-for-term approach leads to false-positive results in these cases, in that the determined overrepresentation results from the structure of the GO DAG and the number of annotated genes rather than truly reflecting the biology of the experiment at hand [98].

8.7 Parent-Child Analysis: An Extended Example

We will now perform parent-child intersection analysis of the same data that was examined above using the term-for-term approach. The study set and the population set are defined in exactly the same way. The analysis procedure is performed analogously to Algorithm 8.1, except that for each term n_t, m_t, $n_{pa(t)}$, and $m_{pa(t)}$ are calculated and Equation (8.4) rather than Equation (8.2) is used to calculate the nominal p-value for the term.

ID	Name	Adj p-value	n_t	m_t
O:0005634	nucleus	6.041×10^{-13}	455 (30.5%)	3791 (23.2%)
GO:0016043	cellular component organization	3.262×10^{-12}	237 (15.9%)	1548 (9.5%)
GO:0031012	extracellular matrix	6.450×10^{-12}	71 (4.8%)	269 (1.6%)
GO:0032502	developmental process	7.633×10^{-8}	311 (20.8%)	2373 (14.5%)
GO:0005488	binding	2.430×10^{-7}	1011 (67.7%)	9890 (60.5%)
GO:0043170	macromolecule metabolic process	3.949×10^{-6}	535 (35.8%)	4809 (29.4%)
GO:0005515	protein binding	2.210×10^{-5}	581 (38.9%)	4847 (29.6%)
GO:0007167	enzyme linked receptor protein signaling pathway	6.171×10^{-4}	50 (3.3%)	265 (1.6%)
GO:0010926	anatomical structure formation	8.653×10^{-4}	103 (6.9%)	672 (4.1%)
GO:0022610	biological adhesion	8.890×10^{-4}	85 (5.7%)	524 (3.2%)
GO:0016055	Wnt receptor signaling pathway	0.00298	30 (2.0%)	129 (0.8%)
GO:0065007	biological regulation	0.00420	651 (43.6%)	6098 (37.3%)
GO:0051301	cell division	0.00563	48 (3.2%)	242 (1.5%)
GO:0007049	cell cycle	0.00819	97 (6.5%)	621 (3.8%)
GO:0048519	negative regulation of biological process	0.02201	150 (10.0%)	1009 (6.2%)
GO:0006807	nitrogen compound metabolic process	0.06495	358 (24.0%)	3126 (19.1%)

TABLE 8.2: Parent-child intersection analysis of the significantly downregulated genes in the aorta experiments. 16,359 genes annotated to at least one GO term were in the population set, and 1494 annotated genes were in the study set. A total of only 16 significant GO terms were identified by the analysis at a significance level of $\alpha = 0.05$. The column *Adj. p-value* shows the Bonferroni-adjusted p-value. Analysis was performed using the Ontologizer [24].

Table 8.2 shows the results of analysis with the parent-child intersection procedure. In contrast to the results with term-for-term analysis, only 16 GO terms were found to be significant at a Bonferroni-corrected significance level of 0.05 instead of 126 with term-for-term analysis. There is now only one GO term related to the extracellular matrix instead of two. Whereas the term-for-term approach returned the terms *developmental process, anatomical structure morphogenesis, multicellular organismal development, anatomical structure development, system development,* all in some way related to the biological process of development, parent-child intersection analysis returned only *developmental process* and *anatomical structure formation.* Arguably, parent-child analysis has returned a smaller list of terms with less redundancies than term-for-term analysis.

In general, parent-child union and parent-child intersection produce different results. Usually, parent-child union flags more terms as significant than parent-child intersection. With the dataset presented here, parent-child union flagged 36 GO terms as significant, including some of those also flagged by the term-for-term procedure. In the simulations described in [98], the parent-child intersection approach gives better results than the parent-child union approach.

8.8   Topology-Based Algorithms

As we have seen, the *propagation problem* represents a major drawback of the *term-for-term* approach because of statistical dependencies related to inferred annotations (see Section 8.5). The parent-child approaches addressed the *propagation problem* by conditioning a term's relevance on the annotations of its parental term or terms. In contrast, topology-based algorithms represent a different approach in that they calculate the significance of a term in light of the annotations to the children (*elim*) or the most significant neighbors (*weight*) of the term [4].

The assumptions of topology-based algorithms are different from those of the parent-child procedure in that the topology-based algorithms seek to capture the most specific overrepresented terms, which are thus regarded as being the most biologically interesting, and simultaneously down-weight annotations to the parents of these terms in order to reduce statistical dependencies. Two distinct topology-based algorithms have been presented to accomplish this objective, called *elim* and *weight* [4].

8.8.1   Elim

As before, we understand the top of the graph as the root of the ontology, while the bottom of the graph consists of the most specific terms. The idea of the *elim* algorithm is to traverse the graph representation of the ontology in bottom-up fashion, which, for instance, can be accomplished by utilizing the backtrack phase of a depth-first search [57].

When the depth-first search (DFS) reaches a leaf node t in the ontology, it performs a Fisher's exact test using Equation (8.2), exactly as in the term-for-term procedure. If the result of this test is significant, that is, $p_t < \alpha$, then all of the genes annotated to the term are marked. In the pseudocode that is shown in Algorithm 8.2, the set of genes annotated to a term t is given as $genes[t]$, and the set of marked genes that apply to a term t is given as $markedGenes[t]$.

When the DFS-elim shown in Algorithm 8.2 returns, the marked genes for term t are returned to the parents of t in a recursive fashion, so that if any term is significant, all of the genes annotated to the term are removed from the significance calculation for all ancestors of t. Algorithm 8.3 sets up the recursion used to implement the DFS, and the recursive function shown in Algorithm 8.2 performs the calculations of statistical significance.[5]

```
 1 DFS-elim(Term t, children(t))
 2 markedGenes[t] ← ∅;
 3 seen[t] ← TRUE ;
 4 foreach u ∈ Children(t) do
 5 │   if seen[u] = FALSE then
 6 │   │   markedGenes[u] ←= DFS − elim(u);
 7 │   end
 8 │   markedGenes[t] ← markedGenes[t] ∪ markedGenes[u];
 9 end
10 Count annotations to t using A and is a/part of relations in G;
11 m_t ← number of genes annotated to t ∈ P;
12 n_t ← number of genes annotated to t ∈ S;
13 Adjust m_t and n_t by subtracting all genes in markedGenes[t];
14 if n_t < 1 then
15 │   jump to next t;
16 end
17 p_t ← p-value calculated by equation 8.2;
18 j ← j + 1;
19 if p_t < α then
20 │   markedGenes[t] ← markedGenes[t] ∪ genes[t];
21 end
22 return markedGenes[t];
```

Algorithm 8.2: DFS-elim(t). The global variables S, P, G, A, m, n, j, and *seen* from Algorithm 8.3 are used by this procedure.

The computational complexity of DFS is linear in the number of terms (nodes) in the ontology, and the overall complexity of the algorithm is the same as that of the term-for-term and parent-child procedures, which also visit all terms in the ontology.

Example 8.5 (continuation of Example 8.4) *The p-value of term t matches the p-value of term t of the* term-for-term *approach, i.e.,* $p_t^{\text{elim}} = p_t^{\text{tft}} = 0.044$. *As this is a significant result, at least, if correction for multiple testing is omitted, all genes annotated to term t are removed in the computation of upper terms.*

---

[5]We note that the original implementation of the *elim* algorithm was based on an iteration over the levels of the GO DAG [4]. The algorithm as shown here yields an equivalent result without needing to explicitly keep track of the DAG levels (distance from the root of the ontology).

Data: study set S, population set P, Gene Ontology G, annotations A
1 $m \leftarrow$ number of unique genes in population set;
2 $n \leftarrow$ number of unique genes in study set;
3 $j \leftarrow 0$;
4 **foreach** $t \in G$ **do**
5 $\quad$ | $\quad seen[t] \leftarrow$ FALSE;
6 **end**
7 $r \leftarrow$ root-term of G ;
8 DFS(r);
9 Perform multiple testing correction for the j nominal p-values;
10 **return** *corrected p-values*

Algorithm 8.3: Depth-First Search for Topology Algorithm (DFS is DFS-elim for the *elim* algorithm and DFS-weight for the *weight* algorithm).

8.8.2 Weight

An equivalent characterization of the *elim* method is the following: If a term t is identified as significant, all genes that are annotated to t are no longer considered in the computation of potential overrepresentation of the ancestors of t. This means that more significant nodes on higher levels in the graph thus can be missed because of the gene removal process [4] (see Example 8.5).

This concern is addressed by the *weight* method, which compares significance scores of connected nodes (a parent and his child) to identify the locally most significant terms in the GO graph and to down-weight genes in less significant neighbors. In order to do this, the *weight* method treats the genes as weighted sets.

Definition 8.1 A weighted set W *is a pair* (A, w). A *is the* underlying set *of* W. *A member of* A *is also a* member *of* W. *Furthermore,* $w : A \to \mathbb{R}$ *is the weight function of* W, *which assigns a weight to each member of* W. *The cardinality of* W, *denoted by* $|W|$, *is a sum over all weights of each member, i.e.,*

$$|W| = \sum_{a \in A} w(a).$$

The *weight* method assigns a weighted set to each GO term t, whereby the underlying set A corresponds to the genes that are annotated to t and the weight function is initially set to 1 for all members. For each term t in the ontology graph, a weighted Fisher's exact test is performed with the cardinalities of the involved weighted sets (which are rounded up to the next higher integer if necessary). Therefore, if all genes had a weight of 1, the calculation would be the same as that for the *term-for-term* approach.

The Fisher's exact test is defined by Equation (8.2) but it is often written as a 2×2 contingency table that records the relation between the numbers of genes belonging to the study set or not and the numbers of genes annotated to some GO term t or not (see Exercise 8.3 for an explanation and example). In the weight algorithm, each of the four counts is replaced with a weighted count. That is, instead of the count a for the count of genes that are both in the study set and annotated to t, the following weighted count is used (where A is used to denote the set of genes that are both in the study set and annotated to t).

$$a^w = \left\lceil \sum_{i \in A} \texttt{weight}(i) \right\rceil$$

Similarly, the counts for b, c, and d in the contingency test for Fisher's exact test are replaced by weighted counts. We will refer to the procedure of computing the weighted cardinality and calculating Fisher's exact test based on the weighted counts in our description of the algorithm as a function called wFisher.

As the *elim* method, *weight* processes the graph representation of the ontology in a bottom-up manner. Initially, the weights of all genes associated with a term are set to 1. Then, the function computeTermSig is called in an iterative manner for each term a. A weighting function sigRatio is used to adjust the weights of the genes; sigRatio is defined for the P-values of two terms a and b, which are denoted $p(a)$ and $p(b)$, as follows.

$$\texttt{sigRatio}(a,b) = \frac{\log p(a)}{\log p(b)}$$

Thus if $p(a) = 1.0 \times 10^{-3}$ and $p(b) = 1.0 \times 10^{-6}$, then (using decadic logarithms for simplicity)

$$\texttt{sigRatio}(a,b) = \frac{\log p(a)}{\log p(b)} = \frac{\log_{10} 1.0 \times 10^{-3}}{\log_{10} 1.0 \times 10^{-6}} = \frac{-3}{-6} = \frac{1}{2}$$

The logarithm represents a relatively soft weighting function. An arbitrary monotonically increasing function can be used in place of the logarithm. For instance, $\texttt{sigRatio}(s,t) = \frac{p(t)}{p(s)}$ will result in a stronger weighting than the logarithm. Considering the previous example, we have:

$$\texttt{sigRatio}(a,b) = \frac{p(b)}{p(a)} = \frac{1.0 \times 10^{-6}}{1.0 \times 10^{-3}} = 10^{-3}$$

Note that with either definition, $\texttt{sigRatio}(a,b) < 1$, if $p(a) > p(b)$.

The core of the *weight* algorithm is the recursive function computeTermSig. For each term, this function is called on a term s. Effectively, the procedure yields a decorrelation of the p-values. The algorithm begins by setting up some variables. In lines 1–3 of Algorithm 8.4, nodeW$[t]$ contains a slot for the weight of each of the genes annotated to GO term t, each of which is set to 1. Note that the weight for a given gene is not necessarily the same for nodeW[t] and nodeW[w] if the gene is annotated to both t and w. In line 4, an array called nodeSig is initialized; this array will contain the p-values coming from the weighted Fisher's exact test for each term. *weight* iteratively goes through all of the levels of the GO DAG, whereby each term is assigned a level corresponding to the length of the longest path from the root (line 5). Then, in lines 6-10, the function computeTermSig is called for each term, whereby the algorithm works its way up level by level from the leaves of the DAG to the root.

Data: study set, population set, Gene Ontology $\mathcal{G}$, annotations $\mathcal{A}$
1 **for** $t \in \mathcal{G}$ **do**
2 $\quad$ nodeW[t] $\leftarrow$ 1;
3 **end**
4 nodeSig $\leftarrow \varnothing$;
5 dagLevels $\leftarrow$ get level of each term in DAG of $\mathcal{G}$;
6 **for** i *from* max(dagLevels) *to 1* **do**
7 $\quad$ **for** $t \in$ terms(dagLevels, i) **do**
8 $\quad\quad$ computeTermSig(t, children(t));
9 $\quad$ **end**
10 **end**

Algorithm 8.4: **Weight algorithm.** This portion of the algorithm sets up the recursion. The function computeTermSig is defined in Algorithm 8.5.

The function computeTermSig, shown in Algorithm 8.5, is the core of the procedure. computeTermSig receives a given GO term t and an array of all of the children of t in the DAG as arguments. It first computes a P-value for t using the weighted Fisher's exact test (wFisher) and the weights for the genes annotated to t that have been stored in nodeW[t]. If term t is a leaf node (i.e., t has no children), then we are finished and the function returns (lines 3–5). Otherwise, the set of significant children is initialized to the empty set in line 6. In lines 7–11, the weight of each child of t is determined by comparing its current P-value with that of t (line 8). If a child ch is more significant than t, then its weight will be greater than 1, and ch is added to sigChildren (line 10). There are now two cases, depending on whether t has at least one child with a lower p-value (i.e., "more significant"). Lines 13–19 cover the case in which there are no such children. Then, the weight calculated in line 8 for each child ch (stored in weights[ch]) will be **less than one**. Line 15 adjusts the current node weight of ch by multiplying it with weights[ch].[6] Following this the weighted Fisher's exact test is run again to calculate the current P-value

---

[6]Note that we use $\otimes$ to indicate an elementwise multiplication of all of the genes in nodeW[t] by all of the corresponding weights in weights[t].

for *ch* in line 16. Following this, we are finished and the function returns in line 18.

Lines 17–25 cover the case in which there is at least one child of t that is more significant than t. For each such child *ch*, weights[*ch*] will be **greater than one**. The upper induced graph of a term t includes term t itself and all of its ancestors in the graph.[7] The weight of the genes annotated to each term in the upper induced graph of t is then multiplied by $\frac{1}{\text{weights}[ch]}$. Since this quantity is less than one, this will have the effect of down-weighting these genes for the calculation of the significance of the terms in the upper induced graph of t. Finally, since the new calculation of the P-value of t in line 22 can affect the significance ratio between t and its other (non-significant) children, computeTermSig is called using the new weights for t and the non-significant children in line 25.

The rationale of the down-weighting is to decorrelate the p-values of the related terms by amplifying their difference while maintaining the signal of the most significant terms. It was shown by the authors of the topology methods that they tend to reduce false-positive results and improve the ability to detect biologically relevant terms.

We conclude this section by noting that although it is possible to use the Benjamini Yekutiele procedure to control the false discovery rate (as the authors of the original publication propose [4]) or a simple Bonferroni correction for multiple testing, the P-values returned by the topology algorithms are not the result of a single test but are potentially adjusted multiple times during the course of the algorithm (*weight*), so that the topology P-values are perhaps better regarded as scores rather than P-values in the classic statistical sense. If desired, empirical P-values could be generated by permutation procedures such as the Westfall Young procedure described in Chapter 3, but this would demand a relatively large amount of computational time.

8.9 Topology-elim: An Extended Example

We will now perform an analysis on the same dataset as we did in the other examples in this chapter, but this time we apply the topology algorithms. We will use a Bonferroni correction based on the number n of GO terms included in the analysis. Within the result list we identify that only nine terms are reported as significant. An inspection of the graph representation of the analysis (which readers will be asked to perform in the exercises) shows that in all cases, if a term is called significant than none of its parent

---

[7]At the time the topology algorithms were published by Adrian Alexa and coworkers [4], GO only had is_a and part_of relations. We recommend that only these relations be used for the *weight* algorithm and not other relations such as the regulates relations.

```
1 computeTermSig(Term t, children(t))
2 nodeSig[t] ← wFisher(genes[t], nodeW[t]);
3 if children[t] = ∅ then
4     return;
5 end
6 sigChildren ← ∅;
7 for ch ∈ children[t] do
8     weights[ch] ← sigRatio(nodeSig[ch], nodeSig[t]);
9     if weights[ch] > 1 then
10         add ch to sigChildren;
11     end
12 end
13 if sigChildren = ∅ then
14     for ch ∈ children[t] do
15         nodeW[ch] ← nodeW[ch] ⊗ weights[ch];
16         nodeSig[ch] ← wFisher(genes[ch], nodeW[ch]);
17     end
18     return;
19 end
20 for ch ∈ sigChildren do
21     for w ∈ upperInducedGraph(t) do
22         nodeW[w] ← nodeW[w] ⊗ (1/weights[ch]);
23     end
24 end
25 computeTermSig(t, children \ sigChildren);
```

Algorithm 8.5: computeTermSig.

terms are called significant, as expected. Table 8.3 presents a summary of the results. It is noteworthy that there is much overlap with the results of the other procedures. For instance, the terms *proteinaceous extracellular matrix, collagen fibril organization,* and *extracellular matrix structural constituent* all point to the extracellular matrix, which is quite similar to the results of the other procedures.

The most significant term, *phosphate transport,* was not seen in the results of the parent-child analysis or in the top hits from the term-for-term analysis. It is interesting that there are indications that vascular smooth muscle cells in the aorta are known to be involved in phosphate transport [185], so this result could lead the researcher to consider the hypothesis that genes involved in phosphate transport processes are more highly expressed during early stages of aortic development than at 6 weeks. Readers will be asked to investigate the results of the *weight* algorithm on the same dataset in an exercise.

ID	Name	Adj p-value	n_t	m_t
GO:0006817	phosphate transport	7.097×10^{-5}	24 (1.6%)	74 (0.5%)
GO:0051301	cell division	7.726×10^{-4}	48 (3.2%)	242 (1.5%)
GO:0005578	proteinaceous extracellular matrix	0.00251	68 (4.6%)	265 (1.6%)
GO:0045944	positive regulation of transcription from RNA polymerase II promoter	0.00427	50 (3.3%)	271 (1.7%)
GO:0030199	collagen fibril organization	0.00497	9 (0.6%)	15 (0.1%)
GO:0016055	Wnt receptor signaling pathway	0.00499	30 (2.0%)	129 (0.8%)
GO:0006468	protein amino acid phosphorylation	0.00981	95 (6.4%)	635 (3.9%)
GO:0007411	axon guidance	0.01560	23 (1.5%)	90 (0.6%)
GO:0005201	extracellular matrix structural constituent	0.02418	13 (0.9%)	35 (0.2%)

TABLE 8.3: **Topology-elim analysis of the significantly downregulated genes in the aorta experiments**. 16,359 genes annotated to at least one GO term were in the population set, and 1494 annotated genes were in the study set. A total of only 9 significant GO terms were identified by the analysis at a significance level of $\alpha = 0.05$. The column p-value (Adj) shows the Bonferroni-adjusted p-value. Analysis was performed using the Ontologizer [24].

8.10 Other Approaches

In addition to approaches that take a fixed subset of the population as input, procedures that take the measurements of the genes into account are also widely in use. This is attractive as it frees the investigator to define a sometimes arbitrary chosen cut off that is used to construct the study set.

A first version of the so-called Gene Set Enrichment Analysis (GSEA) that receives much attention of the scientific community was presented by Mootha and colleagues [166]. Genes were ranked according to an interesting feature (e.g., the difference of the mean of their expression values for two experimental conditions). The null hypothesis is that the genes of the interesting set (e.g., genes annotated to a term) have no association with that list, in which case they would be randomly ordered. The alternative hypothesis is that the genes of the interesting set have an association. For instance, if the genes of the set are grouped together on the top of the list we would assume that there is such an association.

To capture the association via statistical means, the authors propose to use a normalized Kolmogorov-Smirnov (KS) test statistic. Let $r_i \in M$ be the gene of the population M that has rank i in the gene list that is sorted according to the interesting gene feature. Using the previously established notation, i.e., that m is the total number of genes and that N_t is the set of cardinality n_t

which contains only genes that are annotated to t, the score is defined as:

$$ES(N_t) = \max_{i \in \{1,\dots,m\}} \sum_{j=1}^{i} X_j \text{ with } X_j = \begin{cases} -\sqrt{\frac{n_t}{m-n_t}}, & \text{if } r_i \notin N_t \\ \sqrt{\frac{m-n_t}{n_t}}, & \text{otherwise} \end{cases}$$

Thus, the score is the maximum of a running sum that is increased if the gene is annotated to t and decreased if the gene is not annotated to t. In order to check if the obtained score is significant, the calculation is repeated for k randomly chosen sets $N_t^1, \dots, N_t^k$, which all are subsets of M with size n_t. The p-value for a term t is calculated as

$$p_t = \frac{\left| \{ i | ES(N_t^i) \geq ES(N_t) \} \right|}{k}.\text{[8]}$$

The GSEA method has undergone a slight revision [247], where ad-hoc modifications are implemented that are supposed to countervail the well-known lack of sensitivity of the KS test [133, 162].

With *Whole Transcriptome Shotgun Sequencing*, which is also called RNA-seq, advances in sequencing technologies are bringing in new opportunities to the field of transcript expression profiling, since it is now possible to measure levels of vast amounts of transcripts at very high resolution [168, 259]. However, this new technique is also occasioning many new challenges for downstream analysis including the kind of gene enrichment analysis that is the topic of this chapter.

Following the analysis of Oshlack and colleagues, in which a relation between the length of a transcript and the ability to detect this transcript as differential expressed was described [191], Young and colleagues developed an approach that aims to account for this effect within the gene enrichment setting [267]. After determining n genes as differential expressed with any applicable method (e.g., [169] or [208]), the authors propose to fit a probability weighting function that quantifies the likelihood of a gene being differentially expressed by means of the transcript length. That way, any trend, i.e., whether longer transcripts increase or decrease the power of the differential expression test, is included in the statistical test that is used to assesses whether a term is significantly enriched or not.

While the test statistic continues to be the number of differentially expressed genes, i.e., n_t, the null distribution no longer matches the hypergeometric distribution. Therefore, the authors propose a resampling strategy to estimate it. For this purpose, k sets $N^1, \dots, N^k$ all of cardinality n are randomly drawn from the population without replacement, whereby the probability of a gene being included in this set is determined by the pmf. The numbers of genes of these sets that are annotated to term t, i.e., $n_t^1, \dots, n_t^k$

---

[8]Note that according to the formula p_t could become 0. This can be fixed by assuming that the observed test statistic is always a part of the random samples.

determine the null distribution. The p-value is calculated as:

$$p_t = \frac{\left|\{i|n_t^i \geq n_t\}\right|}{k}$$

In addition to the resampling strategy the authors also explain how one can approximate the null distribution using the Wallenius distribution [79].

Note that both procedures that were briefly described here need to be applied for each term t if they are used to generate new hypotheses. As there can easily be more than 10,000 terms that need to be tested, the number of resampling steps needs to be rather large in order to deal with the multiple testing problem.

8.11   Summary

In this chapter, we have discussed several of the most well-known algorithms that have been developed to perform GO overrepresentation analysis. It can be seen that the results of the analysis of the algorithms can be substantially different from one another. There is no one "right" answer in GO analysis, which should be regarded as an exploratory procedure that can help researchers to understand the results of an experiment and generate new hypotheses rather than as a result of the experiment in and of itself. It can be helpful to analyze the results of an experiment with more than one of the analysis procedures outlined in this chapter, but researchers should be cautioned not to assume that a statistically significant GO term does not always represent the key to understanding the biology of the experiment. The results of GO analysis need careful interpretation against the background of current biological knowledge and the experimental setting. In the next chapter, we will introduce two model-based GO analysis procedures that offer a completely different approach to GO analysis.

8.12   Exercises and Further Reading

For readers interested in the bioinformatics of microarray analysis, the book *Statistics for Microarrays: Design, Analysis and Inference* by Wit and McClure [261] is a highly recommended starting place. A special issue of *Nature Biotechnology* on next-generation sequencing technologies from October, 2008[9] and a special issue of *Nature Methods* on next-generation sequencing

---

[9]http://www.nature.com/nbt/focus/sequencing/index.html.

data analysis from November 2009[10] provide valuable introductions to the technology and analysis procedures for next-generation sequence data. The field of next-generation sequencing is moving very rapidly, so readers should additionally consult more recent articles. Rhee and coworkers have published a useful review on GO overrepresentation analysis and how to avoid common pitfalls [206].

Exercises

The first several exercises in this chapter will use R. Consult Appendix A for an introduction to R.

8.1 Use R to implement a function that calculates the probability according to the hypergeometric distribution of observing three genes annotated to t $(n_t = 3)$ if $m_t = 4$ genes in a population of $m = 18$ genes are annotated to t and the size of the study set is $n = 5$ (see this chapter's Example 8.1 on page 185). First implement an R function called ch (for choose, the binomial coefficient) using the builtin function factorial. Then implement a function hypergeometric using the function ch.

8.2 Check your answer from the previous problem using the built-in R function dhyper. Read the documentation for the function by entering >?dhyper at the R prompt.

8.3 Now use the built-in R function fisher.test to calculate the probability with the Fisher exact test (which is identical to the above). In order to use this function, we have to construct a matrix representing the counts as a contingency table (consult the online R documentation for details). You may start with the following contingency table:

	study (n)	not in study $(m - n)$	total (m)
Annotated to t	a	b	m_t
Not annotated to t	c	d	$m - m_t$
	n	$m - n$	m

First, consider what values a, b, c, and d must have in terms of m, n, m_t, and n_t. To calculate the p-value using the Fisher exact test in R, enter the following command using the appropriate values for a, b, c, and d.

```
1  fisher.test(matrix(c(a,b,c,d),nrow=2),alternative="g")
```

The option "alternative" specifies the alternative hypothesis for the statistical test. Since we are interested in overrepresentation, the argument "g" for "greater" is appropriate (see page 184). Compare the results with the results of the previous exercise. Now compare the p-value obtained for all possible different values of k.

---

[10]http://www.nature.com/nmeth/journal/v6/n11s/index.html.

8.4 An equivalent result can be obtained using the function `phypergeom`. Note that `lower.tail=F` implies that the function will take the sum over the upper tail, i.e., $X > x$, but that the definition of the Fisher's exact test implies $X \geq x$. Using the quantities for n_t, m_t, m, and n from the previous exercise, check that you get the same result with the following command. Read the R documentation if anything is unclear.

```
1 phyper(n.t - 1,m.t,m-m.t,n,lower.tail=F)
```

8.5 In this exercise, which will be structured as a tutorial, we will learn how to use the Robo R package from the book Web site to perform term-for-term analysis using the aorta dataset and GO term annotations to mouse genes. See Appendix A for information on how to use R and to install Robo. A version of the GO ontology definition file and the gene association file for the mouse is available in the book data archive (they are gzip-compressed, and the R script will uncompress them behind the scenes). Alternatively, you can download the current ontology definition file and the gene association file for mouse genes (`gene_association.mgi`) from the GO Web site. Start an R session and enter the following commands to initialize the Robo package and input the OBO and association file. The script is written as if all input files are located in the current working directory. If necessary, alter the paths to these files to the locations on your file system.

```
1  ## Aorta data
2  studyset <- "DE_down"
3  popset   <- "Population"
4
5  ## GO Ontology definition file
6  gofile <- "gene_ontology.1_2.obo.gz"
7  ## Gene association file for mouse (MGI)
8  assocfile <- "gene_association.mgi.gz"
9
10 library(robo)
11 session <- openRoboSession(gofile, assocfile)
12 GOterms <- getRoboTerms(session)
13 genes <- getRoboItems(session)
14
15 study.ids <- read.table(studyset,header=F)
16 pop.ids <- read.table(popset,header=F)
17 colnames(study.ids) <- "name"
18 colnames(pop.ids) <- "name"
```

Line 10 loads the Robo package. Robo creates a *session* in which an OBO definition file and a gene association file are linked. `GOterms` is an R list, each of whose members is itself an R list representing a single GO term. For instance, the following command shows the ninth list element. The final two lines assign a name to the column of study.ids and pop.ids (which are R data frames).

```
> GOterms[[9]]
$roboId
[1] 8

$id
[1] "GO:0000010"

$name
[1] "trans-hexaprenyltranstransferase activity"

$parents
[1] 10730

$items
[1] 26163 26164
```

Each GO term is assigned an integer index (roboId) for more efficient computation (indexing begins with zero rather than one as is usual in R because most of the work in Robo is done in the C programming language). id and name show the corresponding fields of the OBO record for the GO term. parents contains one or more integers representing the indices (i.e., roboId of the parents of the current term). Finally, items contains the integer indices assigned to the genes annotated to this GO term in the gene association file.

Lines 15–16 read the study set and population set from the files provided in the book data archive in the directory named Aorta.

```
19  study.mapped <-
20      na.omit(unique(mapRoboItems(session, study.ids)))
21  pop.mapped <-
22      na.omit(unique(mapRoboItems(session, pop.ids)))
```

The Robo package includes a function called mapRoboItems that maps the IDs from the input files to the whole-number indices used by the Robo package, i.e., roboId). The function will attempt to map synonyms that are given in the GO ontology definition file (see Chapter 4). The unique command merges duplicate genes (often, microarrays contain multiple probes for a given gene, but GO analysis is performed at the gene level, rather than for individual probes). The na.omit command removes NAs, which are generated by the mapRoboItems function for gene names from the study or population set which are not mentioned in the annotation data (there are still many genes for which no GO annotations are available). Therefore, study.mapped and pop.mapped contain a unique list of genes for which at least one GO annotation is available — this is the starting point of any GO overrepresentation analysis. For this exercise, we will examine overreprentation to the GO term *extracellular matrix*. We first need to identify the idex of this term in the list GOterms.

```
23  idx <- grep("^extracellular matrix$",
24              sapply(GOterms, function(x) {x$name}))
```

If you do not understand this command, read the R documentation on grep and sapply. We can now use this index to access the element of the R list GOterms for *extracellular matrix* as GOterms[[idx]]. The following code extracts the quantities we will need to perform the Fischer's exact test.

```
25  n<-length(study.mapped)
26  m<-length(pop.mapped)
27  n.t<-length(intersect(GOterms[[idx]]$items,study.mapped))
28  m.t<-length(intersect(GOterms[[idx]]$items,pop.mapped))
```

Finally, we can calculate the *P*-value for *extracellular matrix* using the R function phyper as in the previous exercise.

```
25  phyper(n.t - 1,m.t,m-m.t,n,lower.tail=F)
```

To complete this exercise, calculate the *P*-value for some of the other GO terms showed in Table 8.1.

8.6 In this exercise, you are asked to write an R function with which a *p*-value can be calculated for all of the GO terms. Write a function that takes a GO term (i.e., an item from the R list GOterms) as an argument, then extracts values for $n, m, n.t$, and $m.t$, and then calculates the *p*-value as in the previous exercise. Then use the R function sapply to apply this function to each element of GOterms. Now sort the results to obtain the names and values of the top terms. There are many ways to do this in R. One way is to create a data frame with two columns, one of which holds the term name, and one of which holds the *P*-values. You can then use the order function on the column of the data frame with the *P*-values to sort the dataframe. You can then see the top terms using the head function. Use the R documentation from the command line as explained in Appendix A if needed. The following command may be of value to get a list of the GO term names.

```
1  term.names <- sapply(GOterms,function(x) x$name)
```

Check your work against the results of Table 8.1, but note that there may be minor differences in the results if newer versions of the ontology or annotation files are used.

8.7 In this exercise, we will implement the parent-child union procedure in R. As above, we will first perform the analysis for the term *extracellular matrix*. First, we will get a list of all parents of the term.

```
1  idx <- idx <- grep("^extracellular matrix$",
2                      sapply(GOterms, function(x) {x$name}))
3  ECM <-GOterms[[idx]]
4  parents.items.list <- lapply(GOterms[ECM$parents],
5                               function(x) x$items)
```

Now, we will determine which genes are annotated to the parents. We get a list in which each element contains the genes of a parent. We use the command unlist to create an R vector from the list and return the unique elements, which is the union of all items annotated to the parents.

```
5  parents.items.union<-unique(unlist(parents.items.list))
```

The rest of the procedure is similar to the code for term-for-term analysis, except that the quantities used for Fisher's exact test are defined as in Equations (8.4) and (8.5).

```
6   n<-length(intersect(study.ids,parents.items.union))
7   m<-length(intersect(population.ids,parents.items.union))
8   n.t<-length(intersect(intersect(term$items,study.ids),
9                         parents.items.union))
10  m.t<-length(intersect(intersect(term$items,
11                                  population.ids),
12                                  parents.items.union))
13  phyper(n.t - 1,m.t,m-m.t,n,lower.tail=F)
```

8.8 Write an R function with which a *p*-value can be calculated for all of the GO terms for parent-child union analysis (see the previous exercise).

8.9 Use the p.adjust function from the stats package to correct the *p*-values returned in the previous exercise. The basic syntax of this function is as follows:

```
p.adj <- p.adjust(p, method = p.adjust.methods)
```

p.adjust.methods can be set to one of holm, hochberg, hommel, bonferroni, BH, BY, fdr, or none. Explore how many of the adjusted *p*-values remain significant using the corrections presented in Chapter 3 as well as the other methods.

8.10 The authors of this book have developed a Java Web Start application called the Ontologizer, which can be used to perform all of the calculations described in this chapter, as well as the MGSA procedure described in the next chapter [24, 212]. The Ontologizer is freely available.[11] Java Web Start applications allow users to start Java applications directly from a Web browser. In contrast to earlier Java applets, they do not run directly within the browser window, but are essentially independent applications with some restrictions on access to the underlying operating system. If necessary, users will need to install a recent version of Java to run the software.[12] The Ontologizer additionally uses *graphviz*, which is a freely available, open-source graph visualization framework provided by AT&T research.[13]

In this exercise, we will walk readers through the steps needed to perform term-for-term analysis as in Section 8.4 of this chapter. Make sure you have Java and graphviz installed on your computer as described by the Web sites for Java and GraphViz. Now go to the Web page for the Ontologizer mentioned above and simply click on the icon for the Ontologizer. After a few seconds, you should see a window as in Figure 8.6.

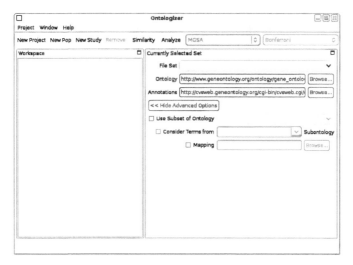

FIGURE 8.6: Ontologizer (1). This is the main window of the Ontologizer [23]. Data for the population set and the study sets are entered in the left panel, while the right panel controls the ontology definition file and the annotation file to be used for the analysis. The **analyze** button is used to specify the type of GO analysis to be performed (in this figure, analysis will be performed with the MGSA method that will be described in the next chapter).

By clicking on the **New Project** button, a "wizard" will be started that

---

[11]http://compbio.charite.de/index.php/ontologizer2.html.

[12]See http://www.oracle.com/technetwork/java/index.html for more information on Java and how to install a version of Java appropriate for your operating system.

[13]See http://www.graphviz.org/ for information about graphviz and how to install it on your system.

will guide you through the process of setting up an analysis. Click on the button, and give the project the name *Aorta* and click on Next. In the window that will appear, you can choose the ontology definition file and the association file (the latter must match the species of the organism being investigated in the experiment). The Ontologizer provides several *File Sets* for commonly investigated organisms that will be downloaded automatically from the GO Web site. For this example, the organism is the common mouse *mus musculus*, so choose "mouse" as File Set and click on next. In the window that appears, you will be asked to choose the *population set*. The data files are included in the directory called "Aorta" in the data archive available from the book Web site. Choose the file called Population and click on Next. In the window that appears, you can choose a study set. Choose the file DE_down from the same directory (it includes a list of differentially expressed genes that were down-regulated in the adult aorta). Name the study set DE-down in the text-entry field. If you click on Next you can add additional study sets, but for now click on Finish. If you let the mouse hover over the name of a gene in the study or population set, you will see a list of the genes it is annotated to (Figure 8.7).

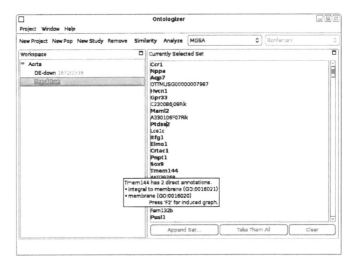

FIGURE 8.7: **Ontologizer (2)**. The right panel displays a list of all genes in the population set. Genes shown in **bold** type are annotated to at least one GO term. We see that the user has let the mouse hover over the gene name *tmem144*, and a dialog appears showing the direct (asserted) annotations of this gene.

Now set the pull-down menu next to the Analyze button to Term-For-Term and the rightmost pull-down menu (which specifies the multiple testing procedure that will be applied) to Bonferroni. Click on the `Analyze` button. A new window will appear shortly with the results of the analysis in tabular form. For instance, the first line in the table should be for the GO term *extracellular matrix*. If a line in the table is marked, the lower panel of the Ontologizer will show more information about the term, including the parents and children of the term and a list of annotated genes. By clicking on the `graph` button (third from right symbol in the menu on the top of the Ontologizer window), users will cause a graph representation of the significant terms and their ancestors to appear in the right panel of the Ontologizer (the graphviz framework is used to generate the visualization of the graph). Especially for the term-for-term algorithm, there will be many significant terms and the graph will be very big. Users can zoom in or out by right-clicking on the graph. Clicking on any particular node will cause the same term to be displayed in the table (Figure 8.8).

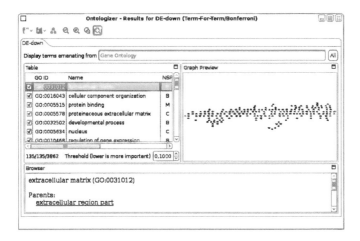

FIGURE 8.8: Ontologizer (3). Here, the results window shows a table of significant GO terms on the left and a graphical representation of the significant terms and their ancestors on the right. The table as well as the graphic can be saved in a variety of formats.

To complete this exercise, explore the some of the significant terms. Use the table browser to examine the lists of annotated genes for the parents and children of a given term. Look at the numbers of genes annotated to a given term in the population and in the study set. Intuitively, what is the relationship between the relative proportions of annotated terms and the p-value? What about the relation between the total count of annotated terms in both sets and the p-value?

Finally, note that the results of the analysis depend on the versions of the ontology definition file and gene annotation file. Therefore, it is possible that you will obtain results that are (presumably slightly) different from the ones shown in this chapter if these files have been substantially updated by the time you hold this book in your hands.

8.11 Use the Ontologizer to perform the analysis with the *parent-child union* and *parent-child intersection* methods as in Section 8.7. What are the main differences in the results of these methods to one another as well as to term-for-term analysis?

8.12 Use the Ontologizer to perform *topology-elim* analysis as in Section 8.9.

8.13 Use the Ontologizer to perform *topology-weight* analysis on the aorta dataset.

8.14 Use the Ontologizer to perform *term-for-term* analysis on the aorta dataset. Use each of the multiple-testing-correction procedures in turn. What effect does this have on the numbers of genes called statistically significant? On the p-values of these genes?

Chapter 9

Model-Based Approaches to GO Analysis

We have seen in the previous chapter that a major difficulty of the standard approach to GO overrepresentation analysis is that each term is analyzed in isolation. Because of the statistical dependencies between terms that are close to one another in the ontology graph, if one term is called significant then commonly one or more related terms are also called significant. A similar problem can affect terms that are distant from one another in the ontology but whose annotations are correlated. The parent-child and the topology algorithms were developed in the attempt to compensate these effects by means of more or less local adjustments to the statistical tests being performed for the GO terms. These procedures are able to reduce false positive results on simulated data, and tend to return smaller lists of terms on real datasets.

This chapter will present a completely different approach to GO analysis that seeks to find the best combination of terms that correspond to an experimental result. The problem reduces to an optimization problem, whereby the choice of active GO terms, and optionally other parameters, are iteratively varied in order to maximize a score or a probability. In contrast to the algorithms in the last section, no statistical test (such as the Fisher exact test or variants thereof) is performed for each term; instead, there is a single score (or probability) that is to be optimized for the entire set of GO terms.

We note that this chapter requires some familiarity with Bayesian statistics. There is a brief introduction to Bayesian statistics and Bayesian networks in Chapter 3, and pointers to more detailed treatments of these subjects are given in the Further Reading section at the end of this chapter.

9.1 A Probabilistic Generative Model for GO Enrichment Analysis

The first algorithm for model-based GO analysis was published by Lu and colleagues in 2008 and was called *GenGO* [155]. The goal of the authors was to create a *generative* model for GO analysis. From a biological point of view, users of GO analysis want to identify a set of biological processes related to

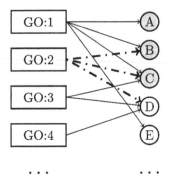

FIGURE 9.1: The GenGO Model. In GenGO [155], GO terms are modeled as active (gray) or inactive (white) nodes that connect to the genes that are annotated to them. The genes are either ON (gray) or OFF (white). Intuitively speaking, the score in GenGO is maximized if a set of GO terms is selected that are connected to as many of the active genes as possible and as few of the inactive genes as possible.

the experiment being analyzed. One way of thinking of this is to imagine that the set of all GO terms can be divided into those that are related to the experimental conditions or results ("active GO terms") and those that are not ("inactive GO terms"). Then, one can model the results of a high-throughput experiment (e.g., a list of differentially expressed genes) by asking which particular set of active GO terms would best "generate" the observed list of differentially expressed genes. The basic assumption of genGO is thus that if a GO term is *active*, then the genes annotated by that term will tend to be differentially expressed (ON). In contrast, genes annotated to *inactive* GO terms will tend not to be differentially expressed (OFF). The genGO algorithm uses an optimization procedure to find the best set of GO terms to "explain" the set of differentially expressed genes resulting from an experiment (Figure 9.1).

The GenGO algorithm can be explained with the use of several categories.

1. A_g: ON gene node that is connected to at least one **active** GO term (e.g., genes B and C in Figure 9.1).

2. A_n: ON gene node that is *not* connected to any **active** GO term (e.g., gene A in Figure 9.1).

3. I: OFF gene node (e.g., genes D and E in Figure 9.1).

4. S_g: edge connecting an **active** GO term with a node in I (e.g., edge from GO:2 to gene D).

5. S_n: edge connecting an **inactive** GO term with a node in I (e.g., edge from GO:4 to gene D).

According to the model of genGO, an `active` GO term does not activate all of the genes it annotated; rather, because of noise or other errors, annotated genes are observed to be `OFF` with a probability of $1 - a$. Similarly, genes that are not annotated by any `active` GO term are observed to be `ON` with a probability of b. The actual values of a and b can be chosen by the user or optimized by the genGO algorithm itself, but representative values are $a = 0.9$ and $b = 0.01$ [155].

One can then define a scoring function that is to be maximized, whereby C is the set of `active` GO terms in the current iteration.

$$\mathcal{L}(C|p, q, G) = |A_g| \log a + |A_n| \log b$$
$$+ |S_g| \log(1 - a) + |S_n| \log(1 - b) - \alpha|C| \quad (9.1)$$

The first four terms of Equation (9.1) are equivalent to the logarithm of the following equation

$$a^{|A_g|} b^{|A_n|} (1 - a)^{|S_g|} (1 - b)^{|S_n|} \quad (9.2)$$

It can be seen that the value of Equation (9.2) is maximized if $|A_g| > |S_g|$ (because $a > 1 - a$) and if $|S_n| > |A_n|$ (because $1 - b > b$). Thus, maximization of Equation (9.2) would tend to identify sets of `active` GO terms that more often than not annotate the `ON` genes ($|A_g| > |S_g|$). On the other hand, the `inactive` GO terms would more often than not annotate `OFF` genes ($|S_n| > |A_n|$).

The final term in Equation (9.1), $-\alpha|C|$ reduces the score linearly in the number of `active` GO terms. The authors of genGO state that a value of $\alpha = 3$ tends to produce good results. The genGO algorithm seeks to optimize the score for the current set of `active` GO terms C. In each iteration, all possible one-step changes of C are considered, both changes that add a term t_M to the current configuration as well as changes that remove a term t_L. In each iteration, the single-step change with the highest improvement of the score is chosen until no further improvement is possible (Algorithm 9.1).

The original description of GenGO includes a procedure for optimizing the values of the parameters p and q. Additionally, a post hoc statistical test for overrepresentation with Fisher's exact test is used to rank and filter the results of the initial analysis. Readers are referred to the publication of Lu and coworkers [155] for further details. The authors showed using simulated datasets that GenGO outperformed the methods described in the previous chapter with respect to false-negative and false-positive results. The idea behind GenGO appears particularly attractive because it represents a novel way of avoiding the statistical dependency problems associated with the overrepresentation methods described in the previous chapter.

Data: study set $\mathcal{S}$, population set $\mathcal{P}$, Gene Ontology $\mathcal{G}$, annotations $\mathcal{A}$

```
1  C ← ∅;
2  repeat
3      foreach t ∈ G do
4          t_L = arg max_{t∈C} L(C_i \ {t});
5          t_M = arg max_{t∈G\C} L(C_i ∪ {t});
6          if L(C_L) > L(Ct_M) then
7              │  C = C \ {t});
8          end
9          else if t_M > t_L then
10             │  C = C ∪ {t});
11         end
12     end
13 until No further improvement of score L(C) ;
14 return C;
```

Algorithm 9.1: GenGO. Note that in each iteration of genGO, the current set of GO terms C is modified by adding or removing a GO term until no further improvement of the score is achieved. In the pseudocode, t_L refers to the term whose removal to the current set C would cause the score to increase the most and $\mathcal{L}(C_L)$ is that score; t_M and $\mathcal{L}(C_M)$ are defined analogously for the term whose addition to the current set would cause the score to increase the most. $\mathcal{L}$ is calculated with Equation (9.1).

9.2 A Bayesian Network Model

Model-based gene set analysis (MGSA) [23] is similar to GenGO in that it seeks to identify an optimal combination of GO terms to "explain" the results of microarray or other high-throughput experiments, but differs from genGO in that it embeds the GO terms and the genes they annotate into a Bayesian network and uses probabilistic methods to search for the optimal combination.

MGSA models gene response in a genome-wide experiment as the result of an "activation" of one or more biological categories. These categories can be pathways as defined by the KEGG database [138], GO terms [22], or any other kind of gene sets [166, 247] that associate genes to potentially overlapping biologically meaningful categories. For simplicity, we will refer to all such categories as *terms* in this chapter, but MGSA is not limited to analysis with GO terms. MGSA does not make use of the graph structure of GO other than utilizing the annotation propagation rule to identify inherited gene annotations (see page 128).

Similar to genGO, MGSA assumes that the experiment attempts to detect genes that have a particular *state* (such as differential expression), which can

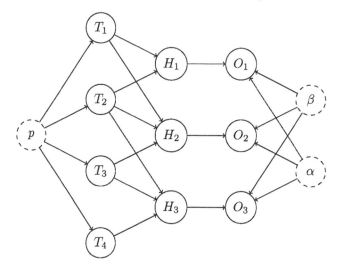

FIGURE 9.2: Structure of the MGSA Network. Gene categories, or terms (T_i) that constitute the first layer can be either `active` or `inactive`. Terms that are `active` activate the hidden state (H_j) of all genes annotated to them, with the other genes remaining `OFF`. The observed states (O_j) of the genes are noisy observations of their true hidden state. The parameters of the model (dashed nodes) are the prior probability of each term to be `active`, p, the false positive rate, α, and the false negative rate, β.

be `ON` or `OFF`. The true state of any gene is hidden. The experiment and its associated analysis provide observations of the gene states that are associated with unknown false positive (α) and false negative rates (β), which we will assume to be identical and independent for all genes.

For instance, in the setting of a microarray experiment, the `ON` state would correspond to differential expression, and the `OFF` state would correspond to a lack of differential expression of a gene. Our model hence assumes that differential expression is the consequence of the annotation to some terms that are `active`.

An additional parameter p represents the prior probability of a term being in the `active` state. The probability p is typically low (less than 0.5), which has the effect of introducing a penalization for increasing the number of active terms. This favors results that identify a relatively low number of `active` terms.

More formally, the model can be described using a Bayesian network with three layers that is augmented with a set of parameters. A simple instance of the model is depicted in Figure 9.2. In more detail, the network consists of:

1. A *term layer* $T = \{T_1, \ldots, T_m\}$ that consists of Boolean nodes corresponding to m terms of the ontology. There is a Boolean variable as-

sociated with each node that can have the state values `active` (1) or `inactive` (0).

2. A *hidden layer* $H = \{H_1, \ldots, H_n\}$ that contains Boolean nodes representing the n annotated genes. There are edges from the terms to the genes they annotate. For instance, if gene H_1 is annotated to terms T_1 and T_2 then there is an edge between T_1 and H_1 and another edge between T_2 and H_1. The state of the nodes reflects the true activation pattern of the genes. Each node can have the state values `ON` (1), or `OFF` (0).

3. An *observed layer* $O = \{O_1, \ldots, O_n\}$ that contains Boolean nodes reflecting the state of all observed genes. The observed gene state nodes are directly connected to the corresponding hidden gene state nodes in a one-to-one fashion.

4. A *parameter set* that contains continuous nodes with values in $[0, 1]$ corresponding to the parameters of the model α, β and p. These parameterize the distributions of the observed and the term layer as detailed below.

For didactic purposes, we will initially explain a simplified version of MGSA in which the parameters α, β and p are considered to have known, fixed values. We will then show how the Bayesian network can be augmented to search for optimal values for α, β and p.

The state propagation of the nodes can be modeled using various *local probability distributions* (LPDs), denoted by P. The joint probability distribution for this Bayesian network can be written as

$$P(T, H, O) = P(T)P(H|T)P(O|H) = P(T) \prod_{i=1}^{n} P(H_i|T)P(O_i|H_i). \quad (9.3)$$

The state of each term $T_j \in T$ is modeled according to a Bernoulli distribution with hyperparameter p, i.e, $P(T_j = 1) = p$. Denoting by $m_{x|T}$ the number of terms that have state x for a given T, i.e., $m_{x|T} = |\{j|T_j = x\}|$, then

$$P(T) = p^{m_{1|T}} (1 - p)^{m_{0|T}}. \quad (9.4)$$

In the following, $T(H_i) \subseteq T$ is used to denote the set of terms to which gene H_i is annotated, i.e., the parents of H_i in the Bayesian network (see Figure 9.2). For the $T \rightarrow H$ links, any node $H_i \in H$ is `ON` ($H_i = 1$) if at least one of its parents is `active`. Otherwise it is `OFF`:

$$P(H_i = 1|T) = \begin{cases} 1, & \text{if } \exists \, T_j \in T(H_i) : T_j = 1 \\ 0, & \text{otherwise.} \end{cases} \quad (9.5)$$

Note that this transition is deterministic. For the $H \rightarrow O$ connection, the following two Bernoulli distributions are used:

$$P(O_i = 1|H_i = 0) = \alpha \tag{9.6}$$

and

$$P(O_i = 0|H_i = 1) = \beta. \tag{9.7}$$

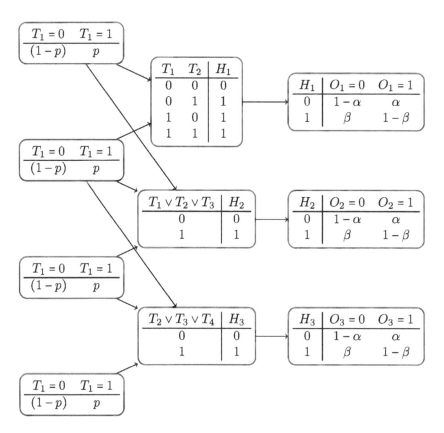

FIGURE 9.3: The Fully Specified MGSA Network from Figure 9.2. Note that the propagation from the term layer to the hidden layer is deterministic.

Therefore, α is the probability that a gene i is observed to be ON (i.e., $O_i = 1$), although its true hidden state is actually OFF (i.e., $H_i = 0$) and thus, none of the terms which annotate the gene are active (such genes can be considered to be false positives according to the logic of the model). Correspondingly, β is the probability of a gene being observed to be OFF although at least one term that annotates it is active (such genes can be considered to be false negatives). Figure 9.3 is a representation of the example network of Figure 9.2 in which the local probability distributions are shown explicitly.

Denote by $n_{xy|T} = |\{i|O_i = x \wedge H_i = y\}|$ the number of genes having observed activation x and true activation y according to the states of T. For instance, $n_{01|T}$ corresponds to the number of genes observed to be not differentially expressed but whose true activation state is *on*. Then, by considering the LPDs of nodes, one gets the following product of Bernoulli distributions for $P(O|T) = \prod_{i=1}^{n} P(H_i|T)P(O_i|H_i)$:

$$P(O|T) = \alpha^{n_{10|T}}(1-\alpha)^{n_{00|T}}(1-\beta)^{n_{11|T}}\beta^{n_{01|T}}. \qquad (9.8)$$

Hence, Equation (9.8) calculates the product over $i = 0, 1$ of the probability of the observed states of the genes given the hidden states of the terms. For instance, for $i = 1$, we need only consider hidden nodes whose parents include `active` terms, because otherwise their probability is zero according to Equation (9.5). Using Equation (9.4) with $p = \beta$, we obtain that $P(H_1|T)P(O_1|H_1) = 1 \times P(O_1|H_1) = (1-\beta)^{n_{11|T}}\beta^{n_{01|T}}$. Similar considerations for $i = 0$ lead to the final expression for Equation (9.8).

9.2.1 Maximum a posteriori

In Bayesian statistics, maximum a posteriori (MAP) estimation is often used to generate an estimate of the maximum value of a probability distribution. That is, if x is used to refer to the data (x can be an arbitrary expression), and θ is used to refer to the parameters of a model, then Bayes' law states that:

$$P(\theta|x) = \frac{P(x|\theta)P(\theta)}{P(x)} \qquad (9.9)$$

The term $P(\theta|x)$ is referred to as the posterior probability, and specifies the probability of the parameters θ given the observed data x. The denominator on the right-hand side can be regarded as a normalizing constant that does not depend on θ, and so it can be disregarded for the maximization of θ. The MAP estimate of θ is defined as:

$$\arg\max_{\theta} P(\theta|x) = \arg\max_{\theta} P(x|\theta)(P(\theta) \qquad (9.10)$$

In the case of MGSA, the parameters comprised by θ would include the set of `active` terms as well as values for α, β, and p.

Although MAP estimation procedures are often relatively simple to implement, they tend to have the disadvantage that they "get stuck" in local maxima (Figure 9.4) without being able to offer a guarantee of finding the global maximum. In complicated networks such as that of MGSA, it is rare to have a single solution that is substantially better than all alternative solutions. Rather, the posterior probability is usually spread over a number of alternative network configurations. This implies that the posterior probability is not adequately represented by a single configuration θ^{MAP}, and it is more

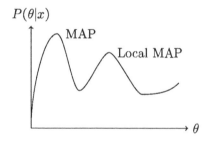

FIGURE 9.4: Global vs. Local Maximum. If the function describing the posterior probability is too complex, it may be difficult to obtain the global maximum. For instance, the optimization problem addressed by genGO and MGSA is known to be NP-complete. An approximation algorithm as the described greedy algorithm may only return a local maximum. In the context of the MGSA algorithm, the parameter vector θ would include the set of `active` terms (and in the full version of MGSA that will be discussed below, also the values for α, β, and p). The data vector x would comprise the lists of ON and OFF genes. Thus, $p(\theta|x)$ would be written $p(T|O)$ for the MGSA algorithm.

appropriate to sample networks from the posterior probability (3), leading to a collection of networks with high posterior probability, each of which offers a good explanation of the data [132]. As we will see in the following section, these considerations motivate the use of the MCMC algorithm to sample from the posterior distribution.

Example 9.1 *Suppose that there are two terms T_1 and T_2. The set of genes to which T_1 is annotated is given by $\{1,2,3\}$. The set of genes to which T_2 is annotated is $\{2,3,4\}$. We observe that gene 2 and 3 are in the ON state (e.g., differentially expressed). This situation is depicted in the left part of Figure 9.5.*

If T_1 were the only `active` term, then the observation could be explained by risking an error of one false-negative. The same can be noticed if T_2 is the only `active` term. Both settings are depicted in middle and in the right of Figure 9.5. MGSA assumes that the false-negative rate is the same for all genes, and thus there is no single optimum solution. A single MAP solution does not account for this.

9.2.2 Monte Carlo Markov Chain Algorithm

A different approach is to calculate the marginal probabilities for each term being in the `active` state. In general, if a joint probability is defined over two random variables X and Y as $P(X,Y) = P(X|Y)P(Y)$, then the marginal probability for $X = x'$ is calculated by summing or integrating over all possible

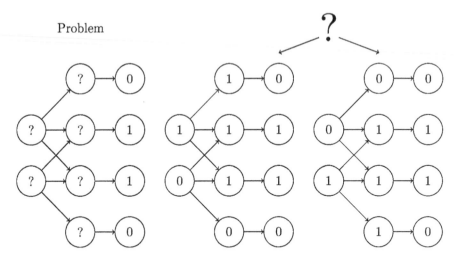

FIGURE 9.5: Two Explanations for the Same Model. The configuration that is shown on the left represents the problem setting of Example 9.1. The configuration that is displayed in the middle explains the observations just as well as the configuration on the right does. A MAP approach would return just one of the two solutions. However, the truth is that we cannot distinguish between both solutions.

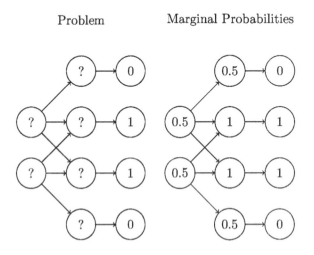

FIGURE 9.6: Marginal Probabilities for the Graph in Figure 9.5.

values of Y: $P(X = x') = \sum_i P(X = x', Y = y_i)$ or $P(X = x') = \int_Y P(X = x', Y)dY$. It is often difficult or impossible to derive marginal probabilities for complicated probability distributions because there are simply too many possible configurations of the variables to be able to calculate each one, as would be required for an analytical solution. For this reason, a number of estimation algorithms have been developed that in essence sample from the distribution of the posterior probability and take the proportion of samples in which X takes on some specific value x' as an estimate of the posterior probability of x' (see Equation 9.14 further on this chapter).

One of the best known and most effective algorithms for this purpose is the Metropolis-Hasting algorithm, which is a Markov chain Monte Carlo (MCMC) method [10, 72, 71]. The MCMC algorithm performs a random walk over the term and parameter configurations, which asymptotically provides a random sampler according to the target distribution $P(T|O)$.

Given the current configuration of the terms denoted by T^t, the algorithm proposes a neighbor state T^p in accordance to a proposal density function $Q_T(\cdot|T^t)$. A value r is sampled uniformly from the range (0,1). Then, if

$$r < P_{accept}(T^t, T^p) = \frac{P(T^p|O)Q_T(T^t|T^p)}{P(T^t|O)Q_T(T^p|T^t)} \tag{9.11}$$

the proposal is accepted, i.e., $T^{t+1} = T^p$, otherwise it is rejected, i.e., $T^{t+1} = T^t$. Using Bayes' law, we have

$$P(T^p|O) = \frac{P(O|T^p)P(T^p)}{P(O)} \tag{9.12}$$

and similarly for T^t. Substituting these expressions for $P(T^p|O)$ and $P(T^t|O)$ cancels out the normalization constant $P(O)$. The acceptance probability is then:

$$P_{accept}(T^t, T^p) = \frac{P(O|T^p)P(T^p)Q_T(T^t|T^p)}{P(O|T^t)P(T^t)Q_T(T^p|T^t)}. \tag{9.13}$$

Equation (9.13) is used iteratively to define a random walk through the space of configurations. A *burn-in period* consisting of a certain number of iterations is used to initialize the MCMC chain (in our implementation of the MGSA algorithm in the Ontologizer [24], the default is 20,000 iterations). Following this, l further iterations (by default, 10^6) are performed. Let $C(T_i)$ be the number of samples in which term T_i was `active`. Then

$$P(T_i|O) \approx \frac{C(T_i)}{l}. \tag{9.14}$$

In order to finish the description of the algorithm, one needs to define classes of operations of which a proposal is chosen, that is, we need to specify $Q_T(T^p|T^t)$. Denote by $T^p \leftrightarrow_T T^t$ the binary relation that states that T^p be constructed from T^t by either

- toggling the `active`/`inactive` state of a single term, or by

- exchanging the state of a pair of terms that contains a single `active` term and a single `inactive` term.

Denote by $N(T)$ the *neighborhood* of a given configuration for T, that is, the number of different operations that can be applied once to T in order to get a new configuration. At first, there are m terms in total, each of which can be toggled. In addition, there are $m_{0|T}m_{1|T}$ possibilities to combine terms that are `active` with terms that are `inactive`. Thus, there are a total of $N(T) = m + m_{0|T}m_{1|T}$ valid state transitions. We would like to sample the valid proposals with equal probability; therefore, the proposal distribution Q_T is determined by

$$Q_T(T^p|T^t) = \begin{cases} \frac{1}{N(T^t)}, & \text{if } T^p \leftrightarrow_T T^t \\ 0, & \text{otherwise.} \end{cases} \tag{9.15}$$

which we can use to rewrite Equation (9.13) to:

$$P_{accept}(T^t, T^p) = \frac{P(O|T^p)P(T^p)N(T^t)}{P(O|T^t)P(T^t)N(T^p)}.$$

The procedure is shown in Algorithm 9.2. For simplicity, the burn-in period is omitted from the pseudocode. In particular, the state space of the situation described in Example 9.1 and the possible transition from one state to another are illustrated in Figure 9.7.

Theorem 9.1 *Algorithm 9.2 converges to the desired stationary distribution.*

Proof 9.1 *It is easy to see that all states of the chain are reachable from any state, as the Markov chain is finite and it is possible to reach an arbitrary state from any other state by a fixed number of operations. This accounts for the irreducibility of the chain. Moreover, the chain is aperiodic as it is always possible to stay in the same state, as any proposal can be rejected. Therefore, the resulting Markov chain is ergodic, which is a sufficient condition for a convergence to a stationary distribution, which matches the desired target distribution[246].*

9.2.3 MGSA Algorithm with Unknown Parameters

In the description of the MGSA algorithm given in the previous section, the values of the parameters α, β, and p were taken as givens. However, these values are not known in advance. Although it is possible to estimate them by running MGSA with many different combinations of values for α, β, and p, the estimation of the parameters α, β, and p can also be easily integrated

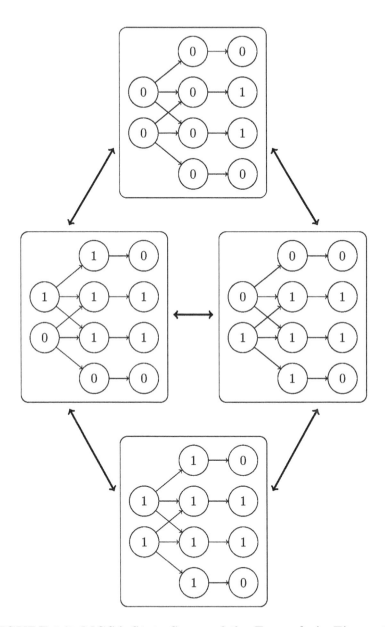

FIGURE 9.7: MGSA State Space of the Example in Figure 9.6.

Data: O, l (number of steps)
Result: $P(T_1 = 1|O), \ldots, P(T_m = 1|O))$
$T^t \leftarrow \underbrace{(0, \ldots, 0)}_{m \text{ times}}$;

1

2 **for** $t \leftarrow 1$ **to** l **do**

3     $T^p \sim Q_T(\cdot|T^t)$, i.e., choose a neighbor candidate by either

- toggling a term

- exchanging an active term with an inactive one

$a \leftarrow \frac{P(O|T^p)P(T^p)N(T^t)}{P(O|T^t)P(T^t)N(T^p)}$
$r \sim U(0,1)$
if $r < a$ **then**
    $T^t \leftarrow T^p$
end

4 **end**

5 **return** $\left(\frac{C(T_1)}{l}, \ldots, \frac{C(T_m)}{l} \right)$

Algorithm 9.2: A Metropolis-Hasting algorithm to estimate $P(T_i = 1|O)$.

directly into the MCMC algorithm. The parameters now must be explicitly considered in the joint probability distribution:

$$P(p,T,H,\alpha,\beta,O) = P(p)P(T|p)P(H|T)P(\alpha)P(\beta)P(O|H,\alpha,\beta), \quad (9.16)$$

where $P(T|p)$ is given by Equation (9.4), $P(H|T)$ is given by Equation (9.5), and $P(O|H,\alpha,\beta)$ corresponds to $P(O|H)$ of the basic model (see Equations 9.6 and 9.7). As p, α, and β are now true random variables, a prior distribution on them must also be defined. A uniform distribution can be used to introduce as little bias as possible.

We are seeking for a scheme to sample from joint posterior distribution

$$P(p,T,\alpha,\beta|O) = \frac{P(p,T,\alpha,\beta,O)}{P(O)}.$$

In order to utilize the Metropolis-Hasting algorithm for this purpose, an efficient calculation for the numerator must be provided. This is straightforward, because the numerator factors to

$$P(p,T,\alpha,\beta,O) = P(p)P(T|p)P(\alpha)P(\beta)P(O|T,\alpha,\beta), \quad (9.17)$$

and moreover, $P(O|T,\alpha,\beta)$ can be determined using Equation (9.8).

In addition to term state transitions, parameter transitions within the proposal density must also be taken into account. The new proposal density can be defined as a mixture of the state transition density Q_T and a parameter transition density Q_Θ. We denote the current realization of the parameters by $\Theta^t = \{\alpha^t, \beta^t, p^t\}$ and by $\Theta^p \leftrightarrow_\Theta \Theta^t$ the relation whether Θ^p can be constructed from Θ^t. The fully specified proposal density is then

$$Q_s(T^p, \Theta^p | T^t, \Theta^t) = \begin{cases} Q_T(T^p | T^t)s & \text{if } T^p \leftrightarrow_T T^t \text{ and } \Theta^p = \Theta^t \\ Q_\Theta(\Theta^p | \Theta^t)(1-s) & \text{if } \Theta^p \leftrightarrow_\Theta \Theta^t \text{ and } T^p = T^t \\ 0 & \text{otherwise.} \end{cases}$$

The parameter $s \in (0,1)$ can be used to balance state transition proposals against parameter proposals. That is to say, depending on the outcome of a Bernoulli process with hyperparameter s, either a new state transition or a new parameter setting is proposed. To achieve an equal balance between both types of proposal, s should be set to 0.5.

The transition $\Theta_p \leftrightarrow_\Theta \Theta_p$ is then defined such that Θ_p differs from Θ_t in the realization of not more than a single variable. In contrast to the configuration space of the terms' activation state, the domain of these new variables is continuous. However, the results of the analysis are not sensitive to very small changes in the values of these parameters, so that it is sufficient to use the values $\alpha, \beta \in \{0.05k | 0 < k < 20\}$ and $p \in \{1/m, .., 20/m\}$, where m is the number of terms.

Finally, we can state the proposal density function for parameter transitions:

$$Q_\Theta(\Theta^p | \Theta^t) = \begin{cases} \frac{1}{|A|+|B|+|P|} & \text{if } \Theta^p \leftrightarrow_\Theta \Theta^t \\ 0 & \text{otherwise,} \end{cases} \quad (9.18)$$

in which A, B, and P stand for the domain of the parameters α, β, and p respectively. Note that Q^Θ is symmetric, i.e., $Q_\Theta(\Theta^t | \Theta^p) = Q_\Theta(\Theta^p | \Theta^t)$.

In the original publication, it was shown that MGSA had a substantially superior performance compared to the term-for-term, parent-child, topology-weight, and GenGO algorithms with respect to precision and recall [23].

9.3 MGSA: An Extended Example

In this section, we will apply MGSA to the same aorta dataset that was examined previously. An important difference between the overrepresentation algorithms and MGSA is that the overrepresentation algorithms essentially are performed as hypothesis tests for each of the GO terms under consideration, whereas MGSA is not a hypothesis test but rather a procedure to find the posterior probability of any GO term being in the active state. In overrepresentation analysis, if there is a statistically significant p-value for a term being overrepresented, then we consider the term to be representative of the results of the experiment. On the other hand, MGSA calculates the probability of each term being in the `active` state. Although the actual cut-off probability can be determined the user, by default a term is found to be `active` if $p > 0.5$.

Another important difference between MGSA and the overrepresentation methods is that a random number generator is used to determine the random walk for the MCMC algorithm in MGSA. This means that the results can differ from run to run, especially if too few iterations are used. Usually, the differences from run to run are very minor, but if the fluctuations are too large, the number of MCMC steps can be increased, in order to promote convergence of the MCMC run.

When actually applied to the same input data that was used for the GO overrepresentation algorithms presented in the previous chapter (see Section 8.4 on page 188), the number of terms decrease to four. The results of the analysis is given in Table 9.1.

ID	Name	Marginal	n_t	m_t
GO:0031012	extracellular matrix	0.974	71 (4.8%)	269 (1.6%)
GO:0040029	regulation of gene expression, epigenetic	0.939	14 (0.9%)	39 (0.2%)
GO:0016055	Wnt receptor signaling pathway	0.859	30 (2.0%)	129 (0.8%)
GO:0017053	transcriptional repressor complex	0.513	7 (0.5%)	16 (0.1%)

TABLE 9.1: Model-based geneset analysis of the significantly down-regulated genes in the aorta experiments. 16,359 genes annotated by at least one GO term were in the population set, and 1494 annotated genes were in the study set. A total of four GO terms were found to have a marginal probability of being in the `active` state of more than 0.5. Analysis was performed using the Ontologizer [24].

9.4 Summary

Data-driven molecular biology experiments can be used to identify a list of genes that respond in the context of a given experiment. With the advent of technologies such as microarray hybridization and next-generation sequencing that enable biologists to generate data reflecting the response profiles of thousands of genes or proteins, gene-category analysis has become ever more important as a means of understanding the salient features of such experiments and for generating new hypotheses. The methods for overrepresentation analysis presented in the previous chapter, together with knowledge bases such as GO, have become a *de facto* standard for molecular biological research.

We suggest that single-term association methods that determine the significance of each term in isolation essentially do "not see the forest for the trees," by which we mean that they tend to return many related terms which are statistically significant if considered individually, but they are not designed to return a set of core terms that together best *explain* the set of genes in the study set.

Modeling requires formulating a generative process of the data. The GenGO [155] and MGSA [23] procedures consider `active` GO terms as the potential *cause* of the gene responses. Fitting the model then enables one to distinguish between the causal categories (according to the model) from the categories merely associated with gene response. Although one cannot conclude that the identified categories are causal in reality (this is only a model and one only has observational data), this feature of model-fitting explains why it provides a better answer to the question *"what is going on?"* than testing for associations on a term-for-term basis.

9.5 Exercises and Further Reading

The article by Lu and coworkers represents the original description of the GenGO algorithm [155]. MGSA was developed by the authors of this book together with Julien Gagneur from the European Molecular Biology Laboratory (EMBL) in Heidelberg [23]. An article by Persi Diaconis provides an excellent overview of the Markov chain Monte Carlo (MCMC) algorithm and some of the applications to which it has been put [71].

Exercises

9.1 Refer to the previous chapter for an introduction on how to use the Ontologizer for GO analysis. Load the data for the aorta dataset as described there or use data from another source (study set, population set). Run MGSA several times. Note that the actual results of MGSA can vary from run to run depending on the numbers provided by the random number generator during the execution of the program. If this is observed, it may be useful to increase the number of iterations (this can be adjusted in the Options menu of the Ontologizer). What are the main differences in the results between MGSA and the procedures that were presented in the previous chapter?

9.2 Download the Yeast dataset that was analyzed for the original publication of the MGSA method.[1] The dataset can be downloaded as part of the Supplementary Data for the article (the second file is the study set, and the third file is the population set). Now compare the results of the term-for-term, parent-child union, and topology-weight algorithms with the results of MGSA. Perform the analysis with the Ontologizer. Now read the discussion in the original publication of MGSA. Do you agree with the comments *of trees and forests?*

9.3 The authors of this book, together with Julien Gagneur (EMBL, Heidelberg), have developed an R package for MGSA analysis with GO and other kinds of metadata. The package is available as a Bioconductor package at `http://www.bioconductor.org`. Perform the MGSA analysis as described above for the Ontologizer using the Bioconductor package by following the instructions in the package vignette. Note that the Bioconductor version of MGSA can be used for arbitrary categories (not just Gene Ontology terms).

---

[1] Available at `http://nar.oxfordjournals.org/content/38/11/3523` or alternatively at the PubMed website under the PubMed ID 20172960 [23]; the article is freely available.

Chapter 10

Semantic Similarity

"... the rest is all semantics."

As we have seen in previous chapters, ontologies consist of well-defined concepts that are connected to one another by semantic relations. This has inspired a number of algorithms that exploit these relationships in order to define similarity measures for terms or groups of terms in ontologies. Semantic similarity measures have been used in computational biology as a way of validating results of gene expression clustering, prediction of molecular interactions, disease gene prioritization, and for clinical diagnostics, among other things. A large number of different methods for calculating semantic similarity have been developed since the seminal paper of Resnik [204]. In this chapter we will review a number of well-known semantic similarity measures, and will explain how they can be used in applications.

10.1 Information Content in Ontologies

In 1995, Resnik introduced a method for evaluating the semantic similarity between two concepts in an ontology with is_a relations [204, 205]. Resnik's idea was to associate probabilities with the concepts of the ontology. Let $C = \{c_1, c_2, \ldots, c_n\}$ be the set of concepts in the ontology, permitting multiple inheritance. Define a function $p : C \longrightarrow [0, 1]$, such that $p(c_i)$ is the probability of encountering an instance of class $c_i \in C$ (Figure 10.1).

Recall from Chapter 6 that x instance_of A and A is_a B implies that x instance_of B. This means that p is a monotonically increasing function as we move up the ontology to more general concepts: if c_i is_a c_j, then $p(c_i) \leq p(c_j)$. Moreover, if the ontology has a unique root node r, then $p(r) = 1$. Resnik then defined the *information content* of a concept (ontology term) as the negative log likelihood of its probability:

$$\mathrm{IC}(t) = -\log p(t). \tag{10.1}$$

This definition makes intuitive sense. As the probability of a concept increases, its information content decreases (we gain less new information from

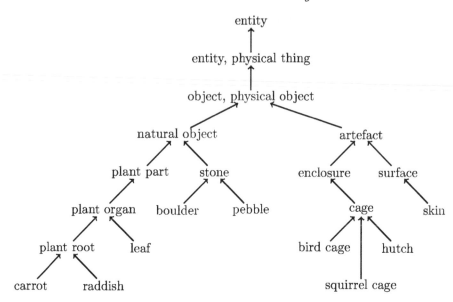

FIGURE 10.1: WordNet. Resnik's work defined semantic similarity in is_a taxonomies using the WordNet as an example. WordNet is a semantic lexicon for the English language that is used extensively by computational linguists and cognitive scientists. WordNet currently contains 155,287 English words. WordNet groups words into sets of synonyms called *synsets* and describes semantic relationships between them. One such relationship is the is_a relationship, which connects a hyponym (more specific synset) to a hypernym (more general synset) [165]. This figure shows a very small excerpt of the WordNet that demonstrates some of the is_a (subclass) relations. For instance, every **pebble** is a **stone**, and every **stone** is a **natural object**. Every concept in WordNet is a descendent of the root term **entity**; therefore, **entity** has a probability of 1, and an information content of log 1 = 0 (i.e., whatever concept we choose from WordNet, it has to be either **entity** or a subclass of **entity**; therefore, there is no added information from knowing that a random term is a descendent of **entity**). On the other hand, the probability of choosing a more specific term such as **pebble** is much lower and the information content is correspondingly higher.

an observation that something common has happened than from an observation that something rare has happened). Moreover, the information content associated with the root term, which subsumes all concepts in the ontology and thus has a probability of 1, is zero because log 1 = 0. Further explanations about the connections between this definition of information content and the definition of entropy in Claude Shannon's information theory are presented in Appendix B.

In the setting of the Gene Ontology (GO), the probability of a GO term *t* is taken to be the probability that a randomly chosen protein is annotated to t, if we choose the protein from the set of all proteins under consideration.[1] Thus, GO terms that are used to annotate many genes have a low information content. For instance, assuming that all genes are annotated to the root (the most general term) of the ontology, the information content of the root is $-\log_2(1) = 0$. Intuitively, this means that if we choose a gene at random and discover that it is annotated to the root term of GO, this is not at all surprising because all genes are annotated to the root. On the other hand, assuming there are 256 annotated genes, the information content of a term used to annotate only one gene is $-\log_2(1/256) = 8$ (recall that $2^8 = 256$), the information content of a term used to annotate two genes is $-\log_2(2/256) = 7$, and so on. Figure 10.2 illustrates the relationship between information content and annotation frequency of ontology terms. Note that although Shannon's definition of entropy and information content used base 2 logarithms, the base of the logarithm is not important for the analysis of semantic similarity in ontologies, logarithms to any base can be used for semantic similarity calculations, and in the rest of this chapter we will use the natural logarithm for simplicity.

Resnik used his definition of information content to define the semantic similarity between two terms in an ontology. The more information two terms share in common, the more similar they are. The information shared between two terms is indicated by the information content of their *most informative common ancestor* (MICA). This is defined using the function $\text{Anc}(t)$, which returns all of the ancestors of the term t in the ontology, including the term t itself.

$$t_{MICA(t_1,t_2)} = \underset{t \in \text{Anc}(t_1) \cap \text{Anc}(t_2)}{\arg\max} -\log p(t). \tag{10.2}$$

Equation (10.2) identifies the term t with the maximum information content that is an ancestor of both t_1 and t_2. We can now define the similarity of two terms t_1 and t_2 as being equal to the information content of their MICA.

$$\text{sim}(t_1, t_2) = -\log p\left(t_{MICA(t_1,t_2)}\right) = \underset{t \in \text{Anc}(t_1) \cap \text{Anc}(t_2)}{\max} IC(t). \tag{10.3}$$

For example, the MICA of terms *cheetah* and *lion* in the ontology of Figure 10.2 is the term *big cat*. Therefore, noting that the term *big cat* has the highest information content of any of the common ancestors of the terms *cheetah* and *lion* (Figure 10.3), we conclude that:

$$\text{sim}(cheetah, lion) = \text{IC}(big\ cat) = 1.386$$

On the other hand, the MICA (and the only common ancestor) of the terms *wildcat* and *beagle* is the root of the ontology (Figure 10.4), meaning that they have a semantic similarity of zero.

---

[1] Usually all proteins or all annotated proteins of some organism.

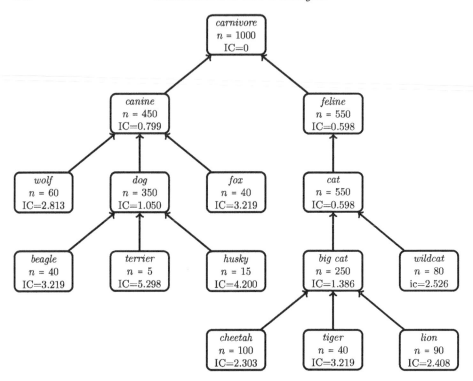

FIGURE 10.2: Information Content of Ontology Terms. The figure shows a carnivore ontology. Imagine the ontology terms have been used to annotate 1,000 documents about carnivores. If we pick a document at random and learn it is annotated to the term *canine*, this gives us some information about the contents of the document, as reflected in the positive information content of $-\log(450/100) = 0.799$. We get much more information if we learn the document is annotated to a more specific term such as *terrier*, with its information content of $-\log(5/1000) = 5.298$. Note that the term dog has 350 annotations. Assuming the annotations to the three child terms are disjoint, the $40+5+15 = 60$ annotations of beagle, terrier, and husky are propagated to the term dog, which thus must have 290 direct annotations. On the other hand, the number of annotations of the term canine is equal to the sum of annotations of its children, meaning that canine has only propagated, but no direct annotations.

A number of modifications of Equation (10.3) have been developed that implicitly take characteristics of the ontology graph and the paths between the terms being compared into account. The motivation is that the measure of Resnik does not take the distance of the terms being compared to one another into account, but only the information content of their MICA. This means for instance that the semantic similarity between the immediate children of the

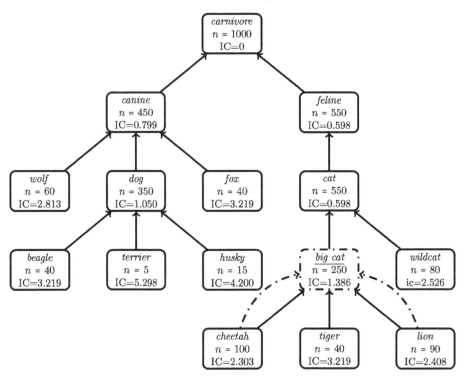

FIGURE 10.3: **Node-Based Semantic Similarity of Ontology Terms.**
The semantic similarity of the terms *cheetah* and *lion* of the ontology of Figure 10.2 is calculated by finding their MICA and calculating the information content of the MICA.

root term, *canine* and *feline* is the same as that of the terms *beagle* and *lion* according to Equation (10.3). As noted by Resnik, a natural way of estimating the similarity of two terms is merely to count the number of edges along the path of one node to another. In the case of multiple paths, the length of the shortest path is taken to be the distance. One problem with this is that the distance between two terms does not always seem to span the same semantic distance. In many ontologies, the semantic distance between terms (as judged by humans) generally gets smaller the further away one gets from the root. Consider for example the following two paths of length two from GO: (1) *cornification* $\xrightarrow{part\ of}$ *keratinization* $\xrightarrow{is\ a}$ *development*, where *cornification* is but one of myriad processes contributing to development; and (2) *serine-type endopeptidase activity* $\xrightarrow{is\ a}$ *serine-type peptidase activity* $\xrightarrow{is\ a}$ *serine hydrolase activity*, which describe serine hydrolase catalytic activities and seem to be

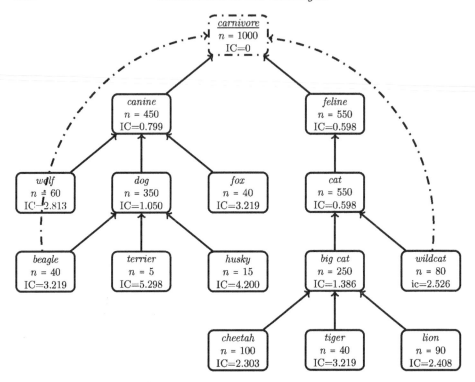

FIGURE 10.4: **Node-Based Semantic Similarity of Ontology Terms.** The semantic similarity of the terms *wildcat* and *beagle* of the ontology of Figure 10.2 is calculated by finding their MICA, which in this case is the root of the ontology. The information content of the root of an is_a ontology (which annotates all items) is $-\log 1 = 0$.

much closer together in meaning. Thus, the path length by itself does not seem to be a good measure of semantic distance

Two measures have been widely used in bioinformatics applications that extend Resnik's approach to take the above considerations into account. Lin proposed a similarity measure that is motivated by three intuitions about what should be considered similar in an ontology [151]:

1. $\text{sim}(A, B)$ is higher, the more A and B have in common (the commonality of two terms corresponds to their common information, which is usually taken to be the information content of their MICA).

2. $\text{sim}(A, B)$ is lower, the more differences exist between A and B.

3. The maximum similarity between A and B is reached when A and B are identical, no matter how much commonality they share.

These considerations lead to a definition of similarity between two ontology terms A and B as the ratio between the amount of information needed to state the commonality between A and B and the information needed to fully describe A and B:

$$\text{sim}_{Lin}(t_1, t_2) = \frac{2 \times IC(t_{MICA(t_1,t_2)})}{IC(t_1) + IC(t_2)}. \tag{10.4}$$

Since the information content of terms in an ontology decreases monotonically as we move up towards the root, $IC(t_{MICA(t_1,t_2)}) \leq \min(IC(t_1), IC(t_2))$. This implies that $0 \leq \text{sim}_{Lin} \leq 1$. In contrast, the similarity measure of Resnik is bounded only by the rarity of the most infrequent term in the ontology, say t_s. Then, $0 \leq \text{sim}_{Resnik} \leq -\log p(t_s)$.

Intuitions 2 and 3 are reflected in the differences between sim_{Resnik} and sim_{Lin}. Consider the semantic similarity between the terms *beagle* and *fox*. According to Resnik's formula, this is the information content of their MICA *canine*, or 0.799. According to Lin's formula, this is weighted by the inverse of the sum of the information content of both terms:

$$\text{sim}_{Resnik}(beagle, fox) = IC(t_{MICA(beagle,fox)}) = IC(canine) = 0.799$$

$$\text{sim}_{Lin}(beagle, fox) = \frac{2 \times IC(canine)}{IC(beagle) + IC(fox)} = \frac{0.799}{3.219 + 3.219} = 0.124$$

Consider now the semantic similarity between *dog* and *fox*. According to Resnik's formula, it is the same as between *beagle* and *fox* because the MICA is identical. For Lin's formula, on the other hand, these two terms are more similar, because the difference between them is less (Figure 10.2).

$$\text{sim}_{Resnik}(dog, fox) = IC(t_{MICA(dog,fox)}) = IC(canine) = 0.799$$

$$\text{sim}_{Lin}(dog, fox) = \frac{2 \times IC(t_{MICA(dog,fox)})}{IC(dog) + IC(fox)} = \frac{0.799}{1.050 + 3.219} = 0.187$$

The similarity of a term to itself is defined by its own information content according to Resnik's formula, and is equal to one according to Lin's formula. Thus, according to Resnik's definition, the self similarity of a term is a function of the probability of the term, which is somewhat counterintuitive. For instance,

$$\text{sim}_{Resnik}(fox, fox) = IC(fox) = 3.219$$

$$\text{sim}_{Lin}(fox, fox) = \frac{2 \times IC(t_{MICA(fox,fox)})}{IC(fox) + IC(fox)} = \frac{2 \times IC(fox)}{2 \times IC(fox)} = 1$$

The MICA of (fox, fox) is *fox*, and thus Resnik's measure calculates the similarity of the term *fox* to itself as being equal to the information content of the term *fox*. On the other hand, Lin's measure always defines the similarity of a term with itself to be equal to 1.

Jiang and Conrath proposed a related distance measure [134].

$$\text{dist}_{JC}(t_1, t_2) = IC(t_1) + IC(t_2) - 2 \times IC(t_{MICA(t_1, t_2)}) \tag{10.5}$$

If $t_1 = t_2$, then $\text{dist}_{JC}(t_1, t_2) = 0$. The maximum distance according to this measure is between two very specific terms whose only common ancestor is the root. There are several ways of making a similarity measure using dist_{JC}, including [268]:

$$\text{sim}_{JC}(t_1, t_2) = 1 - \min(1, \text{dist}_{JC}(t_1, t_2)) \tag{10.6}$$

Another alternative transformation of the distance measure of Jiang and Conrath into a similarity measure defines similarity to be the inverse of distance, whereby 1 is added to the denominator to avoid undefined values, since $\text{dist}_{JC}(t, t) = 0$ for any term t [60].

$$\text{sim}_{JC}(t_1, t_2) = \frac{1}{1 + \text{dist}_{JC}(t_1, t_2)} \tag{10.7}$$

Each of these methods (and others not mentioned here) of measuring semantic similarity is based upon different assumptions of what it is to be semantically similar in an ontology. Different measures may perform better for different applications. However, as we will see below, Resnik's measure tends to perform well in most bioinformatics applications, and we will concentrate on it in the following. More information about other semantic similarity measures can be found in the Further Reading section at the end of this chapter.

All of the above methods for defining the semantic similarity of two terms define the similarity using only the single most informative ancestor of the terms. However, because many ontologies including GO allow multiple parentage, a pair of terms may have multiple distinct common ancestors. For instance, consider the ontology shown in Figure 10.5.

The *Graph-based Similarity Measure* (GraSM) takes all disjunctive common ancestors of two terms into account. In GO, multiple parents of an ontology term represent different interpretations of biological concepts. The intention of GraSM is to take all different interpretations of terms into account when calculating the semantic similarity [60]. In order to explain the algorithm, it will be necessary to present some definitions.

$$\texttt{Parents}(t) = \left\{ u \,|\, (t \xrightarrow{\ is\ a\ } u) \vee (t \xrightarrow{\ part\ of\ } u) \right\}$$

That is, $\texttt{Parents}(t)$ returns all of the parent terms of t in the graph that are related to t by an is_a or a part_of relation. $\texttt{Parents}$ can return a single or multiple terms. $\texttt{Parents}(root) = \varnothing$, but every other term in the GO graph has at least one parent term. We can now use the function $\texttt{Parents}$ to define the set of all paths between two terms t_a and t_b.

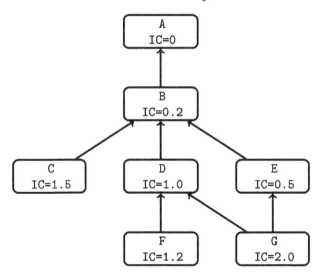

FIGURE 10.5: Graph-Based Similarity Measure: Multiple Ancestors. The Resnik, Lin, and Jiang/Conrath similarity measures consider only the most informative common ancestor. Thus, $sim(F,G) = IC(D) = 1.0$. GraSM takes disjunctive common ancestors into account. In this case, the disjunctive common ancestors of F and G are D and B.

$$\texttt{Paths}(t_a, t_b) = \{\langle t_1, t_2, \ldots, t_n \rangle \,|\, (t_a = t_1) \wedge (t_b = t_n) \wedge (\forall\, i : t_i \in \texttt{Parents}(t_{i+1}))\}$$

That is, if t_a is a more general term in the ontology (closer to the root) and t_b is a more specific term (further from the root), then $Paths(t_a, t_b)$ returns all the paths between the two terms that are connected by direct parent-child edges. In the definition, $1 \le i < n$. For instance, for the ontology shown in Figure 10.5, $Paths(B,G)$ returns the two paths $\langle B, D, G \rangle$ and $\langle B, E, G \rangle$. We can now define the set of Ancestors of a term t as all more general terms for which there is a path to t.

$$\texttt{Ancestors}(t) = \{u \,|\, \texttt{Paths}(u, t) \ne \varnothing\}$$

Note that by definition $\langle t \rangle \in \texttt{Paths}(t, t)$ so that a term is defined to be among the set of its ancestors: $t \in \texttt{Ancestors}(t)$. The set of all common ancestors of two terms t_i and t_j can now be defined as the intersection of the ancestors of t_i and the ancestors of t_j.

$$\texttt{CommonAnc}(t_i, t_j) = \texttt{Ancestors}(t_i) \cap \texttt{Ancestors}(t_j)$$

The GraSM algorithm considers two terms a_i and a_j to be *disjunctive ancestors* of t if there is a path from a_i to t that does not contain a_j and also a path from a_j to t that does not contain a_i:

$$\texttt{DisjAnc}(t) = \{(a_i, a_j) | \exists\, p : p \in \texttt{Paths}(a_i, t) \wedge (a_j \notin p) \wedge$$
$$\exists\, q : q \in \texttt{Paths}(a_j, t) \wedge (a_i \notin q)\}$$

Note that if $a_i \notin \texttt{Ancestors}(a_j)$ and $a_j \notin \texttt{Ancestors}(a_i)$ then a_i and a_j are certainly disjunctive ancestors of t, but there are other possibilities for a_i and a_j to be disjunctive ancestors even if this is not true. Consider the ontology shown in Figure 10.5. B and D are disjunctive ancestors of G because the path $\langle D, G \rangle$ does not include B and the path $\langle B, E, G \rangle$ does not include D. Note that for similar reasons, A and D are disjunctive ancestors of G, as are A and E. GraSM now defines the *disjoint common ancestors* (DCA) of two terms t_i and t_j as the most informative common ancestors of the disjunctive ancestors of t_i and t_j.

$$\texttt{DCA}(t_1, t_2) = \{a_i | a_i \in \texttt{CommonAnc}(t_1, t_2) \wedge$$
$$\forall\, a_j : [a_j \in \texttt{CommonAnc}(t_1, t_2) \wedge IC(a_i) \leq IC(a_j) \wedge a_i \neq a_j]$$
$$\Rightarrow [(a_i, a_j) \in \texttt{DisjAnc}(t_1) \cup \texttt{DisjAnc}(t_2)]\} \quad (10.8)$$

Consider how this definition works for finding the set of common disjunctive ancestors of terms F and G for the ontology shown in Figure 10.5. There is only a single path from F to the root, and thus $\texttt{DisjAnc}(F) = \varnothing$. On the other hand $\texttt{DisjAnc}(G) = \{(D, E), (D, B), (B, E)\}$. It can be seen that $D \in \texttt{DCA}(F, G)$ because $D \in \texttt{CommonAnc}(F, G)$. Letting $a_i = D$, it can be seen that there is no a_j such that $a_j \in \texttt{CommonAnc}(F, G)$ and $IC(D) \leq IC(a_j)$, and the requirements of Equation (10.8) are fulfilled (recall from Chapter 3 that the implication $x \Rightarrow y$ is true if x is false).

Furthermore, $B \in \texttt{DCA}(F, G)$. Firstly, $B \in \texttt{CommonAnc}(F, G)$. Secondly, there are two terms in the ontology with a higher information content than B, i.e., $IC(B) \leq IC(D)$ and $IC(B) \leq IC(E)$. Both (B, D) and (B, E) are in $\texttt{DisjAnc}(F) \cup \texttt{DisjAnc}(G) = \varnothing \cup \{(D, E), (D, B), (B, E)\}$. On the other hand, $E \notin \texttt{DCA}(F, G)$ because E is not an ancestor of F, and $A \notin \texttt{DCA}(F, G)$ because although $A \in \texttt{CommonAnc}(F, G)$, there is no other term a_j such that $(A, a_j) \in \texttt{DisjAnc}(F) \cup \texttt{DisjAnc}(G)$.

We can now use this definition to calculate the GraSM similarity between two terms. The GraSM modification of the Resnik similarity of Equation (10.3) is defined as the average of the information content of the common disjunctive ancestors of the terms.

$$\text{sim}_{ResnikGraSM}(t_1, t_2) = \frac{1}{|\texttt{DCA}(t_1, t_2)|} \sum_{a \in \texttt{DCA}(t_1, t_2)} IC(a) \quad (10.9)$$

Thus, in our example, we would calculate the Resnik GraSM similarity between terms F and G as follows.

$$\text{sim}_{ResnikGraSM}(F, G) = \frac{1}{2} [IC(D) + IC(B)] = 0.6$$

The measures of Lin (Equation 10.4) and Jiang and Conrath (Equation 10.5) can be modified in analogous ways to obtain GraSM versions [60]. Figure 10.6 illustrates how the GraSM similarity between terms F and G is calculated in the ontology of Figure 10.5.

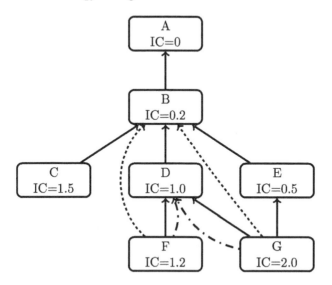

FIGURE 10.6: Calculating GraSM. To calculate the GraSM similarity measure $\text{sim}_{ResnikGraSM}(F, G)$, the common disjunctive ancestors of F and G are identified (symbolized by the two sets of dashed lines), and the average information content of these ancestor terms is calculated as $0.5 \times [IC(B) + IC(D)] = 0.5 \times (0.2 + 1.0) = 0.6$.

10.2   Semantic Similarity of Genes and Other Items Annotated by Ontology Terms

In practice, one is generally more interested in semantic similarity between items that are annotated by one or more ontology terms than in the semantic similarity of the terms themselves. In this section, we will discuss similarity between genes or proteins annotated by GO terms, but the items can be any objects annotated by ontology terms. For simplicity, we will refer to genes or proteins as *genes* in the following. Most variants of semantic similarity measures can be seen as special cases of similarity measures, which are defined in this work as follows:

Definition 10.1 *A similarity measure sim over a finite set* $\mathcal{G} = \{g_1, g_2, \ldots, g_n\}$

of genes is a function $\text{sim} : \mathcal{G} \times \mathcal{G} \to \mathbb{R}$ *with* $\text{sim}(g_i, g_j) \leq \text{sim}(g_i, g_i)$ *and* $\text{sim}(g_i, g_j) \geq 0$ *for all* $g_i, g_j \in \mathcal{G}$. *Additionally, if* $\text{sim}(g_i, g_j) = \text{sim}(g_j, g_i)$ *for all* $g_i, g_j \in \mathcal{G}$ *holds,* sim *is said to be a* symmetric similarity measure.

There are now several ways of harnessing the term similarity measures presented in the preceding section to define a gene similarity measure. For the most part, the different measures vary in the way they combine the term similarities. We will present the gene similarity measures using Resnik's term similarity measure, but gene similarity measures can also be defined using the other similarity measures (with or without GraSM) discussed in the previous section.

In the following equations, we define I_i to be the set of GO annotations for gene g_i; thus $t \in I_i$ returns the set of all annotations for g_i. The simplest similarity measure for two genes g_1 and g_2 is defined as the maximum similarity score for any pair of terms that annotate both g_1 and g_2.

$$\text{sim}^{\text{max}}(g_1, g_2) = \max_{t_1 \in I_1, t_2 \in I_2} \text{sim}(t_1, t_2) \tag{10.10}$$

Clearly, sim^{max} is symmetric. Another commonly used measure iterates over each GO term t that annotated g_i and finds the best matching GO term that annotated gene g_j, taking the average of the similarity scores of these matches:

$$\text{sim}^{\text{avgomax}}(g_1, g_2) = \frac{1}{|I_1|} \sum_{t_1 \in I_1} \max_{t_2 \in I_2} \text{sim}(t_1, t_2) \tag{10.11}$$

In contrast to Equation (10.10), $\text{sim}^{\text{avgomax}}$ is not symmetric, but can depend on the order of the arguments. It is easy to create a symmetric version of this similarity measure by taking the average of both directions: $1/2 \times \text{sim}^{\text{avgomax}}(g_1, g_2) + 1/2 \times \text{sim}^{\text{avgomax}}(g_2, g_1)$. A third commonly used measure simply takes the average of all pairs of terms. This measure is again clearly symmetric.

$$\text{sim}^{\text{avg}}(g_1, g_2) = \frac{1}{|I_1||I_2|} \sum_{t_1 \in I_1} \sum_{t_2 \in I_2} \text{sim}(t_1, t_2). \tag{10.12}$$

10.2.1   Graph-Based and Set-Based Measures of Semantic Similarity

The above measures of semantic similarity all are node-based in that they define similarity using functions of the information content of nodes. In contrast, edge-based similarity measures rely on counting the number of edges between nodes. There are two main strategies. The simplest approach counts the length of the shortest path (number of edges) between two terms in order to determine the distance between them (alternatively, the average of all paths can be calculated). It is also possible to count the number of edges between the MICA of two concepts and the root. Graph-based measures such as this do not perform well in bio-ontologies, because in most bio-ontologies, terms at the same depth do not necessarily have the same specificity, and paths of the same length in different parts of an ontology do not necessarily represent the same semantic distance [198].

Another class of similarity measures makes use of notions of set union and intersection (see Section 2.4, page 25). To illustrate these methods, let us consider how the similarity between two proteins X and Y, each annotated to multiple Gene Ontology terms, would be calculated. Let $GO(X)$ be the set of all GO terms to which protein X is annotated, including implicit annotations resulting from the annotation propagation rule (see Section 5.4.3, page 130). The simplest set-based similarity measure is based on the Jaccard index, which calculates the similarity of two sets as the number of elements in the intersection divided by the number of elements in the union of the two sets.

$$\text{sim}_{Jaccard} = \frac{|GO(X) \cap GO(Y)|}{|GO(X) \cup GO(Y)|} \tag{10.13}$$

A related measures is defined as the sum of the IC values of each term in the intersection divided by the sum of the IC values of each term in the union.

$$\text{sim}_{GIC} = \frac{sum_{t \in GO(X) \cap GO(Y)} IC(t)}{sum_{t' \in GO(X) \cup GO(Y)} IC(t')} \tag{10.14}$$

The sim_{GIC} had the best overall performance in the identification of functional similarity between proteins annotated by GO terms [199].

10.2.2 Applications of Semantic Similarity in Bioinformatics

The semantic similarity measures of the previous section can be used to investigate the similarity between items of any domain which has been annotated by ontology terms. In the field of bioinformatics, this has been done most often for genes or proteins annotated by GO terms, for which it has been shown that semantic similarity correlates with sequence similarity in proteins [153] as well as with gene co-expression [229] and functional relatedness [219]. Figure 10.7 demonstrates how the measures presented above are used to calculate the semantic similarity between genes (or arbitrary items annotated to the terms of an ontology).

There is a statistically significant correlation between the semantic similarity between two proteins and the similarity of their amino acid sequences [153]. This is to be expected because the function of a protein is determined by its sequence. Nonetheless, there are many exceptions to the rule. For instance, if multiple proteins are involved in the same process (e.g., pairs of proteins that heterodimerize), they may share no sequence homology. On the other hand, the fact that a protein pair displays a high degree of sequence homology but a low degree of semantic similarity may indicate that one of the members of the

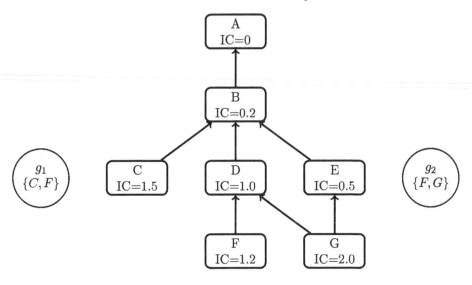

FIGURE 10.7: Calculating Semantic Similarity between Genes. We consider the similarity between gene g_1, which is annotated to terms C and F, and gene g_2, which is annotated to terms F and G. $\text{sim}^{\text{max}}(g_1, g_2) = IC(F) = 1.2$ because $\sim (F, F) = IC(F) = 1.2$, $\sim (F, G) = IC(D) = 1.0$, and $\sim (C, F) = \sim (C, G) = IC(B) = 0.2$. Similarly, $\text{sim}^{\text{avg}}(g_1, g_2) = 0.65$ and $\text{sim}^{\text{avgomax}}(g_1, g_2) = 0.7$. $\text{sim}^{\text{avgomax}}$ is not necessarily symmetric, but in this case $\text{sim}^{\text{avgomax}}(g_1, g_2) = \text{sim}^{\text{avgomax}}(g_2, g_1)$.

pair is underannotated. Lord and colleagues developed a procedure for using semantic similarity to retrieve similar proteins from a database.[153]

Semantic similarity can be used in many settings where an ontology has been used to annotate items of some database in order to retrieve items that are similar to a given query item, and such applications can be substantially better than other ways of searching. For instance, it can be interesting to retrieve a set of all chemicals that are known to or predicted to have some biochemical property such as blood-brain barrier permeability. Many search procedures for computationally comparing chemical compounds have been based entirely on comparisons of chemical structures and physicochemical properties. Recently, Ferreira and Couto [78] were able to substantially improve upon the results of such routines by combining them with an analysis of semantic similarity of the compounds using the ChEBI ontology that was described in Chapter 7. The many other applications of semantic similarity in bioinformatics include the prediction of disease genes [220], improving the detection of differentially expressed gene sets [266], and protein function prediction [195]. A recent article provides a comprehensive review of applications of semantic similarity in computational biology [198].

10.2.3 Applications of Semantic Similarity for Clinical Diagnostics

The above-mentioned applications of semantic similarity were designed to identify similar items in databases. It is also possible to use semantic similarity as a way of processing arbitrary user queries to a database, in which the user enters one or more desired attributes (encoded as ontology terms) and searches for the best matches amongst items in the database that are annotated to the terms of the ontology. This scheme was recently applied to the realm of diagnostics in clinical genetics [145].

Making the correct diagnosis is arguably the most important role of the physician, and is required to plan the correct treatment, to discuss prognosis and natural history of a disease, and to schedule appropriate surveillance examinations to avoid disease complications. Medical diagnostic procedures include a history and physical examination, blood tests and other laboratory investigations, and imaging techniques such as X-rays or computer-assisted tomography scans. The differential diagnostic process attempts to identify candidate diseases that best explain a set of clinical features. This process can be complicated by the fact that the features can have varying degrees of specificity, and by the presence of features unrelated to the disease itself. Depending on the experience of the physician and the availability of laboratory tests, clinical abnormalities may be described in greater or lesser detail.

The Human Phenotype Ontology (HPO), which was described in Chapter 7, was designed to provide a controlled vocabulary for describing human phenotypic abnormalities (signs and symptoms of disease). One obvious advantage of capturing phenotypic information in the form of an ontology is that search routines can be designed to exploit the semantic relationships between terms. For instance, the search procedure can be designed such that a search on *abnormality of the cardiac septa* will not just return all diseases annotated to this term, but also all diseases annotated to related terms such as *ventricular septal defect* or *atrial septal defect*.[2] The search procedure was implemented as a Web-based application called the *Phenomizer*.

Semantic similarity metrics can be adapted to measure phenotypic similarity between queries and hereditary diseases annotated using the HPO. The importance of a clinical finding for the differential diagnosis depends on its specificity. As described above, in ontologies the specificity of a term is reflected by its information content. For medical diagnostics, the physician will enter the various abnormalities observed upon physical and laboratory examination of the patient using terms of the HPO. The information content of each HPO term in this application is defined as $-\log p_t$, where p_t is the frequency of term t among all of the diseases in the database. For instance, if *atrioventricular block* is used to annotate three diseases among a total of 4,813 diseases, its *IC* would be calculated as $-\log(3/4,813) = 7.38$. The more general term

---

[2]The *PhenExplorer* is a search browser for HPO terms with this functionality that is available at the HPO Web site, http://www.human-phenotype-ontology.org.

abnormality of the musculoskeletal system pertains to 2,352 diseases, so its *IC* is $-\log(2,352/4,813) = 0.72$ (see Figure 7.4, page 169).

Any of the term-similarity measures can now be used to calculate a similarity score based on the query terms entered by the physician and the terms used to annotate the diseases in a database. It was found that a measure based on Equation (10.11) had the best performance [145]. That is, for each of the query terms the "best match" among the terms annotated to the disease is found, and the average over all query terms is calculated. This is defined as the similarity:

$$\text{sim}(\mathcal{Q} \to \mathcal{D}) = \text{avg}\left[\sum_{t_1 \in \mathcal{Q}} \max_{t_2 \in \mathcal{D}} IC(MICA(t_1, t_2))\right]. \tag{10.15}$$

Figure 10.8 provides an overview of the approach, whereby the query is made up of the HPO terms *downward slanting palpebral fissures* (a downward slanting of the line defined by the meeting of the eyelids) and *hypertelorism* (widely spaced eyes), both of which are features that can be observed in many different hereditary diseases.

Equation (10.15) will return a high score if a good match is found for each term in the query, but it does not take into account that there could be a number of terms annotated to the syndrome in addition to those used for the maximum match. For instance, this would be the case if a specific query is compared to two syndromes, both of which are annotated by terms that exactly match the query, but one of the syndromes is annotated by a number of additional terms. Using the one-sided formula (Equation 10.15), both syndromes would receive the same score. It is also possible to define a symmetric version of Equation (10.15) in which the similarity of the query to the disease is averaged with the similarity of the disease to the query:

$$\text{sim}_{symmetric}(\mathcal{D}, \mathcal{Q}) = \frac{1}{2}\text{sim}(\mathcal{Q} \to \mathcal{D}) + \frac{1}{2}\text{sim}(\mathcal{D} \to \mathcal{Q}) \tag{10.16}$$

10.3 Statistical Significance of Semantic Similarity Scores

One drawback of the methods presented above is that it is difficult to assign a meaning to any particular semantic similarity score. For instance, even if the phenotypic abnormalities entered by the physician do not correspond to any of the diseases in the database, a result will be returned in which the diseases with the best scores are shown. How does one determine a cutoff below which a semantic similarity score is not meaningful? How trustworthy is any given semantic similarity score?

We have introduced *p*-values in Chapter 3. In the setting of medical diagnostics, the null hypothesis would be that the terms entered by the physician

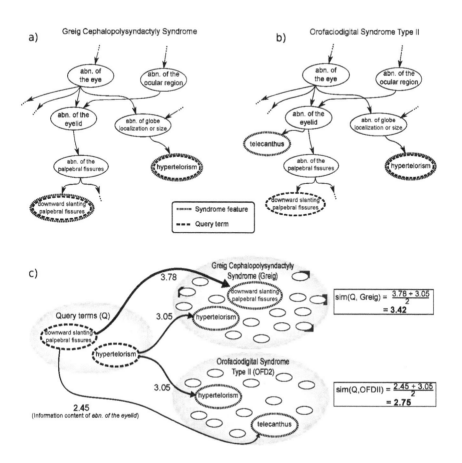

FIGURE 10.8: Searching the HPO with Semantic Similarity. Two of about 5,000 diseases annotated with HPO terms are shown. In **a)**, two of the 33 annotations for Greig Cephalopolysyndactyly syndrome [MIM:175700] are shown. In **b)**, 2 of the 32 annotations for orofacial digital syndrome type II [MIM:252100] are shown. In part **c)**, we see the two query terms entered by the physician: *downward slanting palpebral fissures* and *hypertelorism*. There is an exact match to the terms of Greig Cephalopolysyndactyly syndrome, which is reflected in the similarity score of 3.42; The term *hypertelorism* is also used to annotate the disease orofacial digital syndrome type II, but the term *downward slanting palpebral fissures* is not. Therefore, the information content of the most informative common ancestor of both terms, *abnormality of the eyelid*, is used for the calculation of the semantic similarity. This figure, which was modified from Figure 3 of [209], was kindly provided by Sebastian Köhler.

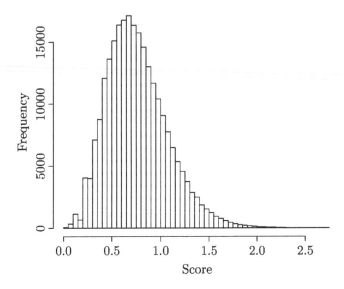

FIGURE 10.9: **Sampled Distribution of Semantic Similarity Scores for the Single Target Set for the Currarino Syndrome.** Random queries of size 10 were generated and their score was determined according to Equation (10.10).

are not related to a disease in the database, meaning that the semantic similarity score obtained for the disease is due only to chance. We can simulate the semantic similarity score distribution for queries of n terms for each of the diseases in the database by drawing n terms at random and calculating the resulting similarity score many times. We can estimate the probability of obtaining a given score by chance by simply calculating the proportion of random scores that are equal to or higher than the score, and use this probability as an empirical p-value. For instance, if we perform 100,000 random queries against the disease *Marfan syndrome* we can store the values obtained for each of these queries. If a real query then has a score that was less than that of only 10 of the random queries, then we can estimate the p-value of this query as $10/100,000 = 10^{-4}$. Figure 10.9 shows the distribution of scores for random queries for one disease.

The score distribution is then calculated for each of the diseases in the database. For each query, the semantic similarity score is calculated and a p-value is generated based on the score distribution. The best matches are returned ranked according to the p-value (which should be corrected for multiple testing against the N diseases in the database), and a threshold can be set against some significance level (usually $\alpha = 0.05$).

This approach now allows a p-value to be assigned to the results of a query. Intuitively, if the highest scoring candidate diagnosis has a significant

p-value, this would indicate to the clinician that this syndrome is a likely differential diagnosis and should be considered further. If on the other hand the highest scoring candidate does not have a significant p-value, this would indicate that the combination of phenotypic abnormalities entered by the physician is not specific enough to allow a diagnosis, or perhaps that the combination of features pertains to a clinical entity that is not present in the database being queried.

Although the idea of estimating the distribution of semantic similarity scores by means of simulations is simple enough, it requires several hundred thousand simulations for each of the diseases in the database and cannot be adjusted dynamically for queries against only a subset of diseases in the database. Newer algorithms are being developed that take advantage of the structure of the ontology graph to calculate the exact score distribution in a fraction of the time required for the simulation approach. Interested readers are referred to the original publication for more details [227].

10.4 Exercises and Further Reading

Appendix B describes concepts of information theory that may be helpful for understanding this chapter, and pointers to books on Information Theory and Entropy are given at the end of that appendix. There is a vast literature on semantic similarity measures and applications thereof. We have not attempted to provide a comprehensive review of similarity measures in this Chapter, but interested readers are advised to consult the review by Pesquita et al. on this subject [198].

Yu and coworkers developed an interesting statistical model of node sharing that counts the number of leaf nodes that share exactly the same set of "higher up" category nodes in comparison to the total number of classified pairs [269].

The algorithms concerning clinical diagnostics were developed within the work group of the authors. The original publication offers more details on the algorithm and on a Web-based application called the *Phenomizer* [145].

Exercises

10.1 In this exercise (which will be structured as a tutorial), we will learn how to use the Robo R package from the Book Web site[3] to calculate the distribution of information content among GO term annotations to

---

[3]http://bio-ontologies-book.org.

mouse genes. See Appendix A for information on how to use R and to install Robo. Download the GO ontology definition file and the gene association file for mouse genes (`gene_association.mgi`) from the GO Web site.[4] Start an R session and enter the following commands to initialize the Robo package and input the OBO and association file.

```
## GO file
gofile <- "gene_ontology.1_2.obo"
## Gene association file for mouse (MGI)
assocfile <- "gene_association.mgi"

## Set up Robo session
library(robo)
session<-openRoboSession(gofile, assocfile)
terms<-getRoboTerms(session)
items<-getRoboItems(session)
```

In order to calculate the information content of each term, we first need to calculate the frequency of annotations to each term amongst all genes of the population. We will use the variable N to hold the number of genes in the population.

```
N <- length(items)
```

The variable `terms` is an R list, each of whose items is a sublist. The following shows the information available for the second item in the list.

```
> terms[[2]]
$roboId
[1] 1

$id
[1] "GO:0000002"

$name
[1] "mitochondrial genome maintenance"

$parents
[1] 5481

$items
[1] 10575 22772 22774 22795 26201 26720 28346 28637 31320
```

---

[4]http://www.geneontology.org.

Thus, the second item in the list represents the GO term GO:0000002, or *mitochondrial genome maintenance*. The Robo package uses a set of integer indices to keep track of the terms, so the index of this term is 1, and the index (also in `terms`) of the parent of this term is 5481. The `items` slot keeps a list of the genes that are annotated to this term also using integer indices. For instance, the gene 10575 is annotated to *mitochondrial genome maintenance*. We can get more information about this gene by entering >`items[[10575]]` at the R prompt. To calculate the information content of *mitochondrial genome maintenance*, we merely calculate its frequency among all annotated genes and take the negative logarithm.

```
m <- length(terms[[2]]$items)
freq <- m/N
IC <- -log(freq)
```

To calculate the information content of all terms at the same time, we have to consider the case where the frequency of a term may be zero, which would lead to a NaN value in R because the logarithm of zero is not defined. We therefore define the following function, which accepts as argument a variable representing a single term. For simplicity, we are using N as a global variable within the function body.

```
inf.content  <- function (term) {
   m <- length(term$items)
   if (m==0){
     IC=0
   } else {
     IC=-log(m/N)
   }
}
```

Finally, we use the R function `lapply` to apply our function in turn to each of the entries of the list `terms`. `lapply` returns a list which is assigned to the variable `ic`, and we use the R function `unlist` to transform `ic` to a vector. Since many of the GO terms are not annotated to any gene in the mouse genome, we remove all of these values by using the `which` command to remove all terms whose information content was calculated to be zero (this will include all terms with no annotations as well as the root, which can be discarded for our purposes). Finally, we use the plotting command `hist` to display the results graphically (Figure 10.10).

```
ic <- lapply(terms,inf.content)
ic <- unlist(ic) ## convert from list to vector
```

```
## Extract just the non-root terms with at least one
## annotation, ie., IC>0
ic <- ic[which(ic>0)]

hist(ic,main="",xlab="Information content",
     ylab="Number of terms",cex.lab=1.5,cex.axis=1.5)
```

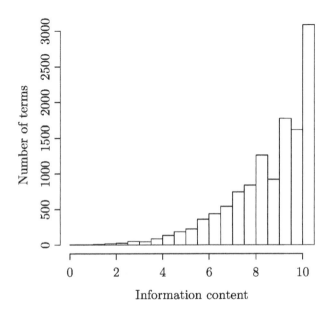

FIGURE 10.10: **Distribution of Information Content of GO Terms Annotating the Mouse Genome**. As can be seen, there are many more specific terms (those with a high information content) than general terms.

To complete this exercise, make sure you understand the script. Now create a similar graphic using a definition of information content with a base 2 logarithm. Investigate the distribution of information content for the genome of the bacteria *E.coli* (the file can also be found at the Gene Ontology Web site). The distribution of scores is different. Why do you think this is so?

10.2 Using the carnivore ontology shown in Figure 10.2, calculate the semantic similarity of the terms *husky* and *fox* using Equations (10.3) and (10.4) and the distance between them using Equation (10.5).

10.3 Using the carnivore ontology shown in Figure 10.2, calculate the semantic similarity of the terms *wild cat* and *cheetah* using Equation (10.4). Repeat your analysis using $\log_2$ instead of the natural logarithm. What do you notice?

10.4 In this exercise (which will be most useful to those with some knowledge of medicine), readers are asked to go to the Web site of the Phenomizer[5] and download the manual via the Help menu. Work through the examples until you understand how to use the program. Enter the features *aortic dissection* and *ectopia lentis* and *arachnodactly*. What differential diagnoses do you see with a significant P-value? Use the improve *differential diagnosis feature* to add more features and explore how this affects the ranking and P-values of the differential diagnoses.

---

[5]The Phenomizer is freely available at `http://compbio.charite.de/phenomizer`.

Chapter 11

Frequency-Aware Bayesian Network Searches in Attribute Ontologies

The last chapter demonstrated how semantic similarity analysis in the Human Phenotype Ontology (HPO) can be used to implement a decision support system for clinical diagnostics. The basic strategy involved finding the diagnosis (disease) that is most similar to the query terms. This approach, however, has a number of drawbacks, including especially that it is not explicitly designed to deal with mistaken or irrelevant query terms, and it does not utilize information about the frequency of a given phenotypic abnormality among all patients with the same disease. These are both clinically important.

First of all, a patient may have signs or symptoms unrelated to the underlying diagnosis. For instance, phenylketonuria, or PKU for short, is a hereditary metabolic disease that is characterized by numerous phenotypic abnormalities in untreated patients. A person with PKU may additionally develop an unrelated disease such as rheumatoid arthritis, abbreviated by RA. However, the examining physician who is trying to make a diagnosis may not recognize that this clinical sign resulting from RA is not related to those resulting from PKU.

Secondly, it is important to recognize that not every person with a given disease necessarily has all of the signs and symptoms that are associated with the disease. For instance, nearly all patients with Marfan syndrome have dilatation (expansion) of the ascending aorta, but only about one half have ectopia lentis (displacement of the lens of the eye). If a feature occurs more frequently in one disease than in another, then, all else being equal, we would tend to believe that the former disease explains the presence of that feature better than the latter disease and therefore can be considered as the more likely candidate.

In this chapter, we introduce a Bayesian approach that takes advantage of these considerations. The algorithm uses a Bayesian approach to deal with false-positive and false-negative search terms and also to take the frequency of individual features into account. For this reason, we will refer to the algorithm as **FABN** (Frequency-Aware Bayesian Network Search). For didactic purposes, we will first present a simplified version of the algorithm that does not take information about the frequencies of the clinical features into account.

11.1   Modeling Queries

FABN is structured in layers, similar to the MGSA model that was introduced in Chapter 9. In MGSA, the goal was to infer the states of the GO terms based on observations of differential expression in genes. With FABN, the Bayesian network consists of a hidden diagnosis layer (each node of this layer corresponds to a single disease, and there is one node for each of the disease[1]) as well as two layers that represent HPO terms. As with MGSA, an optimal configuration of the hidden items is sought in order to explain the potentially noisy observations. In MGSA, the observations corresponded to a list of genes, some of which were in the ON state (e.g., differentially expressed), and some of which were in the OFF state (not differentially expressed). With FABN the observations correspond to HPO terms that represent the clinical observations of the examining physician.

We note that FABN can be used for searches in any database in which items have been annotated with ontology terms, but in this chapter we will describe the algorithm only in the setting of medical differential diagnosis.

11.1.1   High-Level Description of the Model

Figure 11.1 presents an overview of the structure of the FABN network structure. The Bayesian network consists of Boolean variables that represent either a state of a diagnosis (true=1 or false=0) or a state of an HPO term (ON=1 or OFF=0).

We denote the n Boolean variables of the disease layer as $D_1, \ldots, D_n$. They represent the items of interest, e.g., the diseases (diagnoses). If the state of a disease j is true, i.e., the patient has disease j, then $D_j = 1$, otherwise $D_j = 0$. The states of the diseases are jointly described by $D = (D_1, \ldots, D_n)$. The diseases are connected to the variables of the second layer, which represent the hidden state terms of the HPO, which is the ontology that describes the signs and symptoms of the diseases.[2] There are a total of m variables in that layer, which are denoted by $H_1, \ldots, H_m$ and jointly denoted by H. The connections between elements of D and elements of H are made according to the annotations (each of the diseases is annotated to one or usually to multiple terms of the HPO). Edges between the nodes of D and H are only made for asserted annotations, i.e., for the most specific annotations, and not for annotations that can be inferred by the annotation propagation rule.[3]

The hidden layer also contains intra-connections that correspond to the

---

[1] If FABN is used for differential diagnosis in human genetics on the basis of the HPO and annotations to the HPO, then there are approximately 5,000 hereditary diseases in this layer.

[2] The Human Phenotype Ontology, see page 153.

[3] See page 130 for information about the annotation propagation rule.

structure of the ontology, and which implicitly implement the annotation propagation rule within the Bayesian network. Furthermore, the hidden states of the HPO terms are connected to the observed states of the terms, which are the nodes of the third layer of the Bayesian network of FABN. They are denoted as $O_1, \ldots, O_m$ and jointly denoted by O. These nodes correspond to the query terms entered by the physician who is performing the search, and thus represent the signs and symptoms that the physician has observed in his or her patient.

According to the model of FABN, the observed state for a term depends on the corresponding state of the hidden layer, so there are links between elements of H and O in a one-to-one fashion, i.e., H_i is connected to O_i. The propagation between and within the H and O layers accounts for false-positives and false-negatives in FABN.

Example 11.1 *Consider the situation in which a patient has a disease δ that is annotated by the HPO terms 1, 2, 3, and 4. Imagine that a physician has examined the patient and is now entering query terms into a diagnostic program to get some help with the differential diagnosis. Say the physician enters the terms $1, 2, 3$ and 7. Since the disease is also characterized by term 4 but the physician did not enter it, we can consider term 4 to be a false-negative because the physician failed to observe or to enter this term. On the other hand, since the physician entered the term 7 but this term is not associated with the disease, we can consider 7 to be a false-positive because the physician made*

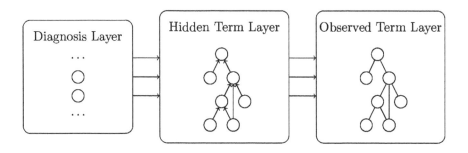

FIGURE 11.1: Structure of FABN. The network is used for modeling searches in ontologies including dependency relations. The diagnosis (item) layer consists solely of links to the next layer, which is the hidden term layer. A node in the disease layer corresponds to a specific disease (diagnosis) and is connected to terms in the hidden layer that represent the signs and symptoms of the disease. The hidden and the observed term layers also contain intra-connections that are used to implement the annotation propagation rule (see page 130). The interplay of the hidden and observed term layers is used to model false-positive and false-negatives, as we will see further below.

some error in the interpretation of the clinical findings, or because the patient has both disease as well as the phenotypic abnormality 7 owing to some other cause.

The observed layer also contains intra-connections according to the structure of the ontology although the dependency relations may have a different direction than those of the hidden layer. The purpose of this is explained in the next section. The joint probability distribution of the network is given by $P(I, H, O)$. Due to links that occur within a single layer, the joint probability distribution (JPD) is not as easily decomposable as it was for MGSA. The next part introduces the notation that is needed to specify a decomposed version of the JPD.

11.1.2   Annotation Propagation Rule for Bayesian Networks

In terms of description logics, the annotation propagation rule is a role inclusion axiom, which passes annotations along other relations such as is_a (see page 130). For diseases annotated to terms of the HPO, it means that if a disease j is annotated to term i then it is also annotated to all subsumers of j (i.e., to all of the ancestors of term j). The process of entering a query is similar to the annotation process, in that it is expected that the person entering the query will use the most specific terms possible. However, if an item in the database matches because of more general terms, it should also be returned. Thus, query annotations are propagated along the is_a relation, which means that a specific query term also implies all of the more general ancestors of the query term, which also describe phenotypic features to be present because of the true path rule.

We will see that the annotation propagation rule requires the propagation of the false-positives and false-negatives to be taken into account. FABN implements this by using the intra-connection in the observed layer.[4] Some additional notation is needed to present the algorithm. We will allow the subscript of a random variable to be a set of indices that refer to the corresponding set of random variables. For instance, $O_{\{1,2\}}$ refers to $\{O_1, O_2\}$. We will denote the set of one or more parents of a term i by $\mathbf{pa}(i)$. In the HPO, this set corresponds to those terms to which annotations are propagated via the is_a relations. We will denote the set of direct children of term i as $\mathrm{ch}(i)$. In the HPO, this comprises the next level of more specific terms. Finally, $\mathbf{a}(i)$ denotes the set of diseases to which the HPO term i is directly annotated.

---

[4]In contrast, in MGSA, the modeling of this propagation was such that the nodes representing the GO terms are connected to nodes representing all of the annotated genes, including both genes that are directly annotated to the term and genes annotated to descendants of the term. This feature allows one to use MGSA for arbitrary gene sets or categories.

Example 11.2 *Consider Figure 11.2 on page 265. For instance, we have:*

$$a(2) = \{\} \qquad\qquad a(3) = \{1\} \qquad\qquad a(4) = \{2\}$$
$$pa(2) = \{1\} \qquad\qquad pa(3) = \{2\} \qquad\qquad pa(4) = \{3\}$$
$$ch(2) = \{3,6\} \qquad\qquad ch(3) = \{4,5\} \qquad\qquad ch(4) = \{\}$$

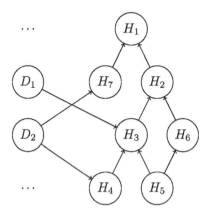

FIGURE 11.2: **Connections between the Item Layer and Hidden Term Layer.** Item D_1 is annotated to term H_3, while item D_2 is annotated to terms H_4 and H_7. The propagation of the ON states within the hidden layer is always directed to the root of the ontology in order to model the effects of the annotation propagation rule within the Bayesian framework.

If X denotes a set of random variables $X_1, \ldots, X_n$ then $X^\vee$ defines a Boolean random variable, such that $X^\vee = 1$ iff there is any $X_i \in X$ with $X_i = 1$, and otherwise $X^\vee = 0$. In other words, $X^\vee$ is a logical disjunction defined by $X^\vee = X_1 \vee X_2 \vee \ldots \vee X_n$. Similarly, we define $X^\wedge$ as the logical conjunction of all variables of X. That is, $X^\wedge = 1$ iff all members of X are 1, otherwise $X^\wedge = 0$.

We will use this notation to express the annotation propagation rule in form of a local probability distribution that can be incorporated into the Bayesian network. Translated to the setting of FABN the rule means that if a term t in H is ON then all terms to which annotations propagate,[5] are also equal to ON. Correspondingly, if a term t is OFF, then all of the descendant terms of t must also be OFF. We will now describe the dependency structure of the Bayesian network as well as the local probability distributions (LPDs) for the various classes of variables.

---

[5] In general, Anc(t), as described on page 239.

11.1.3 LPDs of Hidden Term States

Figure 11.2 displays the connections between the disease and the hidden term layers. Each node in the disease layer represents a specific disease (diagnosis), which is connected to one or more terms in the hidden term layer that represent the clinical features (HPO terms) that characterize the disease. The hidden term layer has interconnections that represent the relations of the terms to one another within the HPO.

For the state propagation between the disease (item) layer and the hidden term layer we specify that the hidden state for term i is ON, if an item that is directly annotated to that term i is true.

Otherwise, if none of the diseases that annotate term i are in the true state, i.e., if all items of $a(i)$ are false, then the state of the hidden term depends on whether a child term of i is in the ON state. In other words, the annotation propagation rule is implemented as an LPD for each variable H_i that assigns ON if at the hidden state of at least one of the child terms of H_i is ON. This specification implements the intra-dependency structure: For the HPO, it corresponds to the propagation of annotations from descendant to ancestor terms over the is_a relations. Formally, the LPD of a single H_i is specified as:

$$P(H_i = 1 | D^\vee_{a(i)}, H^\vee_{ch(i)}) = \max\{D^\vee_{a(i)}, H^\vee_{ch(i)}\} \qquad (11.1)$$

$$P(H_i = 0 | D^\vee_{a(i)}, H^\vee_{ch(i)}) = 1 - \max\{D^\vee_{a(i)}, H^\vee_{ch(i)}\} \qquad (11.2)$$

Equations (11.1) and (11.2) state that if any of the disease nodes are in the true state (i.e., $D^\vee_{a(i)} = 1$) or if any of the children of H_i are in the ON state (i.e., $H^\vee_{ch(i)} = 1$), then the probability of H_i being in the ON state is equal to 1, and otherwise the probability of H_i being in the OFF state is equal to 1. Thus, the states of the nodes in H arise deterministically from the states of the disease nodes in the first layer (Figure 11.3).

For a fixed configuration $D = (D_1, D_2, \ldots, D_n)$ and a combination of hidden states of the m terms $H = (h_1, \ldots, h_m)$ it follows that

$$\prod_i^m P(H_i = h_i | D^\vee_{a(i)}, H^\vee_{ch(i)}) = \begin{cases} 1, & \forall j : D_j = 1 \Leftrightarrow D_j \text{ is annotated to term } i \\ 0, & \text{otherwise.} \end{cases}$$

$$(11.3)$$

Equation (11.3) states that if a disease is in the true state, then all of the terms it annotates (either by asserted annotations or because of the annotation propagation rule) must be in the ON state. Thus, for any given disease, there is only one corresponding configuration of $H = (h_1, \ldots, h_m)$.

11.1.4 LPDs of Observed Term States

What remains to be done is to specify the state propagation to the variables of the observed layer. As in Chapter 9 we model the state propagation between

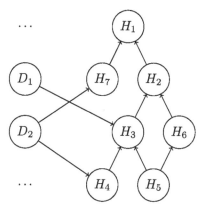

FIGURE 11.3: LPD of Hidden Term States. Item D_1 is in state **true**, i.e., 1. $a(i)$ is the set of terms that are directly annotated by D_1, which in this case comprises only a single term, term 3, which corresponds to the hidden node H_3 as well as to the observed node O_3. $pa(3)$ comprises term 2, and $ch(3)$ comprises terms 4 and 5. $D^v_{a(i)}$ is 1 because item i is **true** and directly annotates term 3. $H^v_{ch(3)} = 0$. Thus, $P(H_3 = 1 | D^v_{a(3)}, H^v_{ch(3)}) = \max\{D^v_{a(3)}, H^v_{ch(3)}\} = \max\{1, 0\} = 1$. Equations (11.1) and (11.2) can be used to calculate the states of all of the hidden nodes in an analogous way. Active (ON or **true**) nodes are shown as dark circles, and inactive nodes as light circles.

the hidden layer and the observed layer probabilistically, whereby global model parameters α and β represent the probability of a false-positive and false-negative observations. The meaning of a false-positive observation for FABN is that a term node in the observed layer, O_i is in the ON state even though the disease that is in the **true** state is not annotated by the term (either directly or by the annotation propagation rule). Similarly, a false-negative observation means that a disease is **true** that is annotated to an HPO term i, but the corresponding term in the observed layer O_i is in the OFF state. FABN assumes that there is one false-positive rate α and one false-negative rate β that apply to all HPO terms.

Example 11.3 *Note that not all configurations of the H and O layers are valid. Suppose that for the network in Figure 11.2 item D_1 is 1. Then H_3 is 1 and following the state propagation, all ancestors of term 3 have state 1 as well. The states of all other terms of the hidden layer are 0. Now suppose that there is a true negative event for the propagation of the hidden state of term 6, which means that state is observed as 0, i.e., $O_6=0$. In addition, there is a false-positive event for term 5 which means that $O_5=1$. This a invalid configuration because the annotation propagation rule is not followed (term 6 is a parent of term 5, and therefore must be ON is term 5 is ON).*

This can be solved by using intra-connections for the observed layer as well, which are in charge of propagating false-positives and false-negatives throughout other observed terms of the ontology independent of their hidden states and therefore may block the state propagation from the hidden layer to the observed layer. In the following, we deal with this problem in two separate cases, in which one case stands for the false-negative propagation and the other one for the false-positive propagation. To simplify the examples, we will demonstrate the propagation of false-positives and false-negatives using a strictly linear ontology in Figure 11.4.

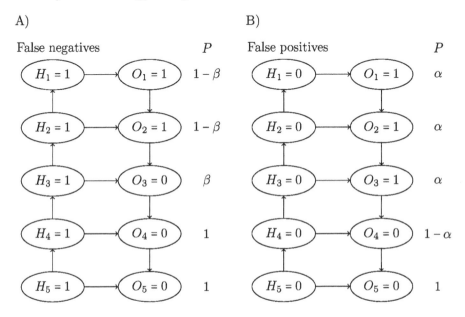

FIGURE 11.4: Propagation of Mistakes. A) False-negative propagation. Here, the 0-case is propagated in a top-down fashion. That means that state 0 of O_4 can be explained by the state 0 of O_3. Therefore, a false-negative is counted only once per branch. **B) False-positive propagation.** A false-positive observation at O_3 is propagated to O_2 and O_1 because of the annotation-propagation rule. In this example, that means that three false positives are counted, one each for O_3, O_2, and O_1.

Propagation of false-negatives

Consider the graph shown in Figure 11.4A. Imagine that a disease is annotated to the (specific) term H_5. Because of the annotation propagation rule, the disease can be inferred to be annotated to the terms $H_4, H_3, H_2,$ and H_1 as well. The observed layer is meant to represent the query. Imagine that a physician is not able to name a certain phenotypic abnormality exactly. Instead, the physician uses a more general term O_2.[6] Thus, terms O_3, O_4, and O_5 are false-negative. This modeled using *top-down propagation*. Thus, the conditional probability $p(O_3 = 0|H_3 = 1, O_2 = 1) = \beta$, where β is the false-negative rate. The remaining false-negative terms can be fully explained due to the fact that at least one single ancestor was already false-negative. Thus, the fact that the state of term 4 is OFF, i.e, $O_4 = 0$ can be fully explained by fact that term 3 has been observed to be OFF, i.e., $O_3 = 0$. Any other explanation in this model would violate the true path rule and therefore, an invalid configuration (with probability 0). Thus, a false-negative is only counted once per branch, where it is first encountered.

Using the notation explained above, the LPD can be defined as follows.

$$P(O_i = 0|H_i = 1, O^\wedge_{\text{pa}(i)} = 0) = 1 \tag{11.4}$$

$$P(O_i = 1|H_i = 1, O^\wedge_{\text{pa}(i)} = 0) = 0 \tag{11.5}$$

$$P(O_i = 0|H_i = 1, O^\wedge_{\text{pa}(i)} = 1) = \beta \tag{11.6}$$

$$P(O_i = 1|H_i = 1, O^\wedge_{\text{pa}(i)} = 1) = 1 - \beta \tag{11.7}$$

In other words, if all parents of an observed term O_i are OFF, then O_i itself also must be OFF, even if H_i is ON (Equations 11.4 and 11.5). The probability that an HPO term i will be observed in the OFF state even though the corresponding hidden term H_i is ON and the parent or parents of O_i are ON is determined by the false-negative rate β (Equations 11.6 and 11.7).

Propagation of false-positives

An example of false-positive propagation is shown Figure 11.4B. FABN models each node with a $H_i = 0 \neq O_i = 1$ mismatch as a false-positive event, i.e., FABN models each O_i that is ON, whose corresponding node $H_i = $ OFF as a false-positive observation. These considerations are reflected more formally by following LPD:

$$P(O_i = 0|H_i = 0, O^\wedge_{\text{pa}(i)} = 0) = 1 \tag{11.8}$$

$$P(O_i = 1|H_i = 0, O^\wedge_{\text{pa}(i)} = 0) = 0 \tag{11.9}$$

$$P(O_i = 0|H_i = 0, O^\wedge_{\text{pa}(i)} = 1) = 1 - \alpha \tag{11.10}$$

$$P(O_i = 1|H_i = 0, O^\wedge_{\text{pa}(i)} = 1) = \alpha \tag{11.11}$$

If term i has at least a single parent that is OFF, then it must be the case that $O_i = 0$ because of the annotation propagation rule. Thus, the probability of an observed term O_i being in the OFF state given that its parent term or terms as well as its corresponding hidden term are in the OFF state is equal to 1 (Equations 11.8 and 11.9). The probability of an observed term O_i being

---

[6]For instance, instead of an exact term such as *Cone-shaped appearance of the epiphysis of the proximal phalanx of the 5th finger*, a physician might simply enter a term such as *Abnormality of the 5th finger*.

in the ON state given that its corresponding hidden term is in the OFF state but that its parents are in the ON state is given by the false-positive rate α (Equations 11.10 and 11.11).

The Joint Probability Distribution for FABN

The joint probability distribution (JPD) of $P(D, H, T)$ for the model is a specified as

$$P(D, H, O) = P(D) \prod_{i=1}^{m} \left[P(O_i | H_i, O^\wedge_{\mathrm{pa}(i)}) P(H_i | D^\vee_{\mathrm{a}(i)}, H^\vee_{\mathrm{ch}(i)}) \right]. \qquad (11.12)$$

Given a particular configuration (H, O) for the variables of the hidden and observed layers, we define

$$m_{xyz|OH} = \left| \left\{ i | O_i = x \wedge H_i = y \wedge O^\wedge_{\mathrm{pa}(i)} = z \right\} \right|$$

to represent the number of cases for a specific combination of values of x, y, and y.[7] Note that

$$m = \sum_{x,y,z \in \{0,1\}} m_{xyz|OH}$$

holds because there is one specific combination for each HPO term i and there are a total of m such terms in the model. We can simplify Equation (11.12) by noting that $m_{110|OH}$ and $m_{100|OH}$ do not occur in the product because they represent invalid configurations of the network. Also, expressions that correspond to $m_{010|OH}$ and $m_{000|OH}$ do not contribute to the product, because they are equal to 1 (cf. Equations 11.4 and 11.8).

Therefore, only 4 of the 8 possible values of $m_{xyz|OH}$ contribute to the condition probability of the O_i:

$$\prod_{i=1}^{m} P(O_i | H_i, O^\wedge_{\mathrm{pa}(i)}) = \beta^{m_{011|OH}} (1 - \beta)^{m_{111|OH}} (1 - \alpha)^{m_{001|OH}} \alpha^{m_{101|OH}} \qquad (11.13)$$

11.2   Probabilistic Inference for the Items

The FABN algorithm was motivated by the goal of providing a decision support system for physicians who would enter a list of signs and symptoms observed in a patient and get back a prioritized list of diseases (a ranked differential diagnosis). Therefore, the interesting quantity is the probability distribution of the activity state of the diseases D given the observations O, which is denoted as $P(D|O)$. After applying the definition of conditional probability

---

[7]Notice that the order of x, y, and z matches the order of the variables within the specifications of the LPDs.

and demarginalizing $P(D,O)$ for H, of which 2^m distinct configurations are possible, we have

$$P(D|O) = \frac{P(D,O)}{P(O)} = \frac{\sum_H P(D,H,O)}{P(O)}. \tag{11.14}$$

The numerator of Equation (11.14) can be calculated by using Equation (11.12):

$$\sum_H P(D,H,O) = P(D) \sum_{H \in \{0,1\}^m} \left[\prod_{i=1}^m P(O_i|H_i, O^\wedge_{\text{pa}(i)}) P(H_i|D^\vee_{\text{a}(i)}, H^\vee_{\text{ch}(i)}) \right]. \tag{11.15}$$

In general, there are 2^m combinations of states of the m HPO terms in the hidden layer H. However, due to our assumption that the state of the hidden terms results deterministically from the state of the disease terms in D, we only need to consider a single configuration $(h_1^D, \ldots, h_m^D)$ for each state of D. Therefore, Equation (11.15) simplifies to

$$\sum_H P(D,H,O) = P(D) \prod_{i=1}^m P(O_i|H_i = h_i^D, O^\wedge_{\text{pa}(i)}). \tag{11.16}$$

In terms of the Bayes Theorem $P(D)$ is the prior, while the product over the probability of the m cases is the likelihood $P(O|D)$, i.e.,

$$\sum_H P(D,H,O) = P(D)P(O|D). \tag{11.17}$$

Finding the configuration of items that best explain the observed data is equivalent to maximizing $P(D|O)$ for D. For this purpose, it is sufficient to maximize the product of the likelihood $P(O|D)$ and the prior $P(D)$ since $P(O)$ is the normalization constant. In general, the optimization problem to maximize this product is NP-complete (see Section 9.2.1). It is tempting to estimate the solution the same way as it was done in Chapter 9 by devising an MCMC algorithm, which would also give access to the marginal posterior probabilities of each disease to be active. In the medical setting, we are typically searching for only a single diagnosis rather than a combination of diseases in any one patient. This is in contrast to the kind of GO analysis done in the MGSA algorithm, in which multiple GO terms are sought to explain the data. This means that it is appropriate to allow only a single disease to be in the true state at once. This makes it easy to iterate through all of the diseases under consideration (typically several thousands) and use Equation (11.16) to calculate the joint probability of the query terms and diseases for each of the diseases in the database and to rank the diagnoses according to these probabilities. This results in an efficient algorithm that runs in time linear to the number of terms and diseases.

In order to implement this, we do not need to leave the Bayesian framework. We realize this model restrictions by defining the prior $P(D)$ as

$$P(D = (D_1, \ldots, D_n)) = \begin{cases} 1, & \text{if } \sum_{j=1}^{n} D_j = 1 \\ 0, & \text{otherwise} \end{cases}.$$

If we agree to restrict ourselves to configurations of D in which just one disease is in the true state at a time and denote the state in which disease i is true as d_i, then we can calculate the marginal probabilities exactly:

$$P(d_i|O) = \frac{P(O|d_i)P(d_i)}{P(O)} = \frac{P(O|d_i)}{\sum_{j=0}^{n} P(O|d_j)}. \tag{11.18}$$

An application implementing this variant of FABN could then rank the diseases according to their marginal probability and apply a threshold (e.g., 50%) beneath which disease are filtered out of the output.

11.3 Parameter-Augmented Network

FABN uses two parameters, α and β, that correspond to the false-positive and false-negative rates. Up to now, we have treated them as constants. In a realistic application, we cannot expect the user to provide them, which means that we have to deal with the parameters within the algorithm. We accomplish this by integrating out α and β. As the integral is not tractable we integrate over a grid of suitable range of different combinations of α and β.

Formally, we augment the Bayesian network with two nodes A and B that represent the respective parameter values, i.e., the realization of A is α while the realization of B is β. The LPD of nodes within the observed layer now also depend on these variables. In the following notation, we represent A and B as a single variable $\Theta = (A, B)$. Thus the LPD is parameterized as:

$$P(O_i|H_i^D, O_{\text{pa}(i)}^{\wedge}, \Theta).$$

The joint probability distribution of the augmented network is factored as:

$$P(D, H, O, \Theta) = P(D) \prod_{i=1}^{m} \left[P(H_i|D_{\text{a}(i)}^{\vee}, H_{\text{ch}(i)}^{\vee}) P(O_i|H_i, O_{\text{pa}(i)}^{\wedge}, \Theta) P(\Theta) \right].$$

The likelihood $P(O|D)$ becomes

$$P(O|D, \Theta) = \sum_{H \in \{0,1\}^m} \prod_{i=1}^{m} \left[P(H_i|D_{\text{a}(i)}^{\vee}, H_{\text{ch}(i)}^{\vee}) \sum_{\Theta} P(\Theta) P(O_i|H_i, O_{\text{pa}(i)}^{\wedge}, \Theta) \right], \tag{11.19}$$

where we assume that A and B and thus Θ are discrete random variables. There are various ways of implementing this, of which perhaps the simplest is to calculate Equation (11.19) for all combinations of appropriate ranges of α and β. For the experiments presented below, 10 values of α and 10 values of β were used for a total of 100 combinations. The average value of Equation (11.19) is then taken as the condition probability of the disease.

11.4   The Frequency-Aware Network

As mentioned above, in many diseases, any given sign or symptom may not occur in all patients but only in a certain proportion of the patients who have the disease. We will refer to this as the *frequency* of a disease feature. The HPO project provides feature frequencies for many of the annotated diseases based upon the original publications and data from OMIM. It is appealing to use this information in a Bayesian framework. Intuitively, a feature that is annotated to diseases X and Y but is two times more common among patients with disease X than among patients with disease Y would tend to provide more evidence for disease X than Y in a patient in whom we are trying to identify the correct diagnosis.

In the probabilistic model, we define the frequency of seeing a certain phenotypic feature represented by term i for disease j as $0 \le f_{j,i} \le 1$. The LPDs of the nodes of the hidden layer given in Equation (11.1) and Equation (11.2) on page 266 must therefore be updated to include the frequencies.

$$P(H_i = 1 | D_{a(i)}, D = d_j, H^{\vee}_{ch(i)} = 0) = f_{j,i} \tag{11.20}$$

$$P(H_i = 0 | D_{a(i)}, D = d_j, H^{\vee}_{ch(i)} = 0) = 1 - f_{j,i} \tag{11.21}$$

$$P(H_i = 1 | D_{a(i)}, D = d_j, H^{\vee}_{ch(i)} = 1) = 1 \tag{11.22}$$

$$P(H_i = 0 | D_{a(i)}, D = d_j, H^{\vee}_{ch(i)} = 1) = 0 \tag{11.23}$$

Note that we have used the notation $D = d_j$ to indicate a configuration of the disease terms in which only disease j is **true**. The probability of the HPO term i in the hidden layer being ON given that disease d_j (and no other disease) is **true** is now determined by the frequency of the feature $f_{j,i}$ (Equation 11.20), rather than by an effect of the global false-negative rate (compare Equation 11.1). As before, if one or more children of a term are in the ON state, then the term must itself be ON because of the annotation propagation rule (Equations 11.22 and 11.23).

These equations imply that the propagation from the disease layer to the hidden term layer is no longer deterministic, but rather depends on the frequencies of the phenotypic features. This in turn affects Equation (11.3) and makes it impossible to simplify the calculations of the likelihood as in Equation (11.13). Instead, 2^k combinations must be considered for the k terms with frequencies less than 1 (Figure 11.5).

Using this definition, the calculation for the likelihood becomes more complex the more annotations with frequencies are available, i.e., the more non-deterministic state propagations are included in the model, because the number of configurations that need to be explored grows exponentially in the number of such annotations. In the search procedure, we therefore restrict the search space to the k annotations with the lowest frequency, and all other annotations always considered as always present (i.e., frequency of 1). As can be seen in the section below, in which k was chosen to be 6, even this heuristic helps to maintain highly precise predictions with a greater recall.

11.5   Benchmark

In order to compare methods, one can conduct a systematic benchmark that is based on simulation. In particular, we simulate patients that have specific diseases according to the annotations of the phenotypic features and their frequencies in the HPO. Furthermore, the signal is distorted somewhat by removing a fixed portion of phenotypic features as well as by adding some unrelated features (this represents a kind of "noise" intended to represent realistic clinical situations in which not all patients have "text book" presentations of disease). Query terms are chosen based on the simulated patients and FABN is performed on the queries. We then ask if the original diagnosis (the one used for the simulation) has gotten the highest marginal probability (i.e., perfect classification) and if not, where it is ranked among all diagnoses. In order to get a reliable estimation of the performance of FABN and other procedures, the simulation is repeated multiple times for each disease and is performed for multiple diseases.

There are different ways to present the final result of benchmarks. In information retrieval scenarios, often so-called precision/recall plots are given in order to graphically compare the methods. The measure *precision* can be defined as the fraction of true-positives among all results that are called as positive (for FABN that is all diseases that are given a marginal probability over 50%, and for the procedure based on semantic similarity presented in Chapter 10, it would be all diseases with a significant p-value). In mathematical terms this can be written as:

$$Precision = \frac{TP}{TP + FP}$$

The measure *recall* refers to the fraction of true positives among all positive

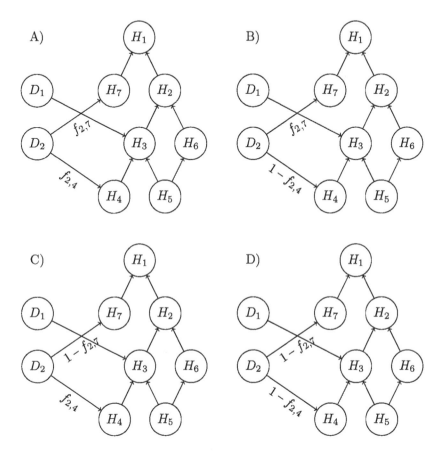

FIGURE 11.5: **Frequency-Aware Propagation.** Here, D_2 is in state true, while D_1 is false. Given that, the probability that H_4 is ON is $f_{2,4}$. The probability that H_7 is ON is $f_{2,7}$. There are four possible configurations of the model. The probability of configuration A) is $f_{2,4}f_{2,7}$, B) is $(1 - f_{2,4})f_{2,7}$, C) is $f_{2,4}(1 - f_{2,7})$, while for D) it is $(1 - f_{2,4})(1 - f_{2,7})$.

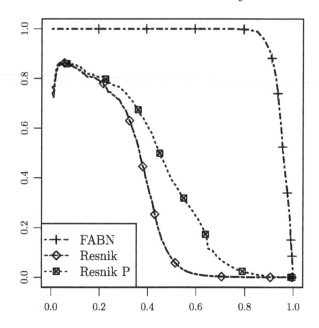

FIGURE 11.6: **Precision/Recall Plots for the Complete Dataset**. For each annotated OMIM disease, of which there were roughly 5000, a patient was generated according to the description in the text using the available frequency information. To disturb the signal, the features were obfuscated with $\alpha = 0.1\%, \beta = 5\%$. Note for this setting, about 7300 terms were considered (the remaining terms had no annotations), thus about seven unrelated terms were added on average, which is a relatively large amount of noise. As of this writing, most of the diseases lack frequency information, there is no substantial improvement noticeable, when frequencies are taken into account.

results, i.e., the proportion of true diagnoses that are successfully retrieved:

$$Recall = \frac{TP}{TP + FN}$$

An optimal classifier would identify all diseases correctly. That is, for a recall of 100% it classifies 100% of the true-positives correctly. We therefore would expect a straight line at the top of the plot.

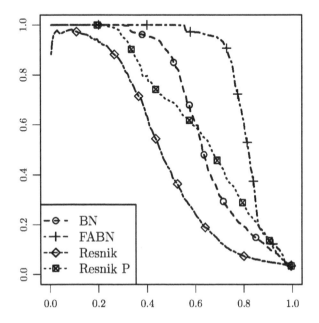

FIGURE 11.7: Benchmarking FABN. 100 patients were generated for each considered disease according to the description in the text. In test setting, we considered only manually annotated OMIM diseases (as per evidence code) that had frequency information, reducing the number of diseases to a total of 30 and the number of terms that had annotations to 740. The resulting term vectors were additionally obfuscated according to $\alpha = 1\%, \beta = 5\%$. In the panel, BN denotes the Bayesian network approach that disregards frequencies. Clearly, the inclusion of the frequencies in the calculation as it is done in FABN improve the performance of the algorithm.

Part IV

Inference in Ontologies

Chapter 12

Inference in the Gene Ontology

At the time the Gene Ontology (GO) was originally introduced in 2000 [14], inference in GO was for the most part limited to exploiting subclass and parthood relations to infer that a gene annotated to a term is also annotated to ancestors of that term connected by is_a and part_of relations on the basis of the annotation propagation rule (see Chapter 5, page 130). Gene Ontology and the OBO ontology language have matured substantially since then. New relations and specifications as to the types of inference that are allowed have been added. This has not only improved the representation of biology in GO, but also facilitates querying and allows GO developers to systematically check for and correct inconsistencies within the GO [86]. This chapter will provide a brief review of inference in the GO ontology and will explain how inferences based on cross-product logical definitions are being used to correct inconsistencies or to add missing links in the GO. We will conclude the chapter with a description of how cross-product logical definitions are being used in the Human Phenotype Ontology (HPO) to link human disease phenotypes with GO and other ontologies such as the Foundational Model of Anatomy ontology of human anatomy.

12.1 Inference over GO Edges

GO currently uses the relations is_a, part_of, has_part, and regulates (which is subdivided into negatively_regulates and positively_regulates). GO has specified a number of rules that define the kinds of inference that can be made. For instance, if a GO term A is a subclass of B, and B is part_of C, then we can infer that A is part_of C:

$$A \xrightarrow{\text{is\_a}} B \wedge B \xrightarrow{\text{part\_of}} C \implies A \xrightarrow{\text{part\_of}} C$$

The typical way relations such as this are encoded in GO is illustrated by the way that mitochondrion and other cellular organelles are defined to be a cytoplasmic part (in the following, the OBO stanzas have been shortened for clarity).

[Term]

```
id: GO:0044444
name: cytoplasmic part
is_a: GO:0044424 ! intracellular part
relationship: part_of GO:0005737 ! cytoplasm

[Term]
id: GO:0005737
name: cytoplasm

[Term]
id: GO:0005739
name: mitochondrion
is_a: GO:0044444 ! cytoplasmic part
```

These assertions allow the inference that mitochondrion is part_of the cytoplasm (Figure 12.1).

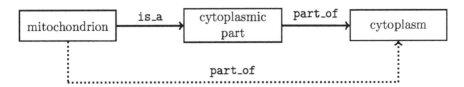

FIGURE 12.1: Inference of part_of Relations in GO. The inferred relation is shown as a dotted edge between *mitochondrion* and *cytoplasm*.

A similar inference is obtained if the order of the assertions is reversed.

$$A \xrightarrow{\text{part\_of}} B \wedge B \xrightarrow{\text{is\_a}} C \implies A \xrightarrow{\text{part\_of}} C$$

As discussed in Chapter 6, the is_a and the part_of relations are transitive. This allows the inference rules:

$$A \xrightarrow{\text{is\_a}} B \wedge B \xrightarrow{\text{is\_a}} C \implies A \xrightarrow{\text{is\_a}} C$$

and

$$A \xrightarrow{\text{part\_of}} B \wedge B \xrightarrow{\text{part\_of}} C \implies A \xrightarrow{\text{part\_of}} C$$

Putting these three rules together allows a chain of inferences. For instance, *ESCRT I complex* $\xrightarrow{\text{part\_of}}$ *endosomal membrane* $\xrightarrow{\text{is\_a}}$ *endosomal part* $\xrightarrow{\text{part\_of}}$ *endosome* $\xrightarrow{\text{is\_a}}$ *intracellular membrane-bounded organelle* $\xrightarrow{\text{is\_a}}$ *intracellular organelle* $\xrightarrow{\text{is\_a}}$ *intracellular part* $\xrightarrow{\text{part\_of}}$ *intracellular* $\xrightarrow{\text{is\_a}}$ *cell part* $\xrightarrow{\text{part\_of}}$ *cell* together imply that *ESCRT I complex* $\xrightarrow{\text{part\_of}}$ *cell*.

Note that this type of inference rule does not hold for is_a relations. For instance, it is not the case that *mitochondrion* is a subclass of *cytoplasm* or that *ESCRT I complex* is a subclass of *cell*.

The two most important relations in GO, is_a and part_of, go from child to parent. Some of the newly introduced relations go from parent to child. The has_part relation sounds at first as if it were the inverse of the part_of relation, but there is actually a subtle difference if the relation is used between types. The part_of relation $A \xrightarrow{\text{part\_of}} B$ implies that all instances of A exist as part of instances of B (see Chapter 6). For instance, the statement

$$\text{mitochondrion} \xrightarrow{\text{part\_of}} \text{cell}$$

implies that every instance of *mitochondrion* exists as part of some instance of *cell*. No statement is made about whether every instance of *cell* has to contain a mitochondrion. On the other hand, the has_part relation implies that the subject of the assertion necessarily must have the parts referred to by the relation. In the following stanza, the *U4/U6 snRNP*, which is a complex involved in RNA splicing, is defined such that it must have the parts *U4 snRNP* and *U6 snRNP*.

```
[Term]
id: GO:0071001
name: U4/U6 snRNP
def: "A ribonucleoprotein complex that contains
      base-paired U4 and U6 small nuclear RNAs."
      [GOC:mah, PMID:14685174]
is_a: GO:0030532 ! small nuclear ribonucleoprotein complex
relationship: has_part GO:0005687 ! U4 snRNP
relationship: has_part GO:0005688 ! U6 snRNP
```

The GO has recently introduced relations to describe biological regulation. The relations are defined using typedef statements in the GO definition file.

```
[Typedef]
id: has_part
name: has_part
xref: OBO_REL:has_part
is_transitive: true

[Typedef]
id: negatively_regulates
name: negatively_regulates
is_a: regulates ! regulates

[Typedef]
id: part_of
name: part_of
xref: OBO_REL:part_of
is_transitive: true
```

```
[Typedef]
id: positively_regulates
name: positively_regulates
is_a: regulates ! regulates

[Typedef]
id: regulates
name: regulates
transitive_over: part_of ! part_of
```

These Typedef statements help define the kind of inference that can be made based on these relations. The regulates relation is used to mean *necessarily regulates*. Thus, a statement such as

$$A \xrightarrow{\text{regulates}} B$$

means that whenever A is present, it always regulates B. On the other hand, no statement is made whether B is always or solely regulated by A.

The regulates relation is transitive over part_of relations. This allows the following inference rule.

$$A \xrightarrow{\text{regulates}} B \wedge B \xrightarrow{\text{part\_of}} C \implies A \xrightarrow{\text{regulates}} C$$

The following stanzas show how the regulates relation is used in practice and demonstrate how inference over the regulates relation works. The term *cyclin-dependent protein kinase activity* is part_of the *cell cycle*. The term *regulation of cyclin-dependent protein kinase activity* is defined using the regulates relation as regulating *cyclin-dependent protein kinase activity*. This allows the inference that a protein[1] annotated to *regulation of cyclin-dependent protein kinase activity* also regulates *cell cycle* (the following stanzas have been shortened for clarity).

```
[Term]
id: GO:0004693
name: cyclin-dependent protein kinase activity
is_a: GO:0004674 ! protein serine/threonine kinase activity
relationship: part_of GO:0007049 ! cell cycle

[Term]
id: GO:0007049
name: cell cycle
is_a: GO:0009987 ! cellular process
```

---

[1] Such as anaphase-promoting complex subunit 2 (ANAPC2).

```
[Term]
id: GO:0000079
name: regulation of cyclin-dependent protein kinase activity
is_a: GO:0045859 ! regulation of protein kinase activity
is_a: GO:0051726 ! regulation of cell cycle
relationship: regulates GO:0004693 ! cyclin-dependent protein
                                     kinase activity

[Term]
id: GO:0051726
name: regulation of cell cycle
is_a: GO:0050794 ! regulation of cellular process
relationship: regulates GO:0007049 ! cell cycle
```

Figure 12.2 shows the inference made on the basis of the stanzas shown above and the inference rules for `regulates` and `part_of`. For instance, the protein ANAPC2 has been annotated with the GO term *regulation of cyclin-dependent protein kinase activity*. We can infer that ANAPC2 is also involved in the regulation of the cell cycle, because *cyclin-dependent protein kinase activity* is part of the cell cycle.

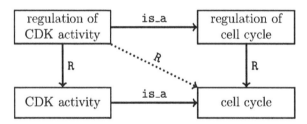

FIGURE 12.2: **Inference with regulates Relations in GO**. The `regulates` relation is shown as an R labeling the edge. The inferred relation is showed using a dashed line. Cyclin-dependent protein kinase has been abbreviated as CDK.

12.2 Cross-Products and Logical Definitions

While it is more or less obvious to humans that the GO term *sodium ion export* refers to the directed movement of sodium ions out of a cell, the definition cannot be "understood" by a computer within the framework of the GO itself because GO does not provide an explicit definition of what a sodium ion is. As described in Chapter 7, the OBO Foundry is aiming to create a suite of orthogonal interoperable reference ontologies in the biomedical domain. This has made it possible to generate computer-readable definitions of many GO terms by referring to ontologies such as ChEBI, which provides a definition for sodium ion, among many others [66]. Cross-product definitions have already proven to be invaluable in ontology maintenance by helping to identify mistakes and point out missing edges in the GO graph; cross-product definitions also provide an ideal entry point for integrative bioinformatics research involving data from multiple domains [177]. Cross-product syntax was introduced briefly in Chapter 4. Here, we will explain how the intersection syntax of OBO corresponds to OWL restrictions and demonstrate how the intersection works both within GO as well as between GO and other bio-ontologies.

12.2.1   Intra-GO Cross-Product Definitions

Consider the term *nuclear ubiquitin ligase complex*. Ubiquitins are small regulatory proteins that bind to other proteins and mark them for degradation by the proteasome, and the ubiquitin ligase complex is a complex of proteins that mediates the attachment of ubiquitin to the proteins that are targeted for degradation. The GO term refers to a class of ubiquitin ligase complex that is localized in the nucleus.

```
[Term]
id: GO:0000152
name: nuclear ubiquitin ligase complex
namespace: cellular_component
def: "A ubiquitin ligase complex found in the nucleus." [GOC:mah]
is_a: GO:0000151 ! ubiquitin ligase complex
is_a: GO:0044428 ! nuclear part
intersection_of: GO:0000151 ! ubiquitin ligase complex
intersection_of: part_of GO:0005634 ! nucleus
```

The two is_a relations place the term within the GO subsumption hierarchy. The two intersection statements provide a definition that can be interpreted as an equivalence axiom between the class specified by the id and the restrictions specified by the intersection_of tags, which provide necessary and sufficient conditions for the class. The following OWL definition is equivalent to the above definition using the intersection syntax.

```
<owl:Class rdf:ID="NuclearUbiquitinLigaseComplex">
  <owl:equivalentClass>
    <owl:Class>
      <owl:intersectionOf rdf:parseType="Collection">
        <rdfs:subClassOf rdf:resource="#UbiquitinLigaseComplex"/>
        <owl:Restriction>
          <owl:onProperty rdf:resource="#part_of"/>
          <owl:someValuesFrom rdf:resource="#Nucleus"/>
        </owl:Restriction>
```

```
        </owl:intersectionOf>
      </owl:Class>
  </owl:equivalentClass>
</owl:Class>
```

It is important to note the Aristotelian design pattern of the definition (see Chapter 6). That is, definitions follow the genus-differentia pattern: an X is a G that D, where G is the more general class (genus), and D is the differentia, i.e., one or more characteristics that *differentiate* (discriminate) instances of the class X from other instances of the class G. In this case, *nuclear ubiquitin ligase complex* (X) is a type of *ubiquitin ligase complex* (G) that is differentiated from other types of *ubiquitin ligase complex* by the fact that its instances are part_of the nucleus (D).

The general pattern is as follows.

```
id:<GO:X>
name:<N>
intersection_of:<G>
intersection_of:<R><D>
```

This defines the class with the id GO:X to be a type of G that is differentiated from other types of G by the fact that instances of GO:X have a relation R to some instance of class D. In RDF/XML syntax, this translates to the following class definition.

```
<owl:Class rdf:ID="GO:X">
  <owl:equivalentClass>
    <owl:Class>
      <owl:intersectionOf rdf:parseType="Collection">
        <rdfs:subClassOf rdf:resource="#G"/>
        <owl:Restriction>
          <owl:onProperty rdf:resource="#R"/>
          <owl:someValuesFrom rdf:resource="#D"/>
        </owl:Restriction>
      </owl:intersectionOf>
    </owl:Class>
  </owl:equivalentClass>
</owl:Class>
```

Note that the differentia can also be a *collection* of characteristics. In the following definition, for instance, there are two statements of the form intersection_of:<R><D>.

```
[Term]
id: GO:0006888 ! ER to Golgi vesicle-mediated transport
intersection_of: GO:0016192 ! vesicle-mediated transport
intersection_of: OBO_REL:results_in_transport_from GO:0005783 !
```

```
                                            endoplasmic reticulum
intersection_of: OBO_REL:results_in_transport_to GO:0005794 !
                                            Golgi apparatus
```

A number of new relations were defined for the GO cross-product definitions [177]. These relations are now candidates for inclusions in future versions of the Relation Ontology described in Chapter 6.

12.2.2 External Cross-Product Definitions

External cross-product definitions make use of terms from OBO ontologies other than the three ontologies that make up the GO. For instance, the following definition is based on a term from the ChEBI ontology (see Chapter 7).

```
[Term]
id: GO:0001510 ! RNA methylation
intersection_of: GO:0008152 ! metabolic process
intersection_of: OBO_REL:results_in_addition_of CHEBI:32875 !
                                            methyl group
intersection_of: OBO_REL:results_in_addition_to CHEBI:33697 !
                                            ribonucleic acid
```

The stanza defines the GO term *RNA methylation* as a metabolic process that is characterized by the addition of a *methyl group* (CHEBI:32875) to a *ribonucleic acid* (CHEBI:33697).

12.2.3 Reasoning with Cross-Product Definitions

One of the main purposes of the cross-product definitions is to allow computer reasoning. It is difficult to maintain a large ontology with multiple links between terms, whereby one term can have multiple parents. Logical definitions of terms allow inference algorithms to detect missing or inconsistent relations to alert curators that attention is needed. Some ontologies such as the fly anatomy ontology (see Chapter 7) maintain a minimal asserted (explicit) hierarchy and use reasoning to generate the full set of relations for release versions of the ontology. Consider the following two GO cross-product definitions.

```
[Term]
id: GO:0030223 ! neutrophil differentiation
intersection_of: GO:0030154 ! cell differentiation
intersection_of: OBO_REL:results_in_acquisition_of_features_of\
                                    CL:0000775 ! neutrophil

[Term]
id: GO:0030851 ! granulocyte differentiation
```

```
intersection_of: GO:0030154 ! cell differentiation
intersection_of: OBO_REL:results_in_acquisition_of_features_of\
                                    CL:0000094 ! granulocyte
```

On the basis of the definition of in the Cell Type Ontology [19], *neutrophil* is defined as a subclass of *granulocyte*:

```
[Term]
id: CL:0000775
name: neutrophil
is_a: CL:0000094 ! granulocyte
```

From this, reasoners can infer that the GO term *neutrophil differentiation* is a subclass of the GO term *granulocyte differentiation*. This kind of analysis has already led to the addition of over 2,000 previously missing relations in the GO [177], and it is to be expected that the quality and consistency of the GO will continue to improve as the cross-product definitions are extended and refined.

12.3   Exercises and Further Reading

The cross-product method for providing computer-readable logical definitions for GO terms was developed by Chris Mungall and colleagues, and their paper in the *Journal of Biomedical Informatics* is a good introduction to the subject [177].

The exercises for this chapter will begin with some questions that ask the reader to infer something about an annotated gene or gene product based on a set of OBO stanzas from the GO.

Exercises

12.1 Consider the following GO term definitions (which have been shortened slightly for clarity).

```
[Term]
id: GO:0007519
name: skeletal muscle tissue development
namespace: biological_process
is_a: GO:0014706 ! striated muscle tissue development
relationship: part_of GO:0060538 ! skeletal muscle organ\
                                    development
```

```
[Term]
id: GO:0014706
name: striated muscle tissue development
namespace: biological_process
is_a: GO:0060537 ! muscle tissue development
```

```
[Term]
id: GO:0060537
name: muscle tissue development
namespace: biological_process
is_a: GO:0009888 ! tissue development
relationship: part_of GO:0007517 ! muscle organ development
```

The protein Cysteine and glycine-rich protein 3, which is encoded by the *CSRP3* gene, is annotated to the GO term *skeletal muscle tissue development*. What other annotations can we infer for this protein and why?

12.2 Consider the following GO term definitions.

```
[Term]
id: GO:0005678
name: chromatin assembly complex
namespace: cellular_component
is_a: GO:0043234 ! protein complex
is_a: GO:0044428 ! nuclear part
```

```
[Term]
id: GO:0044428
name: nuclear part
namespace: cellular_component
is_a: GO:0044446 ! intracellular organelle part
relationship: part_of GO:0005634 ! nucleus
```

```
[Term]
id: GO:0005634
name: nucleus
namespace: cellular_component
is_a: GO:0043231 ! intracellular membrane-bounded organelle
```

The protein Nucleosome assembly protein 1-like 4, which is encoded by the *NAP1L4* gene, is annotated to the GO term *chromatin assembly complex*. What other annotations can we infer for this protein and why?

12.3 Consider the following GO term definitions.

```
[Term]
id: GO:0006310
name: DNA recombination
namespace: biological_process
is_a: GO:0006259 ! DNA metabolic process

[Term]
id: GO:0042148
name: strand invasion
namespace: biological_process
is_a: GO:0006259 ! DNA metabolic process
relationship: part_of GO:0006310 ! DNA recombination

[Term]
id: GO:0060542
name: regulation of strand invasion
namespace: biological_process
is_a: GO:0000018 ! regulation of DNA recombination
relationship: regulates GO:0042148 ! strand invasion

[Term]
id: GO:0060543
name: negative regulation of strand invasion
namespace: biological_process
is_a: GO:0045910 ! negative regulation of DNA recombination
is_a: GO:0060542 ! regulation of strand invasion
relationship: negatively_regulates GO:0042148 ! strand invasion
```

The *Saccharomyces cerevisiae* MPH1 protein is annotated to the GO term *negative regulation of strand invasion*. What other annotations can we infer for this protein and why?

12.4 Consider the following GO cross-product term definition.

```
[Term]
id: GO:0005964 ! phosphorylase kinase complex
intersection_of: GO:0043234 ! protein complex
intersection_of: capable_of GO:0004689 ! phosphorylase kinase\
                                                    activity
```

Explain the meaning of this term in English using the genus-differentia pattern. What is the genus? What is the differentia? Now show the corresponding OWL Class definition using the RDF/XML syntax as described in Section 12.2. Alternatively, use the simplified Manchester syntax [123, 177].

12.5 Consider the following GO cross-product term definition.

```
[Term]
id: GO:0000266 ! mitochondrial fission
intersection_of: GO:0048285 ! organelle fission
intersection_of: OBO_REL:results_in_division_of GO:0005739 !\
                                              mitochondrion
```

Explain the meaning of this term in English using the genus-differentia pattern. What is the genus? What is the differentia? Now show the corresponding OWL Class definition using the RDF/XML or Manchester syntax.

12.6 Consider the following GO cross-product term definition.

```
[Term]
id: GO:0004778 ! succinyl-CoA hydrolase activity
intersection_of: GO:0003824 ! catalytic activity
intersection_of: OBO_REL:has_input CHEBI:15380 ! succinyl-CoA
intersection_of: OBO_REL:has_input CHEBI:15377 ! water
intersection_of: OBO_REL:has_output CHEBI:15346 ! coenzyme A
intersection_of: OBO_REL:has_output CHEBI:30031 ! succinate(2-)
def: "Catalysis of the reaction: succinyl-CoA + H2O \
                       = CoA + succinate." [EC:3.1.2.3]
```

Try to understand the reaction based on the textual definition. Consult the ChEBI [67] Web site as needed. Now explain the meaning of this term in English using the genus-differentia pattern, noting that there are multiple differentiae in this definition. Finally, show the corresponding OWL Class definition using the RDF/XML or Manchester syntax.

Chapter 13

RDFS Semantics and Inference

Formal semantics gives a rigorous definition of "meaning" that can be directly related to computational methods of inference. The W3C has defined the syntax and model theoretic semantics of RDF and RDFS, and these semantics are used to define inference rules: An RDFS graph G entails H iff H is true whenever G is. This chapter will provide an intuitive explanation of RDF semantics based on the specifications of the W3C [108]. See appendix C for background on the syntax of RDF. In this chapter, a simplified version of RDF syntax based on Turtle syntax will be used. The QName prefixes rdf, rdfs and xsd are defined in standard fashion, and we will additionally use the name space ex for an imaginary URI used in our examples.

```
@prefix rdf: <http://www.w3.org/1999/02/22-rdf-syntax-ns#> .
@prefix rdfs: <http://www.w3.org/2000/01/rdf-schema#> .
@prefix xsd: <http://www.w3.org/2001/XMLSchema#> .
@prefix ex: <http://www.example.org/example.rdf#> .
```

This allows us to abbreviate the RDF triple

```
<http://www.example.org/example.rdf#a>
    <http://www.w3.org/1999/02/22-rdf-syntax-ns#type>
        <http://www.w3.org/2000/01/rdf-schema#Class> .
```

as

```
ex:a rdf:type rdfs:Class .
```

This will be sufficient to follow the examples presented in this chapter. Appendix C presents more information about the Turtle syntax for RDF.

13.1 Definitions

Recall from Chapter 4 that RDF graphs contain one or more statements in the form of triples. The elements of an RDF triple are drawn from a set of terms $\mathcal{V}$ that is partitioned into three pairwise disjoint sets. $\mathcal{U}$ is a set of URI references, $\mathcal{B}$ is a set of blank nodes (bnodes), and $\mathcal{L}$ a set of literals. $\mathcal{L}$

in turn is divided into the set $\mathcal{L}_P$ of plain (untyped) literals and the set $\mathcal{L}_T$ of typed literals. An RDF triple is comprised of a subject, predicate, and object $\langle s, p, o \rangle$, where $s \in \mathcal{U} \cup \mathcal{B}$, $p \in \mathcal{U}$ and $o \in \mathcal{U} \cup \mathcal{B} \cup \mathcal{L}$. $\langle s, p, o \rangle$ asserts that there is a relation of type p whose subject is the entity s and whose object is the entity o.

An *RDF graph* is a set of RDF triples. A *ground graph* is an RDF graph with no bnodes. A *name* is a URI reference $(\mathcal{U})$ or a literal $(\mathcal{L})$ (names refer to, but are not identical to, entities in the domain of interest that is being modeled by the RDF graph). A set of names is referred to as a *vocabulary V*. The vocabulary of an RDF graph, $V(G)$, is the set of names that occur as the subject, predicate, or object of at least one triple of the graph.

Denote the set of bnodes of an RDF graph as N. Then suppose that $M(N)$ is a mapping from N to some set of literals, bnodes, and URI references. Any graph obtained from G by replacing some or all of the bnodes N in G by $M(N)$ is an *instance* of G (Figure 13.1). A *proper instance* is an instance in which at least one bnode has been replaced by a name or two bnodes in the graph have been mapped to the same node in the instance.

A *merge* of a set of RDF graphs is merely the union of all of the RDF triples in both graphs, if the graphs do not have bnodes. If the graphs have bnodes, care has to be taken not to accidentally combine bnodes with the same identifiers from the different graphs. A bnode with the identifier _:id1 from one RDF graph does not necessarily have anything in common with a bnode with the identifier _:id1 from another RDF graph. If two different RDF graphs are to be merged that each contains a bnode with the same identifier, then the merge must first replace the nodeID in the different graphs with unique nodeIDs before merging the documents.

13.2   Interpretations

An RDF triple $\langle s, p, o \rangle$ makes a claim about the world by stating that there is a relation p whose subject is s and whose object is o. This assertion is a kind of constraint on the possible ways the world can be. The more RDF triples there are, that is, the larger the RDF graph is, the smaller the set of interpretations that an assertion of the graph allows to be true [108]. Interpretations are made relative to the names of the RDF graph. The W3C defines three levels of interpretation, simple interpretation, rdf-interpretation, and rdfs-interpretation, that make up three levels that place an increasing number of constraints on the vocabulary.

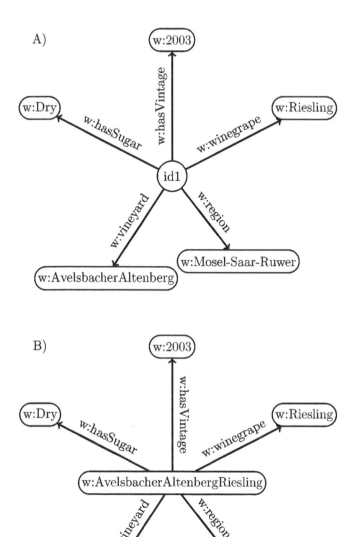

FIGURE 13.1: Instance of an RDF Graph. The RDF graph in part **A)** shows 5 RDF triples that describe an excellent, but unnamed German wine that is represented by the bnode in the center of the graph (the subject of all the triples). The RDF graph in **B)** is an *instance* of the graph in **A)** in which the blank node has been replaced by a resource node representing the wine in question, a Riesling from the Mosel-Saar-Ruwer region in southwest Germany.

Simple Interpretations

Let $V(G)$ be the vocabulary of the graph as defined above, i.e., a set of names of URIs (resources) and literals. A *simple interpretation* $\mathcal{I}$ of V is a 6-tuple $\langle IR, IP, I_{EXT}, I_S, I_L, LV \rangle$, where:

1. IR is a non-empty set of resources (URIs) called the domain or universe of $\mathcal{I}$;

2. IP is the set of properties of $\mathcal{I}$;

3. I_{EXT} is a mapping $IP \to 2^{IR \times IR}$, i.e., a mapping from properties to the powerset of $IR \times IR$. $I_{EXT}(p)$, called the *extension* of p, yields a set of pairs $\langle x, y \rangle$ for some property p such that $x, y \in IR$ and the triple $\langle x, p, y \rangle$ holds;

4. I_S is a mapping from URI references in V into the union of resources and properties: $I_S : V \to IR \cup IP$. Intuitively, this function maps the names of entities (URI references) to the things that they are referring to;

5. I_L is a mapping from typed literals in V into IR; and

6. LV is a subset of IR, called the set of literal values, which contains all the plain literals in V.

The *denotation* of a ground graph[1] in $\mathcal{I}$ can now be given recursively. The following rules extend the interpretation mapping defined by the above sets and functions from the names of the vocabulary V to a ground graph. The rules are combined as an interpretation function $\cdot^{\mathcal{I}}$ that first maps all URIs and literals in V to the corresponding resources and properties.

1. If E is a plain literal "aaa" in V, then $E^{\mathcal{I}}$=aaa;

2. If E is a plain literal with language modifier "aaa"@sp in V, then $E^{\mathcal{I}}$=<aaa,sp>;

3. If E is a typed literal in V, then $E^{\mathcal{I}} = I_L(E)$;

4. If E is a URI reference in V, then $E^{\mathcal{I}} = I_S(E)$;

5. If E is a ground triple $\langle s, p, o \rangle$, then $E^{\mathcal{I}} =$ true if

   - s,p,o $\in V$, and
   - $p^{\mathcal{I}} \in IP$, and
   - <s$^{\mathcal{I}}$,o$^{\mathcal{I}}$>$\in I_{EXT}(\text{p}^{\mathcal{I}})$.

   Otherwise, $E^{\mathcal{I}} =$ false; and

---

[1] We will extend this treatment for graphs that include bnodes shortly.

6. If E is a ground RDF graph, then $E^{\mathcal{I}}$ = false if $t^{\mathcal{I}}$ = false for some triple $t \in E$, otherwise $E^{\mathcal{I}}$ = true.

Each of these six conditions must hold in a valid simple interpretation. The semantic conditions determine whether a ground RDF graph is true or not according to a given interpretation $\mathcal{I}$. A graph is true if all of its triples evaluate to true according to rules 1–6. Note that since the empty graph has no triples, it is trivially true.

An Example

Consider the following RDF document.

```
@prefix rdf: <http://www.w3.org/1999/02/22-rdf-syntax-ns#> .
@prefix rdfs: <http://www.w3.org/2000/01/rdf-schema#> .
@prefix xsd: <http://www.w3.org/2001/XMLSchema#> .
@prefix ex: <http://www.example.org/example.rdf#> .


ex:a ex:p ex:b .
ex:b ex:p ex:a .
ex:A ex:p ex:b .
ex:b ex:p ex:A .
```

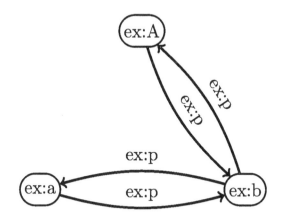

FIGURE 13.2: An RDF Graph. A representation of the example RDF graph.

The document corresponds to the RDF graph shown in Figure 13.2. A simple interpretation of this graph assigns the names {ex:a,ex:A,ex:b,ex:p} to the entities (resources) they denote and fixes the truth value of any given ground RDF triple. The truth value of a triple `<s,p,o>` is determined by the values the property p has for each of the entities in the universe. Any interpretation is made relative to the names of a vocabulary. Consider the following simple interpretation of the RDF graph.

I_R $\quad\{\alpha,\beta,\pi\}$
I_P $\quad\{\pi\}$
I_{EXT} $\quad\pi \Rightarrow \{\langle\alpha,\beta\rangle,\langle\beta,\alpha\rangle\}$
I_S $\quad$ ex:a $\Rightarrow \alpha$, ex:A $\Rightarrow \alpha$, ex:b $\Rightarrow \beta$, ex:p $\Rightarrow \pi$

This interpretation makes all the triples of the graph true. For instance, `<ex:a,ex:p,ex:b>` corresponds to the interpretation $\alpha \xrightarrow{\pi} \beta$ (Figure 13.3).

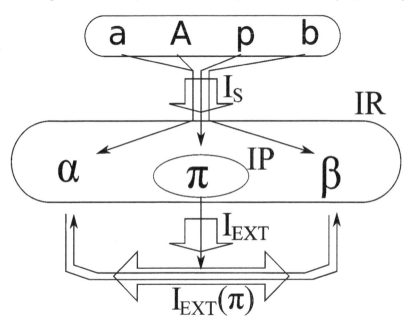

FIGURE 13.3: Simple Interpretation (1). The function I_S maps ex:a and ex:A to the entity α, ex:b to β, and ex:p to π. The three resources (entities) are members of the set of all resources IR. π is a property and thus is also a member of the set of all properties IP. The function I_{EXT} maps the property π to its extension $\langle\alpha,\beta\rangle,\langle\beta,\alpha\rangle$.

Consider now a second interpretation that assigns ex:A to a different entity than ex:a. The interpretation is identical to the first interpretation except that the I_S is defined as ex:a $\Rightarrow \alpha$, ex:A $\Rightarrow \chi$, ex:b $\Rightarrow \beta$, ex:p $\Rightarrow \pi$ (Figure 13.4). According to this interpretation, the following two RDF triples from the graph in Figure 13.2 are false:

```
ex:A ex:p ex:b .
ex:b ex:p ex:A .
```

Thus, the entire graph is false according to the second interpretation.

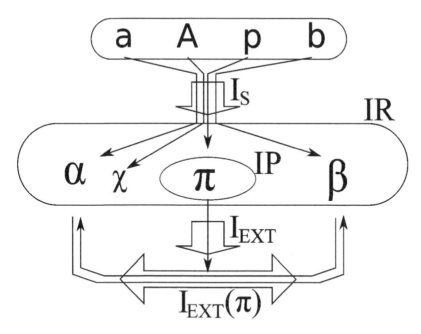

FIGURE 13.4: **Simple Interpretation (2)**. The interpretation is the same as that shown in Figure 13.3 except that the name `ex:A` is mapped to the entity χ.

Interpretations, Models and Truth

Model theory is a formal semantic theory that relates expressions (here, RDF triples) to interpretations. If we denote the set of all expressions $\{e_1, e_2, \ldots\}$ in some RDF graph as E, then an interpretation $\mathcal{I}$ *satisfies* E if $E^{\mathcal{I}}$ is true.

It is the relationship between the interpretation and the triples of the RDF graph that determines the truth value of the graph. For instance, any RDF graph containing the triple `<ex:b,ex:a,ex:p>` is false because `ex:a` is not a property according to the interpretations (and thus, there is no relation $\beta \xrightarrow{\alpha} \pi$ in either interpretation).

An RDF graph F (simply) entails a graph E if every interpretation which satisfies F also satisfies E. This is often written as $F \models E$.

Blank nodes require special treatment. In RDF, a bnode indicates the existence of a thing without naming it. We define an extended interpretation $.^{\mathcal{I}+A}$ such that if E is a bnode, $E^{\mathcal{I}+A}$ maps E according to a mapping A from some set of blank nodes to the universe IR of the simple interpretation $\mathcal{I}$, and if E is a URI or a literal, the mapping is the same as described above:

$$E^{\mathcal{I}+A} = \begin{cases} E^{\mathcal{I}} & E \in \mathcal{U} \cup \mathcal{L} \\ E^{A} & E \in \mathcal{B} \end{cases} \qquad (13.1)$$

Then, if E is an RDF graph, then $E^{\mathcal{I}}$ = true if $E^{\mathcal{I}+A'}$ = true for some mapping A' for the bnodes of the graph, otherwise $E^{\mathcal{I}}$ = false. This procedure essentially just extends the rules for mapping names of the vocabulary to entities of the domain. Thus, let us consider the following RDF graph, in which a triple with a bnode was appended to the graph in Figure 13.2 (Figure 13.5).

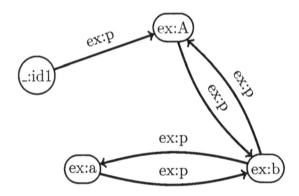

FIGURE 13.5: An RDF Graph with a Blank Node. A blank node has been added to the graph in Figure 13.2.

```
@prefix rdf: <http://www.w3.org/1999/02/22-rdf-syntax-ns#> .
@prefix rdfs: <http://www.w3.org/2000/01/rdf-schema#> .
@prefix xsd: <http://www.w3.org/2001/XMLSchema#> .
@prefix ex: <http://www.example.org/example.rdf#> .

ex:a ex:p ex:b .
ex:b ex:p ex:a .
ex:A ex:p ex:b .
ex:b ex:p ex:A .
_:id1 ex:p ex:A .
```

Thus, the triple <_:id1,ex:p,ex:A> has been added to the graph. This triple will evaluate to true under the first interpretation if the bnode _:id1 is mapped to β. This is the case if the mapping A is defined to include the mapping _:id1 → β.

This effectively treats bnodes as existentially qualified variables within the RDF graph where they occur. In our example, the *meaning* of the bnodes can be given as *"There exists a resource that is related to ex:A by the relation ex:p."* In the language of first-order logic, assuming the existence of a predicate p corresponding to ex:p, we would write

$$p(b, A) \quad \Rightarrow \quad \exists x \ p(x, A)$$

Note that this means that any triple <x,y,z> trivially entails the triple <_:id,y,z>. The latter triple states that there exists some triple with the predicate y and the object z, and the original triple <x,y,z> of course fulfills this description. Similarly, <x,y,z> trivially entails the triple <x,y,_:id2>.

Simple Entailment Between RDF Graphs

Model theory provides a basis for defining entailment rules. Entailment is important for real-world applications as the basis for discovering implicit knowledge on the basis of asserted knowledge. If E entails E' ($E \vDash E'$), then any interpretation that makes E true also makes E' true. For instance, let E be an RDF graph with n asserted triples, and I be an interpretation of E (the simplest interpretation is one that maps each of the names of E to separate entities). Imagine that we use the inference rules that will be explained later in this chapter to deduce a new triple t that would not have been apparent to humans reading the original set of triples. Then as long as the triples in E are true (and presumably the fact that the triples were written down in the RDF file indicates that a human user considered them to be true), then we can assume that t is also true as long as the computational inference rules are correct. Formal semantics provides a mathematical foundation for computational inference rules on the basis of notions of entailment, satisfaction, and validity.

Any process that constructs a graph E' from a graph E is said to be *valid* if E entails E'. Otherwise, the process is *invalid*. Validity of an entailment procedure means that if the procedure is given true inputs, it is guaranteed never to draw false conclusions from them. The main results for simple RDF inference require a small number of lemmas that are easily understood.

Lemma 13.1 *Subgraph Lemma*. *A graph entails all its subgraphs.*

Proof 13.1 *If an RDF graph G is true under an interpretation I, then by definition all of its triples are true under I. Then obviously any subset of its triples, which form a subgraph of G, is also true.*

The subgraph lemma is reminiscent of the rule from logic that $P \wedge Q$ implies P (recall the conjunction elimination rule from Table 2.4).

Lemma 13.2 *Instance lemma. A graph is entailed by any of its instances.*

Proof 13.2 *Recall the definition of graph instance from above, whereby any graph H obtained from G by replacing some or all of the bnodes in G by a mapping from bnodes to literals, bnodes, and URI references, is an instance of G. Suppose an interpretation $\mathcal{I}$ satisfies H and H is an instance of G. Then there exists mapping A on the blank nodes of H such that $\mathcal{I} + A$ satisfies every triple in H. For each blank node b in G, define $b^B = c^{\mathcal{I}+A}$, where c is the blank node or name that is substituted for b in H, or $c = b$ if nothing was substituted for it. Let a be an arbitrary element in the RDF graph H, i.e., a can be a URI reference $(\mathcal{U})$, a literal $(\mathcal{L})$ or a bnode $(\mathcal{B})$. Then we define the following interpretation $\mathcal{I} + B$ for an arbitrary element a of G:*

$$
\begin{cases}
a^{\mathcal{I}+B} = a^{\mathcal{I}} & a \in \mathcal{U} \cup \mathcal{L} \\
a^{\mathcal{I}+B} = a^{A} & a \in \mathcal{B} \text{ and nothing was substituted for } a \text{ in } H \\
a^{\mathcal{I}+B} = c^{B} & a \in \mathcal{B} \text{ and } c \text{ is the name that was substituted for } a \text{ in } H
\end{cases}
$$

Thus, in the first case, a is either a URI reference or a literal and is the same in G and H. In the second case, a is a bnode that is identical in the graph G and the instance graph H. In the third case, a is a bnode in G that was mapped to an element c in the instance graph H, and the interpretation assigns a to whatever c was assigned to by the interpretation of H. Thus, if $H^{\mathcal{I}+B}$ is true, then $G^{\mathcal{I}+A}$ is true. But $\mathcal{I}$, A, and B were arbitrary. Thus, H entails G.

The following lemma means that a set of RDF graphs can be treated as equivalent to a single graph that is formed by merging the graphs as described above. This is clearly important for the purposes of the Semantic Web, in which RDF data from multiple places across the Web may be combined to perform inference. In the remainder of the chapter, we will therefore speak only of "the" graph, but the results apply equally well to "the graph or graphs."

Lemma 13.3 *Merging lemma. The merge of a set S of RDF graphs is entailed by S and entails every member of S.*

Proof 13.3 *If a set of RDF graphs S is true, then all of the triples that make up S are true, and thus all triples in the merge of S are true. Since any subset of the triples in S must also be true, any of the member graphs of S must also be true.*

The main lemma for simple RDF inference provides a complete characterization of simple RDF entailment in syntactic terms. That is, the lemma provides a way of telling whether one RDF graph entails another by checking that some instance of the entailed graph is a subset of the original graph.

Lemma 13.4 *Interpolation lemma. An RDF graph G simply entails another RDF graph H if and only if a subgraph of G is an instance of H.*

Intuitively speaking, the "if" direction of the proof is a simple consequence of the subgraph and instance lemmas: If G is true, then any subgraph of G is also true; Since a graph is entailed by any of its instances, if a subgraph of G is an instance of H, then H must also be true. The proof in the other direction is more technical, and will be omitted here. Interested readers are referred to the documentation of the W3C for the complete proof [108].

13.3 RDF Entailment

Simple interpretations and simple entailment as discussed above apply to RDF graphs when no particular meaning is given to any of the names in the graph (for instance, names such as `rdf:type` have no special significance). Things become more interesting and more useful if the RDF and RDFS vocabulary are taken into account; these forms of interpretation can be regarded as a kind of semantic add-on to simple interpretation. We will begin our discussion with RDF interpretations. RDF interpretations employ the *RDF vocabulary* (**rdfV**), which is a set of URI references in the `rdf` namespace (see Appendix C for more details about RDF and the RDF vocabulary):

```
rdf:type rdf:Property rdf:XMLLiteral rdf:nil
rdf:List rdf:Statement rdf:subject rdf:predicate
rdf:object rdf:first rdf:rest rdf:Seq rdf:Bag
rdf:Alt rdf:_1 rdf:_2 ... rdf:value
```

TABLE 13.1: RDF Vocabulary.

RDF interpretations impose extra semantic conditions on the elements of the RDF vocabulary and on typed literals with the type `rdf:XMLLiteral`.[2] An RDF interpretation of a vocabulary V is a simple interpretation $\mathcal{I}$ of the combined vocabulary $V \cup \mathbf{rdfV}$ which satisfies three additional semantic conditions:

1. $x \in IP \Leftrightarrow <x, \text{rdf:Property}^{\mathcal{I}}> \in I_{EXT}(\text{rdf:type}^{\mathcal{I}})$;

2. If `"a"^^rdf:XMLLiteral` is in V and a is a well-typed XML literal string, then

   - $I_L(\text{"a"^^rdf:XMLLiteral})$ is the XML value of a;

---

[2]We will not cover entailment rules for data types and XML literals in this book; interested readers are referred to the documentation of the W3C [108] for more information.

- $I_L("a"\verb|^^|rdf\verb|:|XMLLiteral) \in LV$; and
- $I_{EXT}(rdf\verb|:|type^\mathcal{I})$ contains
  $<I_L("a"\verb|^^|rdf\verb|:|XMLLiteral), rdf\verb|:|XMLLiteral^\mathcal{I}>;$

and

3. If $"a"\verb|^^|rdf\verb|:|XMLLiteral$ is in V and a is an ill-typed XML literal string, then

   - $I_L("a"\verb|^^|rdf\verb|:|XMLLiteral) \notin LV$; and
   - $I_{EXT}(rdf\verb|:|type^\mathcal{I})$ does not contain
     $<I_L("a"\verb|^^|rdf\verb|:|XMLLiteral), rdf\verb|:|XMLLiteral^\mathcal{I}>.$

The first condition says that if an RDF element x is in the set of properties of $\mathcal{I}$ (IP), then x must be of type rdf:Property. The second condition states that if an element a is a well-formed XML literal then it must be of type rdf:XMLLiteral and be a member of the set of literal values LV. If a is an ill-formed XML literal, then it must be neither of these things. Recall that a string is a well-formed XML literal if it is either well-balanced, well-formed XML content (such as an XML node and all it contains) or if it is a string for which embedding between an arbitrary XML start and end tag yields a valid XML document. Thus, "<foo><bar>hello world</bar></foo>" and "hello world" are valid XML literals, but "<foo" is not.

In addition to the above three semantic conditions, in a valid RDF interpretation, all of the triples in Table 13.2 must evaluate to true. The triples state that the named RDF vocabulary elements must be of type rdf:Property, except for the last triple, which states that the element rdf:nil must be of type rdf:List.

```
rdf:type rdf:type rdf:Property .
rdf:subject rdf:type rdf:Property .
rdf:predicate rdf:type rdf:Property .
rdf:object rdf:type rdf:Property .
rdf:first rdf:type rdf:Property .
rdf:rest rdf:type rdf:Property .
rdf:value rdf:type rdf:Property .
rdf:_1 rdf:type rdf:Property .
rdf:_2 rdf:type rdf:Property .
...
rdf:nil rdf:type rdf:List .
```

TABLE 13.2: RDF axiomatic triples. Each of these triples must be true in a valid RDF interpretation.

An RDF entailment is also a simple entailment. An RDF graph G rdf-entails H if every RDF interpretation which satisfies G also satisfies H. The converse is not true, that is, simple entailment does not necessarily imply RDF entailment. Consider for instance the empty graph. It can be shown that the empty graph is trivially entailed by any graph. Choose an interpretation for the graph such that <rdf:type,rdf:type,rdf:Property> evaluates to false. On the other hand, in any valid interpretation of any RDF graph, <rdf:type,rdf:type,rdf:Property> must evaluate to true. Thus there is an interpretation according to which the empty graph is simply entailed but not RDF-entailed.

13.4 RDFS Entailment

As discussed in Chapter 4, RDFS extends RDF to include a larger vocabulary **rdfsV** (Table 13.3).

```
rdfs:domain rdfs:range rdfs:Resource rdfs:Literal
rdfs:Datatype rdfs:Class rdfs:subClassOf
rdfs:subPropertyOf rdfs:member rdfs:Container
rdfs:ContainerMembershipProperty rdfs:comment
rdfs:seeAlso rdfs:isDefinedBy rdfs:label
```

TABLE 13.3: RDFS vocabulary.

An *RDFS interpretation* of V is an RDF interpretation $\mathcal{I}$ of $V \cup \mathbf{rdfV} \cup \mathbf{rdfsV}$ that satisfies the ten semantic conditions for RDFS.

1. $x \in IC_{EXT}(\mathcal{C}) \Leftrightarrow <x,\mathcal{C}> \in I_{EXT}(\text{rdf:type})$

   - $IC = IC_{EXT}(\text{rdfs:Class}^{\mathcal{I}})$
   - $IR = IC_{EXT}(\text{rdfs:Resource}^{\mathcal{I}})$
   - $LV = IC_{EXT}(\text{rdfs:Literal}^{\mathcal{I}})$

2. If $<p,\mathcal{C}> \in I_{EXT}(\text{rdfs:domain})^{\mathcal{I}})$ and $<u,v> \in I_{EXT}(p)$ then $u \in IC_{EXT}(\mathcal{C})$

3. If $<p,\mathcal{C}> \in I_{EXT}(\text{rdfs:range})^{\mathcal{I}})$ and $<u,v> \in I_{EXT}(p)$ then $v \in IC_{EXT}(\mathcal{C})$

4. $I_{EXT}(\text{rdfs:subPropertyOf}^{\mathcal{I}})$ is transitive and reflexive on IP.

5. If $<p,q> \in I_{EXT}(\text{rdfs:subPropertyOf}^{\mathcal{I}})$ then $p,q \in IP$ and $I_{EXT}(p) \subseteq I_{EXT}(q)$

6. If $\mathcal{C} \in IC$, then $<\mathcal{C}, \text{rdfs:Resource}^{\mathcal{I}}> \in I_{EXT}(\text{rdfs:subClassOf}^{\mathcal{I}})$.

7. If $<\mathcal{C}, \mathcal{D}> \in I_{EXT}(\text{rdfs:subClassOf}^{\mathcal{I}})$ then $\mathcal{C}, \mathcal{D} \in IC$ and $IC_{EXT}(\mathcal{C}) \subseteq IC_{EXT}(\mathcal{D})$.

8. $I_{EXT}(\text{rdfs:subClassOf}^{\mathcal{I}})$ is transitive and reflexive on IC.

9. If $p \in IC_{EXT}(\text{rdfs:ContainerMembershipProperty}^{\mathcal{I}})$, then $<p, \text{rdfs:member}^{\mathcal{I}}> \in I_{EXT}(\text{rdfs:subPropertyOf}^{\mathcal{I}})$

10. If $d \in IC_{EXT}(\text{rdfs:Datatype}^{\mathcal{I}})$, then $<d, \text{rdfs:Datatype}^{\mathcal{I}}> \in I_{EXT}(\text{rdfs:subClassOf}^{\mathcal{I}})$.

RDFS interpretations can be conveniently defined in terms of *classes* (see chapter 4). In RDFS, classes are defined as things that have the rdf:type rdfs:Class. Much of the RDFS vocabulary deals with classes and their properties. We will call the set of all classes in an interpretation IC, and will introduce the function IC_{EXT} for the class extension[3] in the interpretation $\mathcal{I}$. Thus, $IC_{EXT}(\mathcal{C})$ is the set of all individuals in the class $\mathcal{C}$.

In addition to the RDFS semantic conditions, there are a number of triples that need to evaluate to true in an RDFS interpretation. The first set of triples comprises triples of the form <x,rdfs:domain,y>. A triple of the form <r,rdfs:domain,$\mathcal{R}$> allows the inference for a triple in which r is in the predicate position: <p,r,o>, that p is an instance of the class $\mathcal{R}$. For instance, the first axiomatic triple implies for all triples of the form <x, rdf:type, y> that x is an instance of rdfs:Resource (Table 13.4).

The second set of triples comprises triples of the form <x,rdfs:range,y>. A triple of the form <r,rdfs:range,$\mathcal{R}$> allows the inference for a triple in which r is in the predicate position: <p,r,o>, that o is an instance of the class $\mathcal{R}$. For instance, the first axiomatic triple implies for all triples of the form <x rdf:type y> that y is an instance of rdfs:Class (Table 13.5).

Another group of triples deals with containers and properties of list elements. Thus, rdf:Alt, rdf:Bag, and rdf:Seq are defined as subclasses of rdfs:Container; the RDF elements rdf:_1, rdf:_2, (and so on for rdf:_n for $n \geq 1$) are defined to be instances of the property rdfs:ContainerMembershipProperty and the range and domain of the elements is defined as rdfs:Resource (Table 13.6).

Finally, four triples define rdfs:isDefinedBy to be a subproperty of rdfs:seeAlso and provide some specifications for XML literals and data types (Table 13.7).

A number of other triples can be derived from the RDFS axiomatic triples. For instance, consider the two triples

```
rdfs:Datatype rdfs:subClassOf rdfs:Class .
rdfs:subClassOf rdfs:range rdfs:Class .
```

---

[3]In RDFS, each class is associated with a set of individuals, called the *class extension*. The individuals in the class extension are called the *instances* of the class.

```
rdf:type rdfs:domain rdfs:Resource .
rdfs:domain rdfs:domain rdf:Property .
rdfs:range rdfs:domain rdf:Property .
rdfs:subPropertyOf rdfs:domain rdf:Property .
rdfs:subClassOf rdfs:domain rdfs:Class .
rdf:subject rdfs:domain rdf:Statement .
rdf:predicate rdfs:domain rdf:Statement .
rdf:object rdfs:domain rdf:Statement .
rdfs:member rdfs:domain rdfs:Resource .
rdf:first rdfs:domain rdf:List .
rdf:rest rdfs:domain rdf:List .
rdfs:seeAlso rdfs:domain rdfs:Resource .
rdfs:isDefinedBy rdfs:domain rdfs:Resource .
rdfs:comment rdfs:domain rdfs:Resource .
rdfs:label rdfs:domain rdfs:Resource .
rdf:value rdfs:domain rdfs:Resource .
```

TABLE 13.4: RDFS axiomatic triples (1).

According to the RDFS semantic condition 3 (see above), the fact that `<rdfs:subClassOf,rdfs:Class>` $\in I_{EXT}(\text{rdfs:range})^{\mathcal{I}}$ and the fact that `<rdfs:Datatype,rdfs:Class>`$\in I_{EXT}(p)$ implies that `rdfs:Class`$\in IC_{EXT}(\text{rdfs:Class})$. Thus we can infer the following triple

```
rdfs:Class rdf:type rdfs:Class .
```

More details are available in the online documentation of the W3C [108]. Analogous to RDF entailment, we say that an RDF graph G RDFS entails a graph H if every RDFS interpretation of which satisfies every triplet of G also satisfies every triple of H. Also as before, if G RDFS entails H, then G also RDF entails H, but the converse is not necessarily true.

13.5 Entailment Rules

The model theoretic treatment of models and interpretations provides a theory by which to judge whether one RDFS graph is entailed by another, but does not indicate an efficient algorithm for actually doing so in concrete cases. However, a number of inference rules have been characterized which

```
rdf:type rdfs:range rdfs:Class .
rdfs:domain rdfs:range rdfs:Class .
rdfs:range rdfs:range rdfs:Class .
rdfs:subPropertyOf rdfs:range rdf:Property .
rdfs:subClassOf rdfs:range rdfs:Class .
rdf:subject rdfs:range rdfs:Resource .
rdf:predicate rdfs:range rdfs:Resource .
rdf:object rdfs:range rdfs:Resource .
rdfs:member rdfs:range rdfs:Resource .
rdf:first rdfs:range rdfs:Resource .
rdf:rest rdfs:range rdf:List .
rdfs:seeAlso rdfs:range rdfs:Resource .
rdfs:isDefinedBy rdfs:range rdfs:Resource .
rdfs:comment rdfs:range rdfs:Literal .
rdfs:label rdfs:range rdfs:Literal .
rdf:value rdfs:range rdfs:Resource .
```

TABLE 13.5: RDFS axiomatic triples (2).

```
rdf:Alt rdfs:subClassOf rdfs:Container .
rdf:Bag rdfs:subClassOf rdfs:Container .
rdf:Seq rdfs:subClassOf rdfs:Container .
rdfs:ContainerMembershipProperty \
        rdfs:subClassOf rdf:Property .

rdf:_1 rdf:type rdfs:ContainerMembershipProperty .
rdf:_1 rdfs:domain rdfs:Resource .
rdf:_1 rdfs:range rdfs:Resource .
rdf:_2 rdf:type rdfs:ContainerMembershipProperty .
rdf:_2 rdfs:domain rdfs:Resource .
rdf:_2 rdfs:range rdfs:Resource .
```

TABLE 13.6: RDFS axiomatic triples (3).

capture some of the various forms of vocabulary entailment as described in
the previous sections. The rules all have the same basic pattern: If an RDF
graph G has a triple/some triples corresponding to a certain pattern, then

```
rdfs:isDefinedBy rdfs:subPropertyOf rdfs:seeAlso .

rdf:XMLLiteral rdf:type rdfs:Datatype .
rdf:XMLLiteral rdfs:subClassOf rdfs:Literal .
rdfs:Datatype rdfs:subClassOf rdfs:Class .
```

TABLE 13.7: RDFS axiomatic triples (4).

add a new triple to the graph to obtain G'. Then, it can be shown on the basis of model theory that G entails G'. The rules can be used by software as a basis for checking RDF graphs for entailment (see also chapter 15 for more on computational inference algorithms). As with the interpretations, the entailment rules are divided into those for simple entailment, RDF entailment, and RDFS entailment.

Simple Entailment

The interpolation lemma that was explained above pertains to the following two simple entailment rules. The simple entailment rules add bnodes to the original graph, and thus can be thought of as creating a generalization of the original graph, that is, they generate graphs that have the original graph as an instance. By the interpolation lemma, any subgraph of the generalized graph is then entailed by the original graph. No explicit inference rule is required for the subgraph lemma. In the rules, we will use the following conventions. p or q will be used to stand for any URI reference (any possible predicate of an RDF triple). u or v will be used for any URI reference or bnode identifier, i.e., any possible subject of an RDF triple. Finally, x, y, ... will be used to stand for any URI reference, bnode identifier, or literal, i.e., any possible object of an RDF triple. ℓ will be used for any literal, and _:n will be used for bnode identifiers (Table 13.8).

Name	If G contains ...	Then add ...
se1	u p x .	u p _:n .
se2	u p x .	_:n p x .

TABLE 13.8: Simple entailment rules.

If rule se1 is being utilized for the first time on some triple <u,p,x>, then _:n stands for an arbitrary bnode identifier that has not been used before. If x has been allocated a bnode by a previously invocation of rule se1 or se2, then the bnode identifier from the previous allocation is used. Similar comments apply to rule se2. In other words, no more than one bnode identifier is given to a single entity even by multiple invocations of the entailment rules se1 and se2.

Consider for example an RDF graph G with a thousand triples, one of which is `<ex:ArthurDent,ex:hasNationality,ex:English>`. By application of se1, we obtain a graph of 1001 triples including the new triple

`_:1 ex:hasNationality ex:English .`

which is taken to mean, "there exists something (someone) that has the nationality English." Therefore, the graph of 1001 triples entails the graph consisting of just this triple. Therefore a query procedure might be envisaged that would ask if the original graph G entails the triple `<_:1,ex:hasNationality,ex:English>`. If the query procedure applied the rule se1 to G, it would be able to answer in the affirmative because of the interpolation lemma: A subgraph of G (the triple `<ex:ArthurDent,ex:hasNationality,ex:English>`) is an instance of the graph used in the query (the triple `<_:1,ex:hasNationality,ex:English>`). This is related to the logic of SPARQL queries (see Chapter 16), even though SPARQL is not implemented in this fashion.

Applying these rules naively is not an efficient search procedure, and in fact the general problem of determining simple entailment between two arbitrary RDF graphs is known to be NP-complete [108].

The rule 1g is required for RDF entailment. It is a special case of se2. A rule required for RDFS entailment is g1 that can be regarded as the inverse rule (Table 13.9).

Name	If G contains ...	Then add ...
1g	u p ℓ .	u p _:n . where _:n indicates a bnode allocated to the literal ℓ by this rule
g1	u p _:n .	u p ℓ . where _:n indicates a bnode allocated to the literal ℓ by rule 1g

TABLE 13.9: **Entailment rules for literals.** 1g is required for RDF entailment, and g1 for RDFS entailment.

Obviously rule g1 would produce a redundant triple except in cases where it is applied to an allocated bnode that has been introduced as the object of a new triple added by some other inference rule. These rules have the effect of identifying a literal with its allocated bnode.

RDF Entailment

There are two additional RDF entailment rules (Table 13.10). Rule rdf1 states that the predicate of any RDF triple must be an instance of `rdf:Property`. Rule `rdf2` states that a literal (which can only occur in the object position of a triple) must be an instance of `rdf:XMLLiteral`.

Name	If G contains ...	Then add ...
rdf1	u p x .	p rdf:type rdf:Property
rdf2	u p ℓ .	_:n rdf:type rdf:XMLLiteral . where _:n indicates a bnode identifier allocated to ℓ by rule 1g.

TABLE 13.10: RDF entailment rules. Note that rule `rdf2` applies only to well-formed XML literals.

We can now define the criteria for RDF entailment on the basis of simple entailment, the RDF axiomatic triples, and the RDF entailment rules.

Lemma 13.5 *RDF Entailment Lemma G RDF-entails H if and only if there is a graph which can be derived from G plus the axiomatic triples by the application of rules 1g, rdf1, and rdf2 and which simply entails H.*

In other words, the RDF Entailment Lemma states that G RDF-entails H if one can start with G, add all the RDF axiomatic triples to G and use the rules `1g`, `rdf1`, and `rdf2` to create a graph G', which simply entails H.

To prove the "if" direction, one only has to show that the RDF entailment rules are RDF-valid, which is easily done. Consider, for instance, rule `rdf1`. If the graph contains a triple of the form `<u,p,x>`, then we are to add the triple `<p,rdf:type,rdf:Property>`. If we now consider the semantic condition 5 for ground graphs (see page 296), we see that $p^{\mathcal{I}} \in IP$. Furthermore, RDF semantic condition 1 (see page 303) implies that if $p \in IP$, then `<p,`rdf:Property$^{\mathcal{I}}$`>`$\in I_{EXT}(\text{rdf:type}^{\mathcal{I}})$. Thus, `<p,rdf:type,rdf:Property>` is an RDF-valid inference. The other two rules can be proved similarly, as you will be asked to do in the Exercises. The proof in the "only if" direction is more technical and will not be given here. Interested readers are referred to the documentation of the W3C [108].

RDFS Entailment

RDFS offers a number of vocabulary elements to describe classes and the relationships between them for which further entailment rules are necessary (Table 13.11).

Rule `rdfs1` creates a bnode for an untyped literal, which can now take the place of the subject in further triples including the triple created by the rule itself, which states that the bnode (and the literal it stands for) is an instance of `rdfs:Literal`. Rules `rdfs2` and `rdfs3` describe the effects of domain and range constraints in RDFS, as has already been described in Chapter 4: A triple such as `<p,rdfs:domain,x>` allows the inference that the subject of any other triple with `p` as a predicate must be an instance

Name	If G contains ...	Then add ...
rdfs1	u p ℓ .	_:n rdf:type rdfs:Literal Here, ℓ is a plain literal with or without a language tag, and _:n indicates a bnode allocated to ℓ by rule 1g.
rdfs2	p rdfs:domain x . u p v .	u rdf:type x .
rdfs3	p rdfs:range x . u p v .	v rdf:type x .
rdfs4a	u p v .	u rdf:type rdfs:Resource .
rdfs4b	u p v .	v rdf:type rdfs:Resource .
rdfs5	p rdfs:subPropertyOf q . q rdfs:subPropertyOf r .	p rdfs:subPropertyOf r .
rdfs6	p rdf:type rdfs:Property .	p rdfs:subPropertyOf p .
rdfs7	p rdfs:subPropertyOf q . u p v .	u q v .
rdfs8	u rdf:type rdfs:Class .	u rdfs:subClassOf rdfs:Resource .
rdfs9	u rdfs:subClassOf v . y rdf:type u .	y rdf:type v .
rdfs10	u rdf:type rdfs:Class .	u rdfs:subClassOf u .
rdfs11	u rdfs:subClassOf v . v rdfs:subClassOf y .	u rdfs:subClassOf y .
rdfs12	u rdf:type rdfs:CMP .	u rdfs:subPropertyOf rdfs:member .
rdfs13	u rdf:type rdfs:Datatype .	u rdfs:subClassOf rdfs:Literal .

TABLE 13.11: RDFS entailment rules. Abbreviation: rdfs:CMP=rdfs:ContainerMembershipProperty.

of x, and similarly for the range constraint. rdfs4a and rdfs4b state that the subject and object of an arbitrary triple are instances of rdfs:Resource. The rule rdfs5 states that the relation rdfs:subPropertyOf is transitive (rdfs11 does the same for the relation rdfs:subClassOf). rdfs6 states that rdfs:subPropertyOf is reflexive (any property p is a subproperty of itself); rdfs10 does the same thing for the rdfs:subClassOf relation. rdfs9 defines the "inheritance" of subclass/superclass relations, similar to the inheritance of annotations in Gene Ontology (see for instance Figure 5.2 in Chapter 5). The rule rdfs7 defines a constraint for rdfs:subPropertyOf that is similar to the constraint rdfs9 for rdfs:subClassOf. The rule rdfs8 constrains all classes (instances of rdfs:Class to be resources (instances of rdfs:Resource). The rule rdfs12 constrains any property that is a instance of `rdfs:ContainerMembershipProperty` to be a subproperty of `rdfs:member` as well. Finally, `rdfs13` stipulates that any resource that is an `rdfs:Datatype` is a subclass of of `rdfs:Literal`.

The lemma for RDFS entailment mentions a special case for ill-typed XML literals, which are said to cause an `XML clash`. It is possible for a graph to be RDFS-inconsistent because it contains unsatisfiable assertions about an ill-typed XML literal. Consider the following RDFS document.

```
@prefix rdf: <http://www.w3.org/1999/02/22-rdf-syntax-ns#> .
@prefix rdfs: <http://www.w3.org/2000/01/rdf-schema#> .
@prefix xsd: <http://www.w3.org/2001/XMLSchema#> .
@prefix ex: <http://www.example.org/example.rdf#> .


ex:a rdfs:subClassOf rdfs:Literal .
ex:b rdfs:range ex:a .
ex:c ex:b "<:-)"^^rdf:XMLLiteral .
```

We can now use several entailment rules to show that this RDFS graph is unsatisfiable. We apply rule `lg` to the triple `<ex:c,ex:b,"<:-)"^^rdf:XMLLiteral>` to obtain

```
ex:c ex:b _:1 .
```

Note that `_:1` is now assigned to `"<:-)"^^rdf:XMLLiteral>`. Combining the triple `ex:c,ex:b,_:1>` and RDFS entailment rule `rdfs3` allows one to infer the following triple.

```
_:1 rdf:type rdfs:Literal .
```

Thus according to RDFS semantic condition 1 (see page 305), `_:1` as well as the literal value it stands for (`"<:-)"^^rdf:XMLLiteral>`) must be in the class of literal values LV. But according to RDF semantic condition 3, ill-typed XML literal strings must *not* be in LV. Therefore, there is no satisfiable interpretation of this RDFS document. An RDFS-inconsistent graph RDFS-entails all RDF graphs[4]. Of course, this is not an interesting kind of entailment for applications, but it is included in the lemma for reasons of logical correctness.

Lemma 13.6 *RDFS Entailment Lemma.* *A graph G RDFS-entails H if and only if there is a graph which can be derived from G plus the RDF and RDFS axiomatic triples by the applications of rules lg, gl, and the RDF and RDFS entailment rules, and which either simply entails H or contains an XML clash.*

The proof of this lemma is similar to that of the RDF entailment lemma. For the "if" direction, it is sufficient to show that the RDFS entailment rules

---

[4]This is similar to a *vacuous truth* in logic, where if P is false then $P \Rightarrow Q$ is inherently true (see Chapter 2).

are RDFS-valid. Readers will have opportunity to develop some proofs of individual rules in the Exercises, and more technical details of the proof are available in the documentation of the W3C [108].

13.6 Summary

In this chapter, we have discussed the theoretical background of three levels of inference in RDF graphs. The notions of model theory presented in this chapter are used to prove the correctness of a series of entailment rules, which can be used by applications to perform inference. This chapter has intended to provide intuition on how mathematical notions of inference can be translated into entailment rules that then can be used by inference algorithms. Note that the rules themselves do not provide an efficient or practical algorithm for performing inference in ontologies, but rather represent the building blocks which an inference algorithm can use. There are many different algorithms and implementations thereof in current use; Chapter 15 provides an overview of such algorithms and a detailed introduction to one of them.

13.7 Exercises and Further Reading

This chapter was closely based on the documentation of the W3C on RDF semantics [108], and interested readers should study that documentation for further information. A chapter in the excellent book by Hitzler et al. on the *Foundations of the Semantic Web* [117] also provides a useful reading of the W3C documentation about RDF semantics.[5]

Exercises

13.1 Consider the following RDF graph.

```
@prefix rdf: <http://www.w3.org/1999/02/22-rdf-syntax-ns#> .
@prefix rdfs: <http://www.w3.org/2000/01/rdf-schema#> .
@prefix xsd: <http://www.w3.org/2001/XMLSchema#> .
@prefix ex: <http://www.example.org/example.rdf#> .

ex:a ex:p ex:a .
```

---

[5]Readers from German-speaking countries may wish to read the original German version [118].

```
ex:a ex:p ex:b .
ex:b ex:q ex:t .
ex:t ex:q ex:w .
```

Create an *interpretation* of this graph such that the graph evaluates to true. Define IR, IP, I_{EXT}, I_S and show that each of the triples is true (see Figure 13.3 for an example).

13.2 Now create an interpretation of the graph from the previous exercise such that the graph is false.

13.3 Show that all axiomatic RDF triples except the last one can be derived from the RDFS axiomatic triples and the semantic conditions on IC_{EXT}, rdfs:domain, and rdfs:range. (Hint: consider the implication of the triple <rdfs:domain,rdfs:domain,rdfs:Property> and RDFS semantic condition 2).

13.4 Show that the following triples can be derived based on the RDFS axiomatic triples and RDFS semantic conditions.

```
rdfs:Resource rdf:type rdfs:Class .
rdfs:Literal rdf:type rdfs:Class .
rdf:XMLLiteral rdf:type rdfs:Class .
rdfs:Datatype rdf:type rdfs:Class .
rdf:Seq rdf:type rdfs:Class .
rdf:Bag rdf:type rdfs:Class .
rdf:Alt rdf:type rdfs:Class .
rdfs:Container rdf:type rdfs:Class .
rdf:List rdf:type rdfs:Class .
rdfs:ContainerMembershipProperty rdf:type rdfs:Class .
rdf:Property rdf:type rdfs:Class .
rdf:Statement rdf:type rdfs:Class .
```

13.5 Show that the following triples can be derived based on the RDFS axiomatic triples and RDFS semantic conditions.

```
rdfs:domain rdf:type rdf:Property .
rdfs:range rdf:type rdf:Property .
rdfs:subPropertyOf rdf:type rdf:Property .
rdfs:subClassOf rdf:type rdf:Property .
rdfs:member rdf:type rdf:Property .
rdfs:seeAlso rdf:type rdf:Property .
rdfs:isDefinedBy rdf:type rdf:Property .
rdfs:comment rdf:type rdf:Property .
rdfs:label rdf:type rdf:Property .
```

13.6 Show that RDFS rule `rdfs2` is RDFS-valid (see page 311 for an example).

13.7 Show that RDFS rule `rdfs8` is RDFS-valid.

Chapter 14

Inference in OWL Ontologies

In Chapter 13, we saw how model theoretic semantics for RDF and RDFS were used to derive entailment rules. Model-theoretic semantics have been defined for OWL in an analogous fashion [172], but in this chapter we will concentrate instead on how entailment rules relate to inference in an OWL ontology.

We develop the examples based on a toy ontology about the characters from the Harry Potter books in which many of the inference rules defined by the W3C for OWL are demonstrated.[1] Readers may want to construct the ontology using Protégé as they read this chapter.[2]

There are now many different varieties of OWL (see Appendix D for descriptions). For this chapter, we will be using a subset of entailment rules defined for the RL profile of OWL 2, which contains several restrictions on the expressivity of an OWL 2 DL ontology [170, 172]. We will present the ontology using the OWL Manchester syntax that is explained in Appendix D.

14.1 The Semantics of Equality

We will present the entailment rules defined for OWL RL ontologies in a series of tables, followed by illustrative examples from our Harry Potter ontology. Each of the entailment rules has a short name, an **if** condition and a **then** conclusion. If the ontology contains triples satisfying the **if** condition, then the triple or triples under **then** may be inferred. The entailment rules in this chapter are taken from the tables in the W3C documentation of the OWL 2 RL profile, which the reader may wish to consult for further details. Several of the rules, especially those regarding lists and data types, will be omitted from this chapter.

Table 14.1 shows some of the rules about equality. The rules are given as universally quantified first-order implications over a ternary predicate T,

---

[1] We will explain enough of the background along the way for readers who are unfamiliar with these books to understand the inference examples!

[2] For instructions on how to get started with Protégé, see the Exercises of Chapter 4. A file with the complete Harry Potter ontology is available at the book Web site.

Rule	if	then
`eq-ref`	T(?s, ?p, ?o)	T(?s, owl:sameAs, ?s) T(?p, owl:sameAs, ?p) T(?o, owl:sameAs, ?o)
`eq-sym`	T(?x, owl:sameAs, ?y)	T(?y, owl:sameAs, ?x)
`eq-trans`	T(?x, owl:sameAs, ?y) T(?y, owl:sameAs, ?z)	T(?x, owl:sameAs, ?z)
`eq-rep-s`	T(?s, owl:sameAs, ?s') T(?s, ?p, ?o)	T(?s', ?p, ?o)
`eq-rep-p`	T(?p, owl:sameAs, ?p') T(?s, ?p, ?o)	T(?s, ?p', ?o)
`eq-rep-o`	T(?o, owl:sameAs, ?o') T(?s, ?p, ?o)	T(?s, ?p, ?o')
`eq-diff1`	T(?x, owl:sameAs, ?y) T(?x, owl:differentFrom, ?y)	`false`

TABLE 14.1: OWL entailment rules: equality. The table shows a subset of the entailment rules about equality among individuals that hold in the OWL 2 RL profile.

corresponding to a generalized RDF triple[3] with the subject s, the predicate p, and the object o. Variables in the implications (`if` column) are preceded with a question mark, and the corresponding variables in the entailment (`then` column) are bound to the same values. `false` is used as a propositional symbol to indicate a contradiction. If it is derived, then the initial RDF graph was inconsistent.

As previously mentioned, we will be using a toy ontology based on the Harry Potter universe to illustrate the entailment rules. The header of the ontology contains the following information which defines prefixes that will be used as shortcuts. In particular, we define the base IRI of the ontology to be `<http://www.example.org/Potter.owl#>`.

```
Prefix: xsd:   <http://www.w3.org/2001/XMLSchema#>
Prefix: owl:   <http://www.w3.org/2002/07/owl#>
Prefix: xml:   <http://www.w3.org/XML/1998/namespace>
Prefix: rdf:   <http://www.w3.org/1999/02/22-rdf-syntax-ns#>
Prefix: rdfs:  <http://www.w3.org/2000/01/rdf-schema#>
Prefix:   :    <http://www.example.org/Potter.owl#>
```

We can now define two individuals, HarryPotter and Harry. Although it is obvious to us that these two names refer to the same person, remember that for the computer, HarryPotter and Harry have no special meaning. The following declarations state that HarryPotter is an individual. The SameAs

---

[3]In generalized RDF triples, bnodes and literals are allowed in all positions; see [248].

statement declares that *HarryPotter* is the same individual as `Harry`. In all of the examples in this chapter, we will use the pound sign (#) to indicate that names belong to the default namespace for better legibility.

```
Individual: #HarryPotter
  SameAs:
    #Harry

Individual: #Harry
```

The `eq-sym` rule states that if there is a triple of the form `<x, owl:sameAs, y>` for some x and y, then we can infer the triple `<y, owl:sameAs, x>`. Thus, we can infer the following information about *Harry*:

```
Individual: #Harry
  SameAs: #HarryPotter
```

We now define another individual called *Potter* and declare it to be the same as *Harry*.

```
Individual: #Potter
  SameAs: #Harry
```

Then rule `eq-trans` states that if it is asserted that X is the same as Y and that Y is the same as Z, then we can infer that X is the same as Z. Thus, we can infer the following about the individual *Potter*.[4]

```
Individual: #Potter
  SameAs:
    #Harry, #HarryPotter
```

We now define an object property to represent the activity of attending a school.

```
ObjectProperty: #attends
```

We can now declare an individual called *Hogwarts*[5] and revise our declaration about the individual *HarryPotter* to include the assertion that *Harry-Potter* attends *Hogwarts*.

```
Individual: #Hogwarts

Individual: #HarryPotter
```

---

[4]In Protégé 4, open the `Entities` tab and the `Individuals by type` panel. In the `Description` panel, one can see all asserted and inferred `SameAs` relations.

[5]Hogwarts School of Witchcraft and Wizardry was the boarding school for magic in the Harry Potter novels.

```
Facts:
  #attends #Hogwarts
SameAs:
  #Harry, #Potter
```

Rule `eq-rep-s` states that if individual `s` is the same as individual `s'`, and there is a triple `<s,p,o>`, then we can infer the triple `<s',p,o>`. Thus, since *Harry* is known to be the same as *HarryPotter*, we can infer the following fact[6].

```
Individual: #Harry
  Facts:
    #attends #Hogwarts
```

Analogous inferences can be drawn for the rules governing predicates (`eq-rep-p`) and object (`eq-rep-o`).

Finally, the rule `eq-diff1` states that an individual X cannot be both `SameAs` and `DifferentFrom` and individual Y. The propositional variable `false` is shown in the **then** column of Table 14.1 to indicate a logical inconsistency. This would be the case if our ontology contained an assertion of the following sort.

```
Individual: #HarryPotter
  SameAs:
    #Harry
  DifferentFrom:
    #Harry
```

While presumably we are intelligent enough not to explicitly assert such an inconsistency, it is possible that a number of inferred facts that were not explicitly asserted in the ontology file itself lead to the conclusion that some individual is both `SameAs` and `DifferentFrom` another individual. This usually indicates a modeling error.

14.2 The Semantics of Properties

There are many entailment rules based on properties. Intuitively, we know that the subject of the relation `hasWife` is a man (husband), and the object is a woman (wife). The entailment rules for properties formalize this kind of intuition (Table 14.2).

---

[6]In Protégé 4, this is done by invoking the reasoner HermiT from the **Reasoner**. The inferred object property assertion for *Harry* is shown in the **Description** panel when the individual *Harry* is chosen in the **Individuals by type** panel of the **Entities** tab.

Rule	if	then
prp-dom	T(?p, rdfs:domain, ?c) T(?x, ?p, ?y)	T(?x, rdf:type, ?c)
prp-rng	T(?p, rdfs:range, ?c) T(?x, ?p, ?y)	T(?y, rdf:type, ?c)
prp-fp	T(?p, rdf:type, owl:FunctionalProperty) T(?x, ?p, ?y_1) T(?x, ?p, ?y_2)	T(?y_1, owl:sameAs, ?y_2)
prp-ifp	T(?p, rdf:type, owl:InverseFunctionalProperty) T(?x_1, ?p, ?y) T(?x_2, ?p, ?y)	T(?x_1, owl:sameAs, ?x_2)
prp-symp	T(?p, rdf:type, owl:SymmetricProperty) T(?x, ?p, ?y)	T(?y, ?p, ?x)
prp-trp	T(?p, rdf:type, owl:TransitiveProperty) T(?x, ?p, ?y) T(?y, ?p, ?z)	T(?x, ?p, ?z)
prp-spo1	T(?p1, rdfs:subPropertyOf, ?p2) T(?x, ?p1, ?y)	T(?x, ?p2, ?y)
prp-inv1	T(?p_1, owl:inverseOf, ?p2) T(?x, ?p_1, ?y)	T(?y, ?p2, ?x)

TABLE 14.2: OWL entailment rules: properties. The table shows a subset of the entailment rules for properties that hold for the OWL 2 RL profile.

In OWL, every individual is an instance of `owl:thing`, and every user-defined class is implicitly a subclass of `owl:thing`. Up to now, `HarryPotter` was not declared to be an instance of any specific class, and thus `HarryPotter` is an instance only of `owl:thing`. In OWL, it is possible to explicitly assert that an individual is an instance of a class. There are also a number of language constructs that allow the inference that an individual is an instance of a class.

To start off, let us define two classes and use them to specify the domain and range of the property *attends*.

```
Class: #Pupil
  Annotations:  "A person who attends a school."

Class: :School
  Annotations:  "An institution designed for the
                 teaching of pupils."

ObjectProperty: #attends
  Domain: #Pupil
  Range:  #School
```

The rule `prp-dom` states that if the domain of a relation p is class c and there is a triple about an individual x using the predicate p, <x,p,y>, then we can infer that individual x is an instance of class c. Similarly, the rule `prp-rng` states that if the range of a relation p is class d and there is a triple of which an individual y is the object, <x,p,y>, then we can infer that y is an instance of class d. Consider the following assertion.

```
Individual: #HarryPotter
  Facts:
    #attends #Hogwarts
```

This, together with the specifications of the domain and range of the predicate `attends`, allows the following two inferences to be made.

```
#HarryPotter
  Types: #Pupil
#Hogwarts
  Types: #School
```

That is, *HarryPotter* is an instance of *Pupil* (in Manchester syntax, this is expressed with the `Types` keyword), and *Hogwarts* is an instance of *School*.

The rule `prp-fp` states that if a predicate p is a functional property, then a subject can be related to only one specific object by p. This is similar to the definition of a function, which assigns a unique value to any given input. If there are two triples in the ontology of the form <x,p,a> and <x,p,b> for a functional property p, then we can infer that a and b are the same individual. The following statements declare a new individual *SchoolOfWitchcraftAnd-Wizardry* and define the relation `attends` to be functional.

```
Individual: #SchoolOfWitchcraftAndWizardry

ObjectProperty: #attends
  Characteristics:  Functional
  Domain: #Pupil
  Range:  #School
```

We now assert that *HarryPotter* attends both *Hogwarts* and *SchoolOfWitchcraftAnd-Wizardry*.

```
Individual: #HarryPotter
  Facts:
    #attends #Hogwarts,
    #attends #SchoolOfWitchcraftAndWizardry
```

Then rule `prp-fp` allows the inference that *SchoolOfWitchcraftAndWizardry* and *Hogwarts* refer to the same individual.

The rule `prp-ifp` works analogously for inverse functional properties, which are such that a specific object can be related to only one subject by the property. If there are two triples in the ontology of the form `<x,p,z>` and `<y,p,z>` for an inverse functional property p, then we can infer that x and y are the same individual.

The rule `prp-symp` defines the behavior of symmetric properties, stating that if $A \xrightarrow{p} B$ for a symmetric property p, then we can infer that $B \xrightarrow{p} A$. We will define an individual `Ron`, and a symmetric property `isClassmateOf`.

```
ObjectProperty: #isClassmateOf
  Characteristics: Symmetric

Individual: #Ron
  Facts:
    #isClassmateOf #Harry
```

Rule `prp-symp` then allows the inference that *Harry* `isClassmateOf` *Ron*.

```
Individual: #Harry
  Facts:
    #isClassmateOf #Ron
```

The rule `prp-trp` allows inferences about transitive properties to be made. A property is *transitive* if when two triples `<a,p,b>` and `<b,p,c>` are asserted, then it is also the case that `<a,p,c>`. In the following, we declare a new individual *Hermione* and specify that the property `isClassmateOf` is also transitive.

```
ObjectProperty: #isClassmateOf
  Characteristics: Symmetric, Transitive
```

```
Individual: #Hermione
  Facts:
    #isClassmateOf #Ron
```

This allows the inference that Hermione is also a classmate of Harry:

```
Individual: #Hermione
  Facts:
    #isClassmateOf #Ron,
    #isClassmateOf #Harry
```

Just as a class can be a subclass of another class, properties can be related to one another in a hierarchical fashion using the OWL keyword owl:subPropertyOf. A typical example of this sort of property involves family relations; isChildOf is necessarily a subproperty of isRelativeOf. Recalling that Dudley Dursley, Harry Potter's muggle cousin, is the son of Vernon Dursley, we can assert the following facts.

```
ObjectProperty: #isChildOf
  SubPropertyOf: #isRelativeOf
```

```
ObjectProperty: #isRelativeOf
```

```
Individual:  #Vernon
```

```
Individual: #Dudley
  Facts:
    #isChildOf  #Vernon
```

These assertions together with the entailment rule prp-spo1 allow the inference that Dudley is a relative of Vernon.

```
Individual: #Dudley
  Facts:
    #isChildOf  #Vernon,
    #isRelativeOf #Vernon
```

A property q is said to be the inverse of another property p if <x,p,y> necessarily implies <y,q,p> and vice versa. A simple example of this is given by the properties isPupilOf and isTeacherOf.

```
ObjectProperty: #isPupilOf
    SubPropertyOf: owl:topObjectProperty
    InverseOf: #isTeacherOf
```

```
ObjectProperty: #isTeacherOf
    SubPropertyOf: owl:topObjectProperty
    InverseOf: #isPupilOf

Individual: #Snape

Individual: #HarryPotter
  Facts:
    #attends   #Hogwarts,
    #attends   #SchoolOfWitchcraftAndWizardry,
    #isPupilOf #Snape
```

The above assertion that `HarryPotter` is a pupil of `Snape`, together with entailment rule `prp-inv1`, allows the inference that Professor Snape is a teacher of Harry.

```
Individual: #Snape
  Facts:
    #isTeacherOf #Harry
```

14.3 The Semantics of Classes

Much of the power of OWL comes from its constructs for defining classes (Table 14.3). The entailment rule `cls-int1` states that if an individual y is an instance of classes $c_1, c_2, \ldots, c_n$ and some other class C is defined as the intersection of classes $c_1, c_2, \ldots, c_n$, then we can infer that y is also an instance of C.

Let us define the class of Hogwarts wizardry instructors as being equivalent to the intersection of Hogwarts staff, teachers, and wizards.

```
Class: #Wizard
  Annotations:
    rdfs:comment "A person who practices magic."

Class: #HogwartsStaff
  Annotations:
    rdfs:comment "A person under contract to provide
                  services to the Hogwarts School."

Class: #Teacher
  Annotations:
    rdfs:comment "A person who helps children learn reading,
                  writing, arithmetic, and magic."
```

```
Class: #HogwartsWizardryInstructor
  Annotations:
    rdfs:comment "A teacher of magic at the Hogwarts School."
  EquivalentTo:
    #HogwartsEmployee and #Teacher and #Wizard
```

We now modify the declaration of Snape to state that he is a Wizard, and Teacher, and a Hogwarts employee.

```
Individual: #Snape
  Types: #HogwartsStaff, #Teacher, #Wizard
```

These assertions, together with entailment rule `cls-int1`, allow the inference that Snape is also an instance of the class `HogwartsWizardryInstructor`.

The inference also goes in the other direction thanks to entailment rule `cls-int2`. For instance, we can assert that Remus Lupin is an instance of `HogwartsWizardryInstructor`.

```
Individual:  #RemusLupin
  Types: #HogwartsWizardryInstructor
```

The entailment rule `cls-int2` now allows the inference that Remus Lupin is an instance of the classes `HogwartsStaff`, `Teacher`, and `Wizard`.

We can also define a class to be the union of two or more other classes. For the sake of argument, we will assume that any instance of the class Adult is either a Wizard or a Muggle.[7]

```
Class: #Adult
  EquivalentTo: #Muggle or #Wizard
```

We now define the following two individuals:

```
Individual: #Dolores
  Annotations:
    rdfs:comment "Dolores Umbridge, a wizard and
                  instructor at Hogwarts"
  Types: #Wizard
```

```
Individual: #Petunia
  Annotations:
    rdfs:comment "Petunia Dursley, Harry Potter's aunt."
  Types: #Muggle
```

---

[7]In the Harry Potter universe, muggles are people who lack magical abilities.

Rule	if	then
cls-int1	$T(?c, \text{owl:intersectionOf}, ?x)$ $\text{LIST}[?x, ?c_1, ..., ?c_n]$ $T(?y, \text{rdf:type}, ?c_1)$ $T(?y, \text{rdf:type}, ?c_2)$ ... $T(?y, \text{rdf:type}, ?c_n)$	$T(?y, \text{rdf:type}, ?c)$
cls-int2	$T(?c, \text{owl:intersectionOf}, ?x)$ $\text{LIST}[?x, ?c_1, ..., ?c_n]$ $T(?y, \text{rdf:type}, ?c)$	$T(?y, \text{rdf:type}, ?c_1)$ $T(?y, \text{rdf:type}, ?c_2)$ ... $T(?y, \text{rdf:type}, ?c_n)$
cls-uni	$T(?c, \text{owl:unionOf}, ?x)$ $\text{LIST}[?x, ?c_1, ..., ?c_n]$ $T(?y, \text{rdf:type}, ?c_i)$	$T(?y, \text{rdf:type}, ?c)$
cls-svf1	$T(?x, \text{owl:someValuesFrom}, ?y)$ $T(?x, \text{owl:onProperty}, ?p)$ $T(?u, ?p, ?v)$ $T(?v, \text{rdf:type}, ?y)$	$T(?u, \text{rdf:type}, ?x)$
cls-avf	$T(?x, \text{owl:allValuesFrom}, ?y)$ $T(?x, \text{owl:onProperty}, ?p)$ $T(?u, \text{rdf:type}, ?x)$ $T(?u, ?p, ?v)$	$T(?v, \text{rdf:type}, ?y)$
cls-hv1	$T(?x, \text{owl:hasValue}, ?y)$ $T(?x, \text{owl:onProperty}, ?p)$ $T(?u, \text{rdf:type}, ?x)$	$T(?u, ?p, ?y)$
cls-hv2	$T(?x, \text{owl:hasValue}, ?y)$ $T(?x, \text{owl:onProperty}, ?p)$ $T(?u, ?p, ?y)$	$T(?u, \text{rdf:type}, ?x)$

TABLE 14.3: OWL entailment rules: classes. The table shows a subset of the entailment rules for classes that hold for the OWL 2 RL profile. Rule cls-uni holds for all c_i with $1 \leq i \leq n$.

Using entailment rule `cls-uni` and these assertions, we can infer that both `Dolores` and `Petunia` are instances of the class `Adult`. On the other hand, if we assert that some individual is an instance of the class `Adult`, we cannot infer that the individual is also an instance of `Wizard` or `Muggle`. This is an important difference to class definitions that are based on an intersection.

The OWL `someValuesOf` construct is not meant to be used to constrain the possible classes that can be used in assertions, but rather to allow inference. Consider the following example, recalling that all of the pupils at Hogwarts are sorted into one of four houses[8] at the beginning of their school time. We have revised the definition of the class `Pupil` to the effect that every pupil is sorted into some Hogwarts House.

```
Class: #Pupil
  Annotations:
    rdfs:comment "A person who attends a school."
  EquivalentTo: #sortedInto some #HogwartsHouse
  SubClassOf:   #Person

ObjectProperty: #sortedInto

Individual: #Slytherin
    Types: #HogwartsHouse

Individual: #DracoMalfoy
  Annotations:
    rdfs:comment "A pupil at Hogwarts who was sorted
                  into the house of Slytherin."
  Facts: #sortedInto #Slytherin
```

Note that we have not explicitly stated that Draco Malfoy is an instance of the class `Pupil`. The class definition for pupil states that the class is equivalent to the class of all individuals that have been sorted into some `HogwartsHouse`. Since we declare that `DracoMalfoy` has been `sortedInto` the Hogwarts House `Slytherin`, the entailment rule `cls-svf1` allows the inference that `DracoMalfoy` is an instance of the class `Pupil`. The Manchester serialization does not make it easy to see which assertions correspond to the triples of the entailment rule `cls-svf1`. This is perhaps easier to see in the OWL/XML serialization of the class assertion (see Appendix D). The triples $T(?x,owl:someValuesFrom,?y)$ and $T(?x,owl:onProperty,?p)$ correspond to the triples encoded by the `<ObjectSomeValuesFrom>` node in the OWL/XML serialization of the equivalent class axiom:

```
<EquivalentClasses>
  <Class IRI="#Pupil"/>
```

---

[8]Gryffindor, Ravenclaw, Hufflepuff, and Slytherin.

```
  <ObjectSomeValuesFrom>
    <ObjectProperty IRI="#sortedInto"/>
    <Class IRI="#HogwartsHouse"/>
  </ObjectSomeValuesFrom>
</EquivalentClasses>
```

This, together with the assertions that Draco Malfoy was sorted into Slytherin (T(?u, ?p, ?v)) and that Slytherin is an instance of the class HogwartsHouse (T(?v, rdf:type, ?y)) fulfills the four triples for the entailment rule cls-svf1 and thereby allows the inference.

The entailment rule cls-avf defines the semantics of a universal restriction. The owl:allValuesFrom restriction requires that for every instance of the class that has instances of the specified property, the values of the property are all members of the class indicated by the owl:allValuesFrom clause [242]. Consider the following example. In the Harry Potter universe, a magic wand is a tool used by wizards to channel magical powers. Powerful wizards can perform magic without the aid of a wand, but it is quite difficult and requires much skill. If a wizard uses a tool to conjure up magical powers, then the tool must be a magic wand. We will modify the definition of the class *Wizard*, create a class *MagicWand*, and state that a *Wizard* is a subclass of entities that, if they cast spells, do so with a *MagicWand* (the Manchester keyword only is equivalent to *allValuesFrom*).

```
Class: #Wizard
  Annotations:
    rdfs:comment "A person who practices magic."
  SubClassOf: #castsSpellsWith only #MagicWand
  DisjointWith: #Muggle

Class: #MagicWand
  Annotations:
    rdfs:comment "A hand-held stick used
                  to conjure magical powers."

ObjectProperty: #castsSpellsWith

Individual: #DragonHeartstringRod

Individual: #PeterPettigrew
  Types: #Wizard
  Facts:  #castsSpellsWith #DragonHeartstringRod
```

From the above definitions of the individuals *DragonHeartstringRod* and *PeterPettigrew*, we can conclude that *DragonHeartstringRod* is an instance of the class *MagicWand*.

The difference between `someValuesFrom` and `allValuesFrom` is the difference between a universal (`allValuesFrom`) and existential (`someValuesFrom`) quantification. The above existential restriction implies that if a *Wizard* `castsSpellWith` an entity, that entity must be an instance of the class *MagicWand*. It does not imply that a *Wizard* must have any `castsSpellWith` predicate, and thus it is possible that a *Wizard* performs magic without a wand. If we had used `someValuesFrom` instead of `allValuesFrom`, it would have implied that each instance of the class *Wizard* must be related to an instance of the class *MagicWand* by the relation `castsSpellWith`.

In OWL, we can use a `hasValue` restriction to specify classes based on the existence of a particular property value. Recall that in the Harry Potter universe, Goblins speak a language called Gobbledegook, and that they are the only race of creatures that speak this particular language. We can define the class of `Goblin` based on this restriction.

```
ObjectProperty: #speaks

Individual: #Gobbledegook
  Annotations:
    rdfs:comment "The language of the Goblins."

Class: #Goblin
  EquivalentTo: #speaks value #Gobbledegook
```

We now declare an individual Griphook of type `Goblin`

```
Individual: #Griphook
  Types: #Goblin
```

Then, entailment rule `cls-hv1` allows the inference that Griphook speaks Gobbledegook. The entailment rule `cls-hv2` allows inference in the opposite direction based on properties with a `hasValue` restriction. If we assert the following fact about an individual named Ragnok,

```
Individual: #Ragnok
  Facts: #speaks   #Gobbledegook
```

then we can infer that Ragnok is an instance of the type `Goblin`.

Similarly, there are a number of entailment rules that describe the semantics of class axioms (Table 14.4). The entailment rule `cax-sco` implies that if an individual is an instance of some class that is a subclass of another class, the individual is also an instance of that other class. Thus, we have already asserted that `Harry` is an instance of `Pupil` and that `Pupil` is a subclass of `Person`. The rule `cax-eqc1` allows us to infer that `Harry` is also an instance of the class `Person`. Thus, if we declare a class called `NonWizard` and declare it to be equivalent top the class `Muggle`, then we can infer that any instance of the class `Muggle` also must be an instance of the class `NonWizard`.

Rule	if	then
cax-sco	$T(?c_1, \text{rdfs:subClassOf}, ?c_2)$ $T(?x, \text{rdf:type}, ?c_1)$	$T(?x, \text{rdf:type}, ?c_2)$
cax-eqc1	$T(?c_1, \text{owl:equivalentClass}, ?c_2)$ $T(?x, \text{rdf:type}, ?c_1)$	$T(?x, \text{rdf:type}, ?c_2)$
cax-eqc2	$T(?c_1, \text{owl:equivalentClass}, ?c_2)$ $T(?x, \text{rdf:type}, ?c_2)$	$T(?x, \text{rdf:type}, ?c_1)$
cax-dw	$T(?c_1, \text{owl:disjointWith}, ?c_2)$ $T(?x, \text{rdf:type}, ?c_1)$ $T(?x, \text{rdf:type}, ?c_2)$	false

TABLE 14.4: OWL entailment rules: class axioms. The table shows a subset of the entailment rules for class axioms that hold for the OWL 2 RL profile.

```
Class: #NonWizard
  EquivalentTo: #Muggle
```

Based on the above class axiom, we can infer that `Petunia`, who was declared to be an instance of the class `Muggle`, is also an instance of the class `NonWizard`. The rule `cax-eqc2` is entirely analogous.

If we now were to declare that the individual `PomonaSprout` is both a `Wizard` and a `Muggle`, then there is no inconsistency. This is because we have not encoded our background knowledge that these two classes are mutually exclusive. We now add this fact to the class definition of `Muggle`.

```
Class: #Muggle
  EquivalentTo: #NonWizard
  DisjointWith: #Wizard
```

This leads to a logical inconsistency for the individual `PomonaSprout`, because she cannot be an instance of both of these classes.

14.4 The Semantics of the Schema Vocabulary

The entailment rules for the schema vocabulary define the semantics of a number of important OWL keywords (Table 14.5). Let us first consider the following class hierarchy for some of the magical beasts in the Harry Potter universe.

```
Class: #MagicalBeast
```

```
Class: #Dragon
```

Rule	if	then
scm-sco	$T(?c_1, \text{rdfs:subClassOf}, ?c_2)$ $T(?c_2, \text{rdfs:subClassOf}, ?c_3)$	$T(?c_1, \text{rdfs:subClassOf}, ?c_3)$
scm-eqc1	$T(?c_1, \text{owl:equivalentClass}, ?c_2)$	$T(?c_1, \text{rdfs:subClassOf}, ?c_2)$ $T(?c_2, \text{rdfs:subClassOf}, ?c_1)$
scm-eqc2	$T(?c_1, \text{rdfs:subClassOf}, ?c_2)$ $T(?c_2, \text{rdfs:subClassOf}, ?c_1)$	$T(?c_1, \text{owl:equivalentClass}, ?c_2)$
scm-spo	$T(?p_1, \text{rdfs:subPropertyOf}, ?p_2)$ $T(?p_2, \text{rdfs:subPropertyOf}, ?p_3)$	$T(?p_1, \text{rdfs:subPropertyOf}, ?p_3)$
scm-dom1	$T(?p, \text{rdfs:domain}, ?c_1)$ $T(?c_1, \text{rdfs:subClassOf}, ?c_2)$	$T(?p, \text{rdfs:domain}, ?c_2)$
scm-dom2	$T(?p_2, \text{rdfs:domain}, ?c)$ $T(?p_1, \text{rdfs:subPropertyOf}, ?p_2)$	$T(?p_1, \text{rdfs:domain}, ?c)$
scm-rng1	$T(?p, \text{rdfs:range}, ?c_1)$ $T(?c_1, \text{rdfs:subClassOf}, ?c_2)$	$T(?p, \text{rdfs:range}, ?c_2)$
scm-hv	$T(?c_1, \text{owl:hasValue}, ?i)$ $T(?c_1, \text{owl:onProperty}, ?p_1)$ $T(?c_2, \text{owl:hasValue}, ?i)$ $T(?c_2, \text{owl:onProperty}, ?p_2)$ $T(?p_1, \text{rdfs:subPropertyOf}, ?p_2)$	$T(?c_1, \text{rdfs:subClassOf}, ?c_2)$
scm-int	$T(?c, \text{owl:intersectionOf}, ?x)$ $\text{LIST}[?x, ?c_1, \ldots, ?c_n]$	$T(?c, \text{rdfs:subClassOf}, ?c_1)$ $T(?c, \text{rdfs:subClassOf}, ?c_2)$ ... $T(?c, \text{rdfs:subClassOf}, ?c_n)$
scm-uni	$T(?c, \text{owl:unionOf}, ?x)$ $\text{LIST}[?x, ?c_1, \ldots, ?c_n]$	$T(?c_1, \text{rdfs:subClassOf}, ?c)$ $T(?c_2, \text{rdfs:subClassOf}, ?c)$ ... $T(?c_n, \text{rdfs:subClassOf}, ?c)$

TABLE 14.5: OWL entailment rules: schema vocabulary. The table shows a subset of the entailment rules for the schema vocabulary that hold for the OWL 2 RL profile.

```
SubClassOf: #MagicalBeast

Class: #NorwegianRidgebackDragon
  SubClassOf: #Dragon

Class: #ChineseFireball
  SubClassOf: #Dragon
```

We now declare an individual called Norbert to be an instance of the class NorwegianRidgebackDragon.

```
Individual: #Norbert
  Types: #NorwegianRidgebackDragon
```

The entailment rule cm-sco now allows us to infer that Norbert is also an instance of the class Dragon. This in turn allows the inference that Norbert is an instance of the class MagicalBeast.

Readers are asked to provide examples for the other rules of Table 14.5 in the Exercises.

14.5 Conclusions

In this chapter, we have presented a simple ontology based on the Harry Potter universe and have demonstrated some of the kinds of inference that can be performed on the classes and individuals of the ontology. We have not presented nearly all of the inference rules for OWL ontologies, but we have concentrated on the commonly seen rules and ontology design patterns that readers are likely to encounter in real ontologies. Readers are encouraged to develop and extend this ontology themselves in Protégé in the Exercises in order to school their intuition about inference in OWL with an eye towards understanding the reasons why inferences are drawn in OWL ontologies.

The examples of this chapter can all be expressed within the OWL 2 RL profile, which is a syntactic subset of the OWL 2 that allows efficient reasoning without sacrificing too much expressive power. OWL 2 RL is amenable to implementation using rule-based technologies; however, inference in OWL 2 RL ontologies can also be performed using algorithms designed for full OWL 2 DL ontologies, such as the tableau algorithm that will be presented in Chapter 15.

14.6 Exercises and Further Reading

There are a number of useful tutorials on inference in OWL available on the Internet. At the time of this writing, many of them were written in terms of version 1 of OWL, but they are still valid for OWL 2 with at most minor modifications. Particularly recommendable are the tutorials from the Manchester group[9] and the Stanford group.[10] See also Appendix D for references to the W3C documentation on OWL 2.

Several application programming interfaces (APIs) for OWL reasoners are presented in Chapter 15. The exercises in that chapter include a tutorial on using OWL API, which is a Java interface for working with OWL ontologies. In the Exercises, which are also relevant for the material in this chapter, readers will develop a Java program to explore the Harry Potter ontology that was developed in this chapter.

Exercises

14.1 Describe the meaning of the entailment rule scm-eqc1 in English. Provide an example of inference using this rule by extending the Harry Potter ontology or using another ontology of your choice. Use the Protégé ontology editor. See the tutorial in the Exercises of Chapter 4 or consult the online documentation for Protégé[11] if necessary.

14.2 Perform the tasks of Exercise 14.1 for scm-eqc2.

14.3 Perform the tasks of Exercise 14.1 for scm-spo.

14.4 Perform the tasks of Exercise 14.1 for scm-dom1.

14.5 Perform the tasks of Exercise 14.1 for scm-dom2.

14.6 Perform the tasks of Exercise 14.1 for scm-rng1.

14.7 Perform the tasks of Exercise 14.1 for scm-hv.

14.8 Perform the tasks of Exercise 14.1 for scm-int.

14.9 Perform the tasks of Exercise 14.1 for scm-uni.

14.10 For this exercise, use the characters and classifications of a favorite book of yours to create an ontology similar to the one presented in this chapter. Try to find examples to demonstrate at least ten of the entailment rules presented in this chapter.

---

[9]http://owl.man.ac.uk/2003/why/latest/.

[10]http://www-ksl.stanford.edu/software/jtp/doc/owl-reasoning.html.

[11]http://protege.stanford.edu/.

Chapter 15

Algorithmic Foundations of Computational Inference

In previous chapters, we showed how inference is used to address problems but treated it as a kind of "black box." This chapter explains how computational inference can be implemented algorithmically.

In Part III, the algorithms mainly used relations between terms to perform subsumption inference, but did not take advantage of the different sorts of relationships that are part of the ontology. For instance, we have an intuitive understanding of the is_a relation that we encountered many times, i.e., that all instances of concept B that is_a concept A are also instances of concept A (see Figure 5.2 on page 131). However, from the perspective of a computer that particular meaning is not obvious, as long as we don't instruct the machine how these relationships are to be handled.

One way of implementing inference based on is_a relations is to instruct the computer to propagate the annotations to terms linked by the is_a relationship before the overrepresentation analysis is carried out. This means that the semantics of the is_a relationship is encoded in the computer program. There is nothing wrong with that approach, and many of the current programs for GO overrepresentation analysis use graph algorithms to implement inferences based on is_a relations (as well as part_of and the other GO relations).

However, things get more involved as soon as the number of different relationships increases, and it quickly becomes difficult or impossible to implement inference with simple graph algorithms or with a separate function for each inference rule. One of the key features of semantic modeling is that the actual semantics of the relationships can be formally defined such that the computer can also "grasp" their meaning. As a part of the modeling process, a set of facts and relations are asserted as an ontology using a formal language that is equipped with entailment rules. This was covered in many previous chapters. There are several classes of algorithms that can be used to infer new relations not explicitly asserted in the ontology by using inference rules. The history of computational inference algorithms, which goes hand-in-hand with the history of the development of description logics and ontology languages, revolves around attempts to find a good compromise between the expressivity of the ontology and the computational properties of the corresponding inference algorithms. As a general and unsurprising rule, the more expressive an ontology language is, the more difficult it is to perform computational inference in the language.

Rule-based algorithms for reasoning are relatively straightforward. They require a predefined set of rules that follow the *modus ponens*, i.e., if a condition P is fulfilled then Q holds ($P \Rightarrow Q$).[1] In the forward-chaining algorithm, for each iteration, the knowledge base is checked against P for each rule. If P of a rule matches then Q of that rule is added to the knowledge base. This process is repeated until no more rules can be applied or depending on the problem, until the query can be answered. Rule-based algorithms including forward and backward chaining are relatively simple to understand and implement, but are not the primary choice when performing inference in OWL DL ontologies.

In 1991, Schmidt-Strauss and Smolka introduced the relatively expressive description language $\mathcal{ALC}$[221], and also provided a procedure that solved the subsumption problem for this particular language for practical problems in acceptable time bounds. As we have seen in Chapter 2, many different extensions of $\mathcal{ALC}$ were subsequently developed, and corresponding extensions of the inference procedure first presented for $\mathcal{ALC}$ were devised to handle the increased expressivity of these languages.[16] All of those variants were essentially specialized versions of the *tableau calculus* of first order logic, and the term *tableau algorithms* is now commonly used to refer to them.[2]

The tableau algorithm is the basis for most software for inference in OWL DL ontologies, and forms the starting point for inference algorithms used in ontology software such as FaCT++ [252] and Pellet [232]. Tableau algorithms are able to perform complete and sound inference for all of the inference rules covered by OWL DL ontologies. We will present one version of this algorithm without making any attempt to describe the bells and whistles of actual software implementations as found in mature ontology software, or to discuss current developments of the algorithm such as hypertableau reasoning [175]. Nevertheless, readers will come away from this chapter with basic conceptions about how inference is done on a computational level.

15.1   The Tableau Algorithm

In this section we explain the basics of the tableau algorithm, which actually represents a family of related algorithms. For didactic purposes, we will present the tableau algorithm for reasoning in the $\mathcal{ALC}$ description logic,

---

[1]Note that P may be either a single rule or a conjunction of several rules: $P = P_1 \sqcap P_2 \sqcap \dots \sqcap P_n$ for $n \geq 1$.

[2]It turned out that the algorithm of [221] was equivalent to a previously developed tableau algorithm for K, a formal logic language that marks the foundation of many modal logics [218].

which was presented in Chapter 2. The algorithm presented here can easily be extended to handle other description logics as well as OWL DL ontologies. Recall from previous chapters how the language elements of $\mathcal{ALC}$ and other description logics can be combined to form complex class constructors:

Hogwarts Wizardry Instructor	`Wizard ⊓ HogwartsStaff ⊓ Teacher⊓`
who is married to a muggle	`∃ married_to.Muggle ⊓`
and teaches at least 12 pupils	`≥ 12teaches.Pupil ⊓`
all of whom are Wizards	`∀ teaches.Wizard`

Inference is presented in the setting of a knowledge base that contains a number of asserted (explicit) axioms (sentences or statements) from which additional axioms can be inferred. As already mentioned in Chapter 2, a distinction is drawn between axioms in the TBox and axioms in the ABox.

TBox and ABox

The terms "ABox" and "TBox" are used to describe two different kinds of statements in ontologies or knowledge bases. Axioms in the TBox (terminological box) describe the concept hierarchy (subsumption and other relations between classes), while axioms in the ABox (assertional box) specify which classes individuals belong to.

The TBox contains
axioms for defining concepts:
   `HogwartsWizardryInstructor ≡ Wizard ⊓ HogwartsStaff ⊓ Teacher`

The ABox contains axioms that describe the properties of individuals:
Class membership:
   `Wizard(Harry)`

Relations:
   `attends(Harry,Hogwarts)`

In the setting of the tableau algorithm, inference procedures are usually formulated in terms of queries to a knowledge base that contains all of the statements of the ontology. Typical queries include whether two classes are equivalent or whether there is a subsumption relation between them. An overview of different query types including their formal syntax in description logic is given in Table 15.1. Regardless of which form the query is, all of them can be reduced to a test for the (un)satisfiability of the knowledge base that is extended by a transformation of the query that is referred to as the *goal*.

For instance, if we would like to infer that class C is a subclass of class D, we can formulate a query of whether the axiom $(C⊓¬D)(a)$ is satisfiable. Thus, we try to identify an individual a that is an instance of the class C but not an instance of the class D. If it is impossible to find such an individual, we can conclude that whenever an individual is an instance of class C, the individual

Query	Goal	Description
$\bot$	$\top$	Is the knowledge base inconsistent?
$C \sqsubseteq D$	$(C \sqcap \neg D)(a)$	Is class C a subclass of class D?
$C \equiv D$	$(C \sqcap \neg D)(a)$, $(D \sqcap \neg C)(a)$	Are two classes C and D equivalent?
$C \sqcap D \equiv \bot$	$(C \sqcap D)(a)$	Are class C and D disjoint?
$C \equiv \bot$	$C(a)$	Is class C inconsistent?
$C(a)$	$\neg C(a)$	Is a a member of C? (instance checking)

TABLE 15.1: **Typical simple queries against a DL knowledge base.** In order to answer whether a knowledge base entails a given query, we add the goal to the knowledge base, which is then tested for unsatisfiability.

Task	Description
Retrieval problem	Determines all known instances of a class C
Realization problem	Find the most specific concepts to a given individual a
Least common subsumer	Find the most specific concept that subsumes a given set of concepts

TABLE 15.2: **Typical complex queries.** Complex queries can typically be reduced to the application of other queries. For instance, the retrieval problem for a class C can be implemented by applying instance checking for class C for each known individual.

must also be an instance of class D, which is equivalent to saying that C is a subclass of D. More complex queries, such as those given in Table 15.2, can be reduced to procedures involving instance checking but also more optimized approaches may exist.

Formally, given a class description $\mathcal{C}_0$ (such as those in the left column of Table 15.1), the tableau algorithm tries to construct a finite interpretation that satisfies $\mathcal{C}_0$, i.e., that contains an element x_0 such that x_0 is an instance of $\mathcal{C}_0$ according to the interpretation [16]. The tableau algorithm does this by applying a number of consistency-preserving transformation rules that we will meet in the next section. Before the rules can be applied, the axioms of the ontology must be transformed into a standard form.

15.1.1 Negative Normal Form

For simplicity, the tableau algorithm assumes that classes of the knowledge base are given in the *negative normal form* (NNF). A formula is in NNF if the negation symbol $\neg$ is only present before literals. If this is not the case, any knowledge base can be converted into the NNF by applying DeMorgan's laws and the fact that the negation is nullified if it is applied twice consecutively. The NNF of an arbitrary formula F is defined using the following rules.

$$\text{NNF}(A) = A \tag{15.1}$$
$$\text{NNF}(\neg A) = \neg A \tag{15.2}$$
$$\text{NNF}(\neg\neg C) = \text{NNF}(C) \tag{15.3}$$
$$\text{NNF}(C_1 \sqcup C_2) = \text{NNF}(C_1) \sqcup \text{NNF}(C_2) \tag{15.4}$$
$$\text{NNF}(C_1 \sqcap C_2) = \text{NNF}(C_1) \sqcap \text{NNF}(C_2) \tag{15.5}$$
$$\text{NNF}(\neg(C_1 \sqcup C_2)) = \text{NNF}(\neg C_1) \sqcap \text{NNF}(\neg C_2) \tag{15.6}$$
$$\text{NNF}(\neg(C_1 \sqcap C_2)) = \text{NNF}(\neg C_1) \sqcup \text{NNF}(\neg C_2) \tag{15.7}$$
$$\text{NNF}(\forall R.C) = \forall R.\text{NNF}(C) \tag{15.8}$$
$$\text{NNF}(\exists R.C) = \exists R.\text{NNF}(C) \tag{15.9}$$
$$\text{NNF}(\neg\forall R.C) = \exists R.\text{NNF}(\neg C) \tag{15.10}$$
$$\text{NNF}(\neg\exists R.C) = \forall R.\text{NNF}(\neg C) \tag{15.11}$$

The following example demonstrates how the rules can be used to transform a formula that is not in NNF to an equivalent formula that is.

Example 15.1 *The $\mathcal{ALC}$ formula $F = \neg(C \sqcap \neg(D \sqcup E))$ is not in valid NNF syntax, because there are two negations of nonliterals: there are two negation ($\neg$) symbols used directly preceding parentheses. By recursively applying NNF on F we obtain:*

$$\begin{aligned}
\text{NNF}(F) &= \text{NNF}(\neg(C \sqcap \neg(D \sqcup \neg E))) \\
&= \text{NNF}(\neg C) \sqcup \text{NNF}(\neg\neg(D \sqcup \neg E)) && \text{by (15.6)} \\
&= \text{NNF}(\neg C) \sqcup \text{NNF}(D \sqcup \neg E) && \text{by (15.3)} \\
&= \text{NNF}(\neg C) \sqcup \text{NNF}(D) \sqcup \text{NNF}(\neg E) && \text{by (15.4)} \\
&= \neg C \sqcup D \sqcup \neg E && \text{by (15.2)}
\end{aligned}$$

15.1.2 Algorithm for ABox

The idea of the tableau algorithm is to make the implicit explicit by transforming the expressions of the knowledge base step by step. To answer satisfiability, the objective is to find an obvious contradiction (also called *clashes*) within the knowledge base. A knowledge base contains an obvious contradiction if both $C(a)$ and $\neg C(a)$ can be logically derived for at least a single C. To simplify the presentation, we initially assume that all expressions of the knowledge base are assertional axioms.

Tableau

The word tableau (plural form: tableaux) originally comes from a French word meaning "small table," and has several meanings in English including a pause at the end of a scene of a play when all the performers briefly freeze in position. This led to the usage of the word in the field of logic to refer to a diagram constructed to demonstrate the consistency of a set of statements (the various statements of an ontology are thus imagined to be arranged similar to the stage full of actors).

In the current context, a *tableau* refers to a set of branches representing sets of asserted and inferred axioms of an ontology. A *branch of a tableau* is a finite set of expressions of form $C(a)$ and $R(a, b)$, where C is a class given in NNF of DL. A branch is said to be *closed* if it contains contradictions, i.e., both $C(a)$ and $\neg C(a)$. If the tableau contains such a contradiction, it is said to contain a *clash*. If all branches of a tableau are closed, then the tableau is said to be *closed*. Otherwise, the tableau is said to be *open*. A branch of a tableau is *complete* if every formula on it has been processed. A tableau is completed if all of its branches are complete.

The procedure is carried out by a function that we denote as TBL. The procedure applies the rules of Table 15.3 to the axioms of the ontology until no more rules can be applied. The tableau algorithm can be most easily described in terms of two sets, one representing the processed axioms ($\mathcal{P}$), and one represent the axioms that are waiting to be processed ($\mathcal{A}$): TBL($\mathcal{P}, \mathcal{A}$). Before processing starts, $\mathcal{P}$ is empty and $\mathcal{A}$ contains all of the axioms of the ontology. The tableau algorithm uses the rules described in Table 15.3 to transform the axioms in $\mathcal{A}$, and adds transformed axioms to $\mathcal{A}$. Each of the rules (15.13–15.19) has the effect of choosing and removing an axiom from $\mathcal{A}$ and adding the axiom to $\mathcal{P}$. Rules (15.16–15.19) additionally add new axioms to $\mathcal{A}$ for axioms that describe unions or intersections of classes as well as existential or universal restrictions. Rule (15.17) describes the behavior needed to deal with class unions, which essentially introduces a branch into the processing of the axioms that reflects the "either-or" nature of the axiom. $(C \sqcap D)(a)$ implies that either $C(a)$ or $D(a)$ or both are true. If the tableau algorithm processed the ontology until no more rules apply to $\mathcal{A}$, i.e., $F = \varnothing$, then it remains to be tested whether there is a contradiction in the axioms in $\mathcal{P}$. This is done based on Rule (15.12).

The following describes the rules in more detail. Note that the axioms that are a result of the processing rules are only added to $\mathcal{P}$ if they are not already present. This is expressed using set union notation (see Figure 2.1 on page 28). The set subtraction notation is used to express the procedure for removing an axiom to $\mathcal{A}$.

Tail of recursion. Rule (15.12) marks the tail of the recursion. Here, the current expression denoted as F within Table 15.3 is empty because no

Condition: $F \in \mathcal{A}$	Action: call TBL$(\mathcal{P}, \mathcal{A})$
$\varnothing$	$\text{return} \begin{cases} \texttt{true} & \text{if } C(a) \in \mathcal{P} \wedge \neg C(a) \in \mathcal{P} \\ \texttt{false} & \text{otherwise} \end{cases}$ \hfill (15.12)
$A(a)$	TBL$(\mathcal{P} \cup \{F\}, \mathcal{A} \smallsetminus \{F\})$ \hfill (15.13)
$\neg A(a)$	TBL$(\mathcal{P} \cup \{F\}, \mathcal{A} \smallsetminus \{F\})$ \hfill (15.14)
$R(a,b)$	TBL$(\mathcal{P} \cup \{F\}, \mathcal{A} \smallsetminus \{F\})$ \hfill (15.15)
$(C \sqcap D)(a)$	TBL$(\mathcal{P} \cup \{F\}, (\mathcal{A} \cup \{C(a), D(a)\}) \smallsetminus \{F\})$ \hfill (15.16)
$(C \sqcup D)(a)$	TBL$(\mathcal{P} \cup \{F\}, (\mathcal{A} \cup \{C(a)\}) \smallsetminus \{F\})$ \hfill (15.17) and TBL$(\mathcal{P} \cup \{F\}, (\mathcal{A} \cup \{D(a)\}) \smallsetminus \{F\})$
$(\exists R.C)(a)$	TBL$(\mathcal{P} \cup \{F\}, (\mathcal{A} \cup \{R(a,b), C(b)\}) \smallsetminus \{F\})$ \hfill (15.18)
$(\forall R.C)(a)$	TBL$(\mathcal{P} \cup \{F\}, (\mathcal{A} \cup \{C(b)\}) \smallsetminus F)$ \quad $\forall b : R(a,b) \in \mathcal{P} \cup \mathcal{A}$ \hfill (15.19)

TABLE 15.3: Tableau algorithm for ABoxes of $\mathcal{ALC}$. TBL is a recursive function that accepts two sets and returns **true** if the knowledge base is satisfiable and **false** otherwise. The algorithm successively processes axioms in $\mathcal{A}$, adds processed axioms back to $\mathcal{A}$ and the original axioms are added to the processed sets of axioms $\mathcal{P}$. Once an axiom is in $\mathcal{P}$, it is not allowed to apply a rule on it again. Otherwise, the TBL function is called recursively with the arguments shown in the right column. See text for details.

more axioms are available for processing in $\mathcal{A}$. We evaluate whether the set of decomposed expressions in $\mathcal{P}$ contains a contradiction. If that is the case we return **false**, otherwise **true**.

Atomic expression. If the selected expression F is atomic, one of the Rules (15.13-15.15) applies. F is simply added to the processed set as there is nothing that can be decomposed and we recursively call TBL with F being removed from $\mathcal{A}$.

Intersection. If the selected expression F has an instance a of a class that is defined as the intersection of two classes C and D, then a must be an instance of both C and D. Thus, Rule (15.16) causes two new assertions $C(a)$ and $D(a)$ to be inserted into $\mathcal{A}$ and simultaneously causes the expression $F = (C \sqcap D)(a))$ to be added to $\mathcal{P}$ and removed from $\mathcal{A}$.

Union. If an individual a is a member of a union of two class expressions C and D, i.e., Rule (15.17) applies, then $C(a)$ or $D(a)$ is true. We therefore split the current branch of the tableau into two branches by adding the expression $C(a)$ to one and $D(a)$ to the other. This can be implemented by two recursive calls of TBL using the arguments shown in Table 15.3. If either or both of the new branches is satisfiable, then **true** is returned.

Existential quantification. In Rule (15.18) we encounter the expression $(\exists R.C)(a)$. This means that a member of a class is instantiated that is constructed via an existential quantification. From the instantiation, we know that there must be a member b of class C that is the object of relation R with a. Thus, we add $C(b)$ and $R(a, b)$ to the unprocessed set, whereby b denotes an instance of the class C that has not yet been used (named) in the ontology.

Universal quantification. In Rule (15.19) we handle the translation step for the universal quantification $(\forall R.C)(a)$. For each relation $R(a, b)$ in the tableau, we add $C(b)$ to the unprocessed set.

Lemma 15.1 *If the tableau derived from a knowledge base by applying rules of Table 15.3 is closed, the knowledge base is unsatisfiable (inconsistent). On the other hand, if the tableau is complete and open, then the knowledge base is satisfiable (consistent).*

The lemma, of which the proof can be found in [16], implies that the transformation rules can be used to show whether a concept $\mathcal{C}$ is satisfiable in a finite interpretation. We can use the algorithm to ask queries by processing appropriate statements (the goals shown in Table 15.1). The basic strategy is to investigate whether an axiom that contradicts the query can be added to a knowledge base without causing a contradiction. For instance, if we want to know whether a knowledge base entails that an individual a is an instance of a class C, we add the contradictory axiom, $\neg C(a)$ to the knowledge base and use the tableau algorithm to test the consistency of the resulting knowledge base. If the tableau algorithm returns an inconsistency, then we can infer that the knowledge base does in fact entail that a is an instance of C.

Example 15.2 *Consider the ABox* $A_0 = \{(\neg\exists R.(D \sqcup E))(a), R(a,b)\}$ *with two individuals a and b. We want to check whether it follows that b is an instance of D, i.e., $D(b)$. According to Table 15.1 this is the instance checking problem, therefore we add $\neg D(b)$ to the ABox. Then, each axiom is converted to the NNF. This gives* $A_1 = \{(\forall R.(\neg D \sqcap \neg E))(a), R(a,b), \neg D(b)\}$. *We show the trace of an invocation of the recursive function* TBL. *Note how $\mathcal{P}$ is initially empty, and $\mathcal{A}$ initially contains all of the original axioms of the knowledge base. i.e., $\mathcal{A} = A_1$. With each step, axioms are shifted or created until* A *is empty or there is a contradiction in $\mathcal{P}$.*

$$\text{TBL}(\varnothing, \{(\forall\boldsymbol{R}.(\neg\boldsymbol{D} \sqcap \neg\boldsymbol{E}))(\boldsymbol{a}), R(a,b), \neg D(b)\}) \overset{(15.19)}{=}$$

$$\text{TBL}(\{(\forall R.(\neg D \sqcap \neg E))(a)\}, \{\boldsymbol{R(a,b)}, \neg\boldsymbol{D(b)}, (\neg D \sqcap \neg E)(b)\}) \overset{(15.14),(15.15)}{=}$$

$$\text{TBL}(\{(\forall R.(\neg D \sqcap \neg E))(a), R(a,b), \neg D(b)\}, \{(\neg\boldsymbol{D} \sqcap \neg\boldsymbol{E})(\boldsymbol{b})\}) \overset{(15.16)}{=}$$

$$\text{TBL}(\{(\forall R.(\neg D \sqcap \neg E))(a), R(a,b), \neg D(b), (\neg D \sqcap \neg E)(b)\}, \{\neg\boldsymbol{D(b)}, \neg\boldsymbol{E(b)}\}) \overset{(15.14)}{=}$$

$$\text{TBL}(\{(\forall R.(\neg D \sqcap \neg E))(a), R(a,b), \neg D(b), (\neg D \sqcap \neg E)(b), \neg E(b)\}, \varnothing) \overset{(15.12)}{=} 0$$

The rules that were applied in each step are shown in bold. Obviously, no clash can be detected, therefore TBL *returns 0 and b is not an instance of D. It is easy to see that if one would ask explicitly whether b is not an instance of D, i.e., $\neg D(b)$, by adding the goal $D(b)$, then* TBL *would identify an obvious contradiction between $D(b)$ and $\neg D(b)$. Thus, the knowledge base should entail that b cannot be an instance of D.*

A tableau for $\mathcal{ALC}$ and other description languages is often implemented computationally as a set of directed graphs. For each directed graph, each individual a of the knowledge base is represented by a separate node of the graph. Each node a is labeled with $\mathcal{L}(a)$, which is a set that contains all axioms to which a is asserted. Edges between two individuals a and b contain the names of the binary relations in which they participate as subject and object. Using this representation, the tableau algorithm then can be seen as procedure that acts on nodes and edges of the graph. Rules (15.13-15.15) are NOPs.[3] Rule (15.16) relabels the matching node a to add the expression C and D to $\mathcal{L}(a)$. Rule (15.17) relabels the matching node a such that the current graph is duplicated, and $\mathcal{L}(a)$ is extended by expression C in the first graph, and expression D is added to $\mathcal{L}(a)$ of the other graph.[4] Rule (15.18) adds a new node b to the graph and connects a and b via a new edge for R. For node a, rule (15.19) adds C to $\mathcal{L}(b)$ if edge (a,b) contains R. These rules are applied until the tableau is saturated.

---

[3]No operation performed.

[4]In an implementation, one would not duplicate the entire graph. Instead, one would usually decide for one path and then trace back to this point if a clash is found.

Example 15.3 *We take the knowledge base of Example 15.2 and we want to query whether b is not an instance of D, i.e., whether the knowledge base entails ¬D(b). We therefore add the goal D(b) to the knowledge base. Transforming this knowledge base to the graphical representation yields two nodes a and b with $\mathcal{L}(a) = \{\forall R.(\neg D \sqcap \neg E)\}$ and $\mathcal{L}(b) = \{D\}$. We also have an edge from a to b due to R(a,b) being in the knowledge base. We then apply our rules in order to decompose the expressions. This can be visualized as follows:*

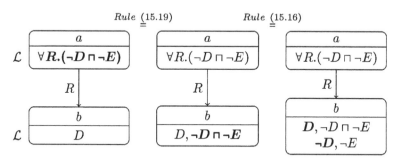

Thus, from the initial representation of the knowledge base, we apply first Rule (15.19) and then Rule (15.16). As expected in the previous example, we get an obvious contradiction for individual b, which means that b is indeed not an instance of D.

15.1.3   Adding Support for the TBox

Recall that the TBox contains expressions of form $C \sqsubseteq D$, and that $C \sqsubseteq D$ can be equivalently written as $\neg C \sqcup D$ (see Chapter 2). We can then convert the terminological axioms to assertional ones by adding $(\neg C \sqcup D)(a)$ to the tableau for each individual a. This is also done for individuals that are newly added to the tableau because of applying the existential quantification rule, i.e., Rule (15.18). Within the graph representation this means that we add $\neg C \sqcup D$ into each existing and newly created node. As described so far, the modified algorithm may lead to an endless loop due to Rule (15.18), as can be seen in the following example.

Example 15.4 *Consider the knowledge base $K_0 = (\mathcal{T}, \mathcal{A})$ with $\mathcal{T} = \{\top \sqsubseteq \exists R.C\}$ and $\mathcal{A} = \{C(a)\}$. First, we bring K_0 into NNF. Thus $K_1 = (\{\neg\top \sqcup \exists R.C\}, \{C(a)\}) = (\{\exists R.C\}, \{C(a)\})$. We then apply the TBox rule for individual a, which yields:*

$$K_2 = (\{\exists R.C\}, \{C(a), (\exists R.C)(a)\})$$

Then we apply Rule (15.18) to get:

$$K_3 = (\{\exists R.C\}, \{C(a), (\exists R.C)(a), R(a, a_1), C(a_1)\})$$

Obviously, we can apply the TBox rule another time for the new individual

a_1. *The resulting expression* $(\exists R.C)(a_1)$ *yields another new individual* a_2, *on which we can apply the TBox rule again. Hence, the algorithm always chooses between the TBox rule and existential quantification rule ad infinitum.*

The example shows that Rule (15.18) needs to be rewritten to avoid infinite loops.

Definition 15.1 *An expression* $F = (\exists R.C)(a)$ *for a given ABox* $\mathcal{A}$ *is blocked, if there is an individual* b *with* $\{C|C(a) \in \mathcal{A}\} \subseteq \{C|C(b) \in \mathcal{A}\}$.

Intuitively, we call the expression $F = (\exists R.C)(a)$ blocked, if the addition of a new member b does not yield any new information. Using this definition, we reformulate Rule (15.18) as follows:

$$\{(\exists R.C)(a)\} = \begin{cases} \text{TBL}(\mathcal{P} \cup \{F\}, \mathcal{A} \setminus \{F\}), & \text{if } F \text{ is blocked} \\ \text{TBL}(\mathcal{P} \cup \{F\}, (\mathcal{A} \cup \{R(a,b), C(b)\}) \setminus \{F\}) & \text{otherwise.} \end{cases}$$

(15.20)

Example 15.5 (Cont. of Example 2.3 from page 32) *Denoting the relation hasDomain(x,y) as $h(x,y)$, suppose that we want to turn Example 2.3 into a knowledge base with further information that a particular individual y is of type DNA binding domain. Another individual x has the domain y. Using the $\mathcal{ALC}$ syntax we can express this via axioms $D(y)$ and $h(x,y)$, which we add to the ABox of the knowledge base. We then have*

$$K = \{T \equiv \exists h.D, D(y), h(x,y)\}$$

for the entire knowledge base.

The hypothesis is that x is of type transcription factor, i.e., $T(x)$. In order to verify that $T(x)$ follows from the knowledge base, we temporarily add the goal $\neg T(x)$ to K and check whether the modified K is unsatisfiable. The TBox axiom $T \equiv \exists h.D$ is the same as to say $T \sqsubseteq \exists h.D$ and $\exists h.D \sqsubseteq T$, which is equivalently expressed by formulas $\neg T \sqcup \exists h.D$ and $\neg\exists h.D \sqcup T$. Denoting all axioms in NNF we have

$$K = \{\neg T \sqcup \exists h.D, \forall h.\neg D \sqcup T, D(y), h(x,y), \neg T(x)\}.$$

We have to assert the two TBox statements with the two existing individuals x and y. The ABox thus is

$$K = \{h(x,y), (\neg T \sqcup \exists h.D)(x), (\forall h.\neg D \sqcup T)(x), \neg T(x),$$
$$(\neg T \sqcup \exists h.D)(y), (\forall h.\neg D \sqcup T)(y), D(y)\}$$

Focusing on x, we can apply Rule (15.17) two times which yields four different branches. We can immediately close two of them, namely those in which we choose $T(x)$ as this is an obvious contradiction to $\neg T(x)$. Thus, for x, *we need to consider* $\{\neg T(x), (\forall h.\neg D)(x)\}$ *and* $\{(\exists h.D)(x), (\forall h.\neg D)(x)\}$. *In both branches, we apply Rule (15.17) on* $(\forall h.\neg D)(x)$ *and* $h(x,y)$, *which gives* $\neg D(y)$. *This is in contradiction to* $D(y)$, *so both branches can be closed as well. As all branches are closed now, the knowledge base is unsatisfiable and thus it entails that* x *is indeed a transcription factor.*

Notice that in the last example we would encounter more branches if we would have started to apply the rules on individual y first. However, as all of them would have been eventually closed with a clash, the result would have been the same. This is generally the case: the algorithm is *complete*[5] regardless of which rule is applied first. However, it is not possible to determine the path that yields the result using the least number of steps in advance.

As it turns out, the test for satisfiability for $\mathcal{ALC}$ has been proven to be PSPACE-complete [221] with respect to combined complexity which includes the size of the ABox and the TBox. This is a worse result than one would usually expect to be practicable. Yet, for real world problems, in which the concept definitions consist only of few other concepts, optimized implementations of the algorithm are employed that are more usable than the general complexity result initially suggests. Also, by reducing the set of possible constructors, which in turn decreases the expressivity of the language, one can obtain algorithms that have much better complexity. For instance, if we derive a language from $\mathcal{ALC}$, in which we disallow the usage of constructors for the univeral quantification ($\mathcal{A}$), complex class negation ($\mathcal{C}$) as well as conecpt union ($\mathcal{U}$), we get $\mathcal{EL}$, which is the base of many bio-ontologies and of which traceable algorithms are known.

The described tableau algorithm is a progenitor of a whole family of algorithms. In order to support more features of description logics, one can extend the procedure by adding new rules. Often also the blocking definition has to be refined in this process. An overview of the complexities of satisfiability algorithms for other description logics is given in Table 15.4.

15.2 Developer Libraries

For integrating inference in software projects there is no need to implement the more involved algorithms such as the described tableau procedure on its own. In fact, there are a bunch of software libraries available that can be used to equip software with features of the Semantic Web.

Jena [50] is Java-based framework for RDF, RDFS, OWL and SPARQL. It comes with a rule-based inference algorithm, but provides necessary interfaces such that other reasoners can be easily hooked in.

---

[5] An inference procedure is complete, if every consequence can be correctly drawn.

DL Language	Complexity
$\mathcal{AL}$	P
$\mathcal{EL}$	P
$\mathcal{EL}^{++}$	P
$\mathcal{ALC}$	$PSPACE$-complete
$\mathcal{SHIF}$	$ExpTime$-complete
$\mathcal{SHOIN}$	$NExpTime$-complete
$\mathcal{SROIQ}$	$NExpTime$-hard and decidable

TABLE 15.4: Complexity of inference algorithms for selected description logics. The given complexities for $\mathcal{ALC}$ and $\mathcal{SHIF}$ only hold if the TBox contains no cycle. Note that $P \subseteq NP \subseteq PSPACE \subseteq ExpTime \subseteq NExpTime$ and the suffix -hard provides a lower bound.

The *OWL API* [122] is a specification of a Java programming interface that helps developers to take advantage of ontologies within their application without the need to replicate common, often boring tasks. This includes the reading and storing of ontologies in different formats (which encompasses OWL but also OBO) as well as manipulating them in-memory. The API also provides a common interface to various popular reasoners, for instance to *FaCT++* or *Pellet*. In the Exercises, we will make use of this versatile API.

FaCT++[6] is an actively developed LGPL-ed reasoner whose core is written in the C++-language.[252] At this writing, version 1.5.0 is current which supports most features of $\mathcal{SROIQ}$ including support for concrete domains for simple data types. In the past, FaCT++ was also used as reference implementation for many recently introduced extension of description logics.

Pellet[7] is a Java-only reasoner based on the tableau algorithm.[232] Like FaCT++ it supports $\mathcal{SROIQ}$ with simple data types. Pellet also supports OWL 2 profiles such as OWL2-EL and DL-safe rules.

Hermite[8] is the first reasoner that is based on the hypertableau calculus [174] that promises reasonable running time also for large-sized ontologies. Like *Pellet* it supports DL-safe rules.

TrOWL [249] is a Java-based reasoning infrastructure for OWL2. It comes with an interface to reasoners such as FaCT++ or Pellet but also supports approximate reasoning which sacrifices completeness for traceability (polynomial running time). Depending on the inference question, OWL2-DL ontologies are transformed into the lighter profiles OWL2-QL or OWL2-EL which then can be processed in an efficient manner.

---

[6]http://owl.man.ac.uk/factplusplus/.

[7]http://clarkparsia.com/pellet/.

[8]http://hermit-reasoner.com/.

15.3   Exercises and Further Reading

Comprehensive coverage of computational inference would require at least another book on its own. In this chapter, we have concentrated on the tableau algorithm, which forms a basis for most of the software currently used for inference in OWL ontologies. An article by Franz Baader reviews the history of the development of inference algorithms and is a good place to start for readers who would like to know what classes of inference algorithms exist and what relationships they have with one another [16].

The Description Logic Handbook: Theory, Implementation, and Applications, edited by Franz Baader and colleagues, not only covers many several introductory chapters for description logics but also includes a comprehensive description of tableau algorithms including their complexity analyses [15]. The *Description Logic Complexity Navigator* is an interactive Web resource that provides very convenient access to complexity results for various description logics and has a large list of references to original works. It can be accessed via http://www.cs.man.ac.uk/~ezolin/dl/.

Exercises

The exercises for this chapter will demonstrate how to use the OWL API, which is an extremely useful application programming interface (API) for working with OWL 2 ontologies. The OWL API is written in Java and can parse and render OWL 2 ontologies in the serializations that are described in Appendix D. OWL API can be used for developing OWL-based applications by means of a set of interfaces for inspecting, manipulating and reasoning with OWL ontologies [122]. One of the most attractive features of OWL API is that it has various interfaces for interacting with OWL reasoners, with which programs can check the consistency of ontologies, check for unsatisfiable classes, compute class and property hierarchies, and check which axioms are entailed by an ontology. OWL API currently has interfaces to reasoners including the before-mentioned FaCT++, Pellet, and HermiT.

The exercises involve modifying a Java program that will be used to analyze the Harry Potter ontology that was used to illustrate the OWL 2 inference in Chapter 14. You will need to have a recent version of Java (1.6 or newer) installed on your computer.[9] We have provided some additional explanations for readers who may be less familiar with the Java programming language.

Getting Started with the OWL API. Go to the OWL API homepage[10] and download the latest binary release (at the time of this writing, version

---

[9]See http://www.oracle.com/technetwork/java/index.html for downloads of the Java SE Development Kit if needed.

[10]http://owlapi.sourceforge.net/.

3.2.2 from February 17, 2011, was the current release). Unzip the archive; we need the jar (Java archive) file `owlapi-bin.jar` for these exercises.

Getting Started with OWL Reasoners: HermiT. Go to the HermiT Web page mentioned above and download the current stable version (at the time of this writing, version 1.3.1 was the current version). Unzip the downloaded archive. We need the file `HermiT.jar` for these exercises. As we will see below, the OWL API provides a simple interface for HermiT that allows the HermiT reasoner to be used in application programs through a standard interface. If desired, you can use any of the other reasoners mentioned in this chapter (a few others are described in the OWL API documentation as well) with very minor modifications to your code (see the OWL API documentation for details).

Compiling and Running the Program

For the exercises in this chapter, we will construct a Java program that makes use of the OWL API and the HermiT reasoner to perform various tasks. We will use the Harry Potter ontology from Chapter 14, but the code can easily be adapted to work with arbitrary ontologies. We will assume that the files `owlapi-bin.jar`, `HermiT.jar`, and `Potter.owl` are all in the current directory; otherwise, the paths will need to be adjusted accordingly. We will write a Java program consisting of a single class in a single file called `OWLAPI.java`. The exercises will ask the reader to create or modify functions to perform various tasks. The code for this program is based on LGPL'ed code by Matthew Horridge, one of the authors of OWL API [122]. The OWL API Web site contains additional code examples.

To compile the program, enter the following at the command line:

```
$javac -cp .:HermiT.jar OWLAPI.java
```

To run the program, enter the following:

```
$java -cp .:HermiT.jar OWLAPI Potter.owl
```

We will now explain the basic framework of the Java code for readers who are less familiar with Java. Java-savvy readers can skip directly to the first exercise. The source code of the Java program for these exercises can be downloaded from the book Web site.[11]

Java programs generally begin with import statements that tell the compiler where to find the class definitions for classes that are not in the default `java.lang` package. `OWLAPI.java` begins with such statements to pull in class definitions from the OWL API and from the HermiT reasoner (these classes

---

[11] http://bio-ontologies-book.org.

are defined in the two jar files mentioned above). In addition, some non-standard Java classes are pulled in for File I/O and to provide an iterator.

```
1  import org.semanticweb.owlapi.apibinding.OWLManager;
2  import org.semanticweb.owlapi.model.*;
3  import org.semanticweb.owlapi.io.*;
4  import org.semanticweb.owlapi.util.*;
5  import org.semanticweb.owlapi.reasoner.*;
6  import org.semanticweb.owlapi.reasoner.impl.OWLClassNode;
7
8  import org.semanticweb.HermiT.Reasoner;
9
10 import java.io.*;
11 import java.util.*;
```

In Java, each file usually contains a single class whose definition follows the import directives and is enclosed in brackets. The class `OWLAPI` contains three class-scope variables that can be used by the functions of the class, as we will see below. Within the body of the class (shown here as ...), there is a `main` method that controls the flow of execution as well as a number of functions.

```
1  public class OWLAPI {
2
3      private OWLOntologyManager manager;
4      private OWLOntology ontol;
5      private OWLReasoner reasoner;
6
7  (...)
8
9  }
```

The main method accepts arguments from the command line via the `args` variable. Here it assigns to the variable `owlfile` the contents of `args[0]`, which will contain the name of the OWL file if the program was called from the command line as shown above. A new instance of the class is created and assigned to the variable `owlapi`. Each of the tasks the program is implemented in a function. For instance, the function `parseOntology` is used to parse the file containing the ontology. The reader will create or modify such functions and should put them where (...) is shown in the following listing. Note that Java uses a system of exceptions to catch run time errors. The catch clauses are called if an error corresponding to the exceptions occurs during program execution. To get more information about such errors, programmers can add `e.getMessage()` to the `println` commands.

```
1  public static void main(String[] args) {
2    String owlfile = args[0];
3    OWLAPI owlapi = new OWLAPI();
4
```

```
5   try {
6     owlapi.parseOntology(owlfile);
7     (...)
8   } catch (UnparsableOntologyException e) {
9     System.out.println("Could not parse the ontology.");
10    usage();
11  }
12  catch (OWLOntologyCreationException e) {
13    System.out.println("Could not load ontology.");
14  }
15  catch(UnsupportedOperationException e) {
16    System.out.println("Unsupported reasoner operation.");
17  }
18 }
```

parseOntology is a typical function in our program. This function creates an OWLOntology object that will be used throughout the program.

```
1  public void parseOntology(String filename)
2    throws UnparsableOntologyException,
3  OWLOntologyCreationException  {
4      manager = OWLManager.createOWLOntologyManager();
5      File file = new File(filename);
6      ontol = manager.loadOntologyFromOntologyDocument(file);
7      System.out.println("Loaded ontology: " + ontol);
8  }
```

We can now come to the first exercise.

15.1 Download and compile the Java program OWLAPI as described above. Note that all of the commands within the try block have been commented out except for the one shown above (owlapi.parseOntology). For this exercise, look through the code and try to understand the general structure of the program.

Now compile and run the program and consider the output of the last command of the function owlapi.parseOntology. How many axioms does the ontology contain? How many logical axioms?

The following code creates a reasoner object using the interface to the HermiT reasoner. Note that the name of the reasoner is not used in the createReasoner command. This is determined instead by the import statement as described above. One would modify the import statement to use a different reasoner. The code additionally creates a simple progress monitor that will print progress messages to the console. We use the command precomputeInferences to have the reasoner perform complete inference over the ontology. The final commands cause the reasoner to output a diagnostic message as to whether the ontology is consistent.

```
1 public void createReasoner () {
2   OWLReasonerFactory reasonerFactory =
3     new Reasoner.ReasonerFactory ();
4   ConsoleProgressMonitor progressMonitor =
5     new ConsoleProgressMonitor ();
6   OWLReasonerConfiguration config =
7     new SimpleConfiguration (progressMonitor );
8   reasoner = reasonerFactory.createReasoner (ontol, config );
9
10   reasoner.precomputeInferences ();
11
12   boolean consistent = reasoner.isConsistent ();
13   System.out.println ("Consistent: " + consistent );
14 }
```

15.2 Go to the file OWLAPI.java that you downloaded from the book Web site, and uncomment the following command.

```
1 // owlapi.createReasoner ();
```

Now compile and run the program again. Is the Harry Potter ontology consistent? The following exercises will ask you to uncomment other functions in the main function of OWLAPI.java. You will also be asked to modify some of the functions or to create your own.

OWLAPI.java is able to print the entire class hierarchy of the ontology to the console using recursive functions. A call to the function printFromRoot sets up the recursion by first getting a reference to the root of the ontology from the OWL API function getTopClassNode, and then passing it to the recursive function print. This function simply returns (i.e., skips) if it is called with the bottom node (which contains owl:Nothing and unsatisfiable classes and would appear as a leaf of every other node). Otherwise, it prints two spaces for every level of the ontology and then uses the printNode function (see below) to print the name of the node. Finally, for each of the subclasses of the current node, the print function is called recursively. Thus, the procedure performs a traversal over the entire ontology graph.

```
1 public void printFromRoot () {
2   Node<OWLClass> topNode = reasoner.getTopClassNode ();
3   print (topNode, 0);
4 }
5
6 private void print (Node<OWLClass> parent, int depth) {
7   if (parent.isBottomNode ()) {
8     return;
9   }
10   // Print an indent
```

```
11  for(int i = 0; i < depth; i++)
12    System.out.print("   ");
13
14  printNode(parent);
15  for(Node<OWLClass> child : reasoner.getSubClasses\
16        (parent.getRepresentativeElement(), true)) {
17    print(child, depth + 1);
18  }
19 }
```

The `print` function calls the following function to actually print the IRIs of the nodes of the ontology. Each function call prints a single IRI within curly brackets.

```
1  private void printNode(Node<OWLClass> node) {
2    System.out.print("{");
3      for(Iterator<OWLClass> it =
4        node.getEntities().iterator(); it.hasNext(); ) {
5          OWLClass cls = it.next();
6          System.out.print(cls);
7          if (it.hasNext()) {
8            System.out.print(" ");
9          }
10       }
11       System.out.println("}");
12  }
```

15.3 Uncomment the `printFromRoot` function, compile and run the program. Note that the class hierarchy of the ontology is printed using the full IRIs of the classes. For this exercise, you will be asked to modify the program code such that QNames are displayed instead. To do so, we will make use of the OWL API class `DefaultPrefixManager`. Go to the Javadoc documentation of this class as available on the OWL API Web page or as a part of the OWL API download. Add the following line (line breaks have been added to make it fit here) to the OWLAPI.java file outside of any other function definitions:

```
1 private static DefaultPrefixManager pm = new
2   DefaultPrefixManager
3     ("http://www.example.org/Potter.owl#");
```

This will set the default prefix to the IRI of the Harry Potter ontology as specified in the ontology file. Study the documentation of the **getShortForm** method of the **DefaultPrefixManager** class and use it to output the abbreviated (QName) form of the classes. You will need to modify the **printNode** method.

It is possible to print out the subhierarchy of all descendants of any given ontology term. To do this, we will use the OWL API method called **getSubClasses**. Note that to do this, we create a reference to the class **Adult** in the Harry Potter ontology and pass this to the function **getSubClasses**. If you do not understand the code, consult the Javadoc documentation for OWL API.

```
1 public void printSubclasses() {
2   OWLDataFactory fac = manager.getOWLDataFactory();
3   OWLClass adult = fac.getOWLClass(IRI.create
4     ("http://www.example.org/Potter.owl#Adult"));
5   NodeSet<OWLClass> subClses =
6     reasoner.getSubClasses(adult, true);
7   System.out.println("Printing subclasses of:\n" + adult);
8   OWLClassNode adultN = new OWLClassNode(adult);
9   print(adultN,0);
10 }
```

15.4 Uncomment the **printSubclasses** function call, compile and run the code, and check whether the results correspond to the ontology hierarchy for the complete ontology. Modify the code to obtain the subtree for the class **MagicalBeast**.

We can also use the OWL API to retrieve a list of all instances of any given class, including both explicitly asserted and inferred instances.

```
1 public void getInstances() {
2   OWLDataFactory fac = manager.getOWLDataFactory();
3   OWLClass goblin = fac.getOWLClass(IRI.create\
4     ("http://www.example.org/Potter.owl#Goblin"));
5   NodeSet<OWLNamedIndividual> individualsNodeSet
6     = reasoner.getInstances(goblin, true);
7
8   Set<OWLNamedIndividual> individuals
9     = individualsNodeSet.getFlattened();
10  System.out.println("Instances of : " + goblin);
11  for(OWLNamedIndividual ind : individuals) {
```

```
12      System.out.println(" " + ind);
13    }
14    System.out.println("\n");
15  }
```

15.5 Uncomment the `getInstances` function call in `main`, compile and run the code, and check whether the results correspond to the inferences made by Protégé about the instance of `Goblin` in Chapter 14. Modify the code to obtain all inferences for the class `Pupil`.

We may want to print out a list of all inferred axioms in order to check the results of inference procedures. OWL API provides a number of functions for doing so.

```
1  public void printInferredOntology()
2    throws OWLOntologyCreationException,
3          OWLOntologyStorageException {
4    List<InferredAxiomGenerator<? extends OWLAxiom>> gens=new
5      ArrayList<InferredAxiomGenerator<? extends OWLAxiom>>();
6    gens.add(new InferredSubClassAxiomGenerator());
7    OWLOntology infOnt = manager.createOntology();
8    InferredOntologyGenerator iog =
9    new InferredOntologyGenerator(reasoner, gens);
10   iog.fillOntology(manager, infOnt);
11
12   Set<OWLAxiom> axioms = infOnt.getAxioms();
13   System.out.println("Inferred Axioms:");
14   for(OWLAxiom ax: axioms) {
15     System.out.println("  " + ax);
16   }
17  }
```

15.6 Uncomment the `printInferredOntology` function call in `main`, compile and run the code. What axioms have been inferred? For this exercise, you will be asked to modify the function to print out a list of all the individuals as well as any inferences that different names represent the same individual. You may wish to use the functions `getIndividualsInSignature` and `getSameIndividualAxioms` for your solution. Consult the Javadoc for the `OWLOntology` Interface of OWL API as needed for information on these functions.

15.7 We trust you have gotten the idea of the basics of OWL API. For this exercise, write a new method for the class `OWLAPI.java` that will retrieve all individuals that are connected to the individual *HarryPotter* by the property `isClassmateOf`. Check the original ontology file. Were all of these relations explicitly asserted? To complete this exercise, the functions `getOWLNamedIndividual`, `getOWLObjectProperty`, and `getObjectPropertyValues` may be helpful. Consult the Javadoc information on these methods if necessary.

Although we have only demonstrated a fraction of the capabilities of OWL API, you should be in a good position to begin to use OWL API to create some interesting software. See the article by Horridge and Bechhofer for further information [122].

Chapter 16

SPARQL

SPARQL stands for *SPARQL Protocol and RDF Query Language.*[1] SPARQL is the W3C recommendation for querying RDF data, providing a querying and search technology designed specifically for the Semantic Web. SQL (Structured Query Language) is a standardized computer language for querying data in relational databases. SPARQL plays a similar role for RDF stores. There are a number of important differences between SQL and SPARQL. SQL is designed to query a single database at a time, whereas SPARQL queries can be distributed in order to search multiple online resources simultaneously, gather the results, and present them to the user, a process known as federated query. In order to generate an SQL query, a user must have knowledge about the structure of the tables of the database in order to formulate JOIN and other SQL statements to retrieve the results. In contrast, SPARQL queries are based on triple patterns, conjunctions, disjunctions, filters, and optional patterns that can be used with any RDF store. SPARQL syntax allows queries to be formulated that are globally unambiguous because of the use of URIs[2]. These features make SPARQL a flexible and extensible framework for searching and representing information about resources on the Web.

SPARQL was made an official W3C recommendation in 2008 [201] and is beginning to be used in the realm of biomedical ontologies. In this chapter, we will present the main features of SPARQL using a small example ontology that can be downloaded from the book's Web site. In the exercises, readers will find a tutorial on how to use the Jena framework to perform SPARQL queries as well as pointers to other SPARQL implementations.

16.1 SPARQL Queries

The SPARQL query language consists of the syntax and semantics for asking and answering queries against RDF graphs (see Chapters 4 and 13 as well

---

[1]SPARQL continues the tradition of recursive acronyms in computer science, such as GNU: GNU's Not Unix.

[2]The SPARQL standard allows Internationalized Resource Identifiers (IRIs) to be used in place of Uniform Resource Identifiers (URIs). IRIs represent a generalization of URIs that allow non-ASCII characters, such as Chinese characters, to be used.

as Appendix C for background on RDF). There is a relatively small number of basic building blocks from which SPARQL queries can be constructed. In this section, we will use a small and simple RDF representation of four proteins to demonstrate how simple SPARQL queries are written.

```xml
<?xml version="1.0"?>
<!DOCTYPE rdf:RDF [
<!ENTITY prot "http://compbio.charite.de/prot#">
]>
<rdf:RDF
xmlns:rdf="http://www.w3.org/1999/02/22-rdf-syntax-ns#"
xmlns:prot="http://compbio.charite.de/prot#">

<rdf:Description rdf:about="&prot;Catalase">
  <prot:name>Catalase</prot:name>
  <prot:length>527</prot:length>
  <prot:EC>1.11.1.6</prot:EC>
  <prot:location>Peroxisome</prot:location>
  <prot:cofactor>Heme</prot:cofactor>
 <prot:cofactor>NADP</prot:cofactor>
</rdf:Description>

<rdf:Description rdf:about="&prot;Acetylcholinesterase">
  <prot:name>Acetylcholinesterase</prot:name>
  <prot:length>614</prot:length>
  <prot:EC>3.1.1.7</prot:EC>
  <prot:location>Synapse</prot:location>
</rdf:Description>

<rdf:Description rdf:about="&prot;Fibrillin-1">
  <prot:name>Fibrillin-1</prot:name>
  <prot:length>2817</prot:length>
  <prot:location>Extracellular matrix</prot:location>
</rdf:Description>

<rdf:Description rdf:about="&prot;SuperoxideDismutase">
  <prot:name>Superoxide dismutase</prot:name>
  <prot:length>154</prot:length>
  <prot:EC>1.15.1.1</prot:EC>
  <prot:location>Extracellular matrix</prot:location>
  <prot:cofactor>Copper</prot:cofactor>
  <prot:cofactor>Zinc</prot:cofactor>
</rdf:Description>

</rdf:RDF>
```

SPARQL statements are built from triple patterns that are like RDF subject-predicate-object triples but have the option of a variable in place of the RDF terms in the subject, predicate or object positions. The following triple pattern has the variable `?x` as the subject. The predicate is enclosed in angled brackets, and the literal "Catalase" is the object.

```
?x  <http://compbio.charite.de/prot#name>  "Catalase"
```

This SPARQL query will try to find triples in the RDF graph that match this triple pattern. The subject `?x` is a variable, and the predicate and object have fixed values. The pattern matches any and all triples in the RDF graph with the corresponding values for the predicate and object.

A complete SPARQL query can use a SELECT, a FROM, and a WHERE clause similar to SQL, as well as a number of SPARQL graph patterns to restrict the answer to the query (the effect of this restriction is roughly similar to that of a JOIN clause in SQL). The following SPARQL query searches for all triples in the RDF graph with the predicate `<http://compbio.charite.de/prot#name>` and the object `"Catalase"`. The FROM clause contains a Uniform Resource Indicator (URI) that indicates a file or other resource that is to be used to form the default graph for the query. In this case, we have simply indicated the name of the file containing the RDF graph for the protein data shown above.

```
SELECT ?x
FROM <proteins.rdf>
WHERE { ?x  <http://compbio.charite.de/prot#name>  "Catalase" }
```

This query returns the following result.

```
-------------------------------------------
| x                                       |
===========================================
| <http://compbio.charite.de/prot/Catalase> |
-------------------------------------------
```

Recall that `rdf:about="http://compbio.charite.de/prot/Catalase"` indicates the subject of the triples represented by the elements enclosed within the `<rdf:Description>` element (Figure 16.1). Therefore, the query has returned the subject of the triple. This works by matching the triple pattern in the WHERE clause against the triples in the RDF graph. The predicate and object of the triple are fixed values so the pattern is going to match only triples with those values. The subject is a variable, and there are no other restrictions on the variable. Thus, the pattern matches any triples with these predicate and object values.

It is possible to use prefix bindings to make the SPARQL queries less cluttered. For instance, the following query is equivalent to the query shown above.

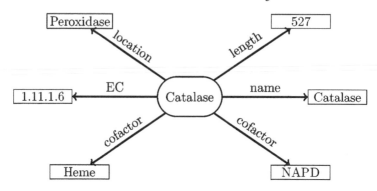

FIGURE 16.1: RDF Triples for Catalase. The figure shows an excerpt of the RDF graph represented by the XML/RDF serialization shown above in this chapter. The URIs have been shorted for legibility. The resource http://compbio.charite.de/prot/Catalase (shown as a gray oval) is the subject of 6 RDF triples, whereby the predicates are represented as edges and the objects as the six nodes pointed to by the edges.

```
PREFIX prot:   <http://compbio.charite.de/prot#>

SELECT ?x
FROM <proteins.rdf>
WHERE { ?x  prot:name  "Catalase" }
```

In order to retrieve the names of all proteins in the file, we modify the SPARQL query.

```
PREFIX prot:   <http://compbio.charite.de/prot#>

SELECT ?name
FROM <proteins.rdf>
WHERE { ?x  prot:name  ?name }
```

This produces the following result.

```
-------------------------
| name                  |
=========================
| "Superoxide dismutase" |
| "Fibrillin-1"          |
| "Acetylcholinesterase" |
| "Catalase"             |
-------------------------
```

SPARQL offers a number of constructs with which to write more complicated queries. Group patterns are interpreted conjunctively (i.e., as AND).

The following query identifies all triples in the RDF graph for which the same subject (?x) forms triples with both the predicates prot.name and prot.length. The query returns one row for all such variable *bindings*.

```
PREFIX prot:  <http://compbio.charite.de/prot#>

SELECT ?name ?len
FROM <proteins.rdf>
WHERE { ?x  prot:name  ?name .
        ?x prot:length ?len  .}
```

This retrieves the following data.

```
------------------------------------
| name                    | len    |
====================================
| "Superoxide dismutase"  | "154"  |
| "Fibrillin-1"           | "2817" |
| "Acetylcholinesterase"  | "614"  |
| "Catalase"              | "527"  |
------------------------------------
```

SPARQL provides mechanisms for filtering the patterns that are retrieved from the RDF graph. One such filter tests strings based on regular expressions. Let's say we would like to retrieve all proteins having the string "fib" with case-insensitive matching ("i").

```
PREFIX prot:  <http://compbio.charite.de/prot#>

SELECT ?name ?len
FROM <proteins.rdf>
WHERE { ?x  prot:name  ?name .
        ?x prot:length ?len  .
        FILTER regex(?name, "fib", "i") }
```

This query retrieves only the entry for Fibrillin-1.

```
---------------------------
| name          | len    |
===========================
| "Fibrillin-1" | "2817" |
---------------------------
```

SPARQL also provides the ability to filter on Boolean conditions. The Boolean expression, which can be much more complex than the one shown here, is enclosed in parentheses following the keyword FILTER.

```
PREFIX prot:  <http://compbio.charite.de/prot#>

SELECT ?n ?len
FROM <proteins.rdf>
WHERE { ?x prot:name ?n .
        ?x prot:length ?len .
        FILTER(?len >= 600) .
      }
```

Note that this query will not produce the expected results with the RDF graph shown above. This is because SPARQL does not "know" that <prot:length>527</prot:length> refers to a number (by default, 527 would be treated as a character string). To fix this problem, we need indicate the datatype (integer). This is done with the XMLSchema specifications [42]. Each of the lines used to indicate the length of the protein needs to be changed to include an attribute specifying the datatype. We will add an entity to the DOCTYPE declaration at the top of the proteins.rdf file.

```
<?xml version="1.0"?>
<!DOCTYPE rdf:RDF [
<!ENTITY prot "http://compbio.charite.de/prot#">
<!ENTITY xsd "http://www.w3.org/2001/XMLSchema#">
]>
```

We can now specify that the value of the <prot:length> element is an integer.

```
<prot:length rdf:datatype="&xsd;integer">614</prot:length>
```

SPARQL now returns only those proteins with a length of at least 600 (amino acids).

```
---------------------------------
| n                     | len  |
=================================
| "Fibrillin-1"         | 2817 |
| "Acetylcholinesterase" | 614  |
---------------------------------
```

Consider now the following query.

```
PREFIX prot:  <http://compbio.charite.de/prot#>

SELECT ?n ?cf
FROM <proteins.rdf>
WHERE { ?x prot:name ?n .
        ?x prot:cofactor ?cf.
      }
```

SPARQL retrieves all triples for the proteins that have cofactors.

```
------------------------------------
| n                      | cf       |
====================================
| "Superoxide dismutase" | "Zinc"   |
| "Superoxide dismutase" | "Copper" |
| "Catalase"             | "NADP"   |
| "Catalase"             | "Heme"   |
------------------------------------
```

In general, it cannot be assumed that RDF graphs have complete information. SPARQL provides a query syntax that enables one to retrieve information if available, but not to reject the query if some part of the query does not match. In our protein RDF graph, some, but not all, of the proteins have cofactors. *Optional graph pattern matching* provides this ability using the keyword OPTIONAL.

```
PREFIX prot:  <http://compbio.charite.de/prot#>

SELECT ?n ?cf
FROM <proteins.rdf>
WHERE { ?x prot:name ?n .
        OPTIONAL {?x prot:cofactor ?cf} .
      }
```

This query now returns a result for Fibrillin-1 and Acetylcholinesterase (which do not have a cofactor), leaving the corresponding column blank.

```
------------------------------------
| n                      | cf       |
====================================
| "Superoxide dismutase" | "Zinc"   |
| "Superoxide dismutase" | "Copper" |
| "Fibrillin-1"          |          |
| "Acetylcholinesterase" |          |
| "Catalase"             | "NADP"   |
| "Catalase"             | "Heme"   |
------------------------------------
```

The UNION keyword forms a disjunction (OR) of two graph patterns. For instance, the following query returns all proteins whose length is more than 1000 or less than 200.

```
PREFIX prot:  <http://compbio.charite.de/prot#>

SELECT ?n ?len
```

```
FROM <proteins.rdf>
WHERE {
        { ?x prot:name ?n .
          ?x prot:length ?len .
          FILTER(?len >= 1000)
        }
        UNION
        { ?x prot:name ?n .
          ?x prot:length ?len .
          FILTER(?len < 200)
        }
      }
```

This query returns a row for Fibrillin-1, which comprises 2817 amino acids, as well as for Superoxide dismutase, which comprises only 154 amino acids.

```
-----------------------------------
| n                      | len  |
===================================
| "Fibrillin-1"          | 2817 |
| "Superoxide dismutase" | 154  |
-----------------------------------
```

There are cases where one is interested only in whether something exists but does not need any more information. The keyword ASK can be used in this situation.

```
PREFIX prot:  <http://compbio.charite.de/prot#>

ASK
FROM <proteins.rdf>
WHERE {    ?x prot:length ?len .
           FILTER(?len >= 10000)
      }
```

This query asks whether the RDF graph has any proteins whose length is at least 10,000. SPARQL answers "no":

```
Ask => No
```

SPARQL provides a method for asking an RDF store for all information surrounding some topic. For instance, if we want to ask for all information regarding "Catalase," we would execute the following query.

```
PREFIX prot:  <http://compbio.charite.de/prot#>

DESCRIBE ?x
WHERE {    ?x prot:name "Catalase" . }
```

This yields the following information.

```
@prefix rdf:      <http://www.w3.org/1999/02/22-rdf-syntax-ns#> .
@prefix prot:     <http://compbio.charite.de/prot#> .


<http://compbio.charite.de/prot/Catalase>
      prot:EC        "1.11.1.6" ;
      prot:cofactor  "NADP" ;
      prot:cofactor  "Heme" ;
      prot:length    527 ;
      prot:location  "Peroxisome" ;
      prot:name      "Catalase" .
```

In this case, we specified the RDF dataset via the command line rather than in the SPARQL query itself. The Exercises will explain how this is done using the Jena framework. This section has by no means covered all the keywords and features of SPARQL. Readers wishing to have more information should consult the sources listed at the end of this chapter.

16.2   Combining RDF Graphs

An important feature of SPARQL is the capability to combine multiple RDF data sources, which is a major step towards making the Web a single, semantically driven database, which is a vision of the Semantic Web. The ability of SPARQL to combine multiple target graphs is thus an effective data integration mechanism.

Several SPARQL syntactic constructs support combining data from different RDF graphs. Here, we will assume there is a second local file called genes.rdf, but RDF stores from anywhere on the Web can be combined[3]

```
<?xml version="1.0"?>
<!DOCTYPE rdf:RDF [
<!ENTITY gene "http://compbio.charite.de/gene#">
<!ENTITY xsd "http://www.w3.org/2001/XMLSchema#">
]>

<rdf:RDF
   xmlns:rdf="http://www.w3.org/1999/02/22-rdf-syntax-ns#"
   xmlns:gene="http://compbio.charite.de/gene#">
```

---

[3]SPARQL implementations can be made to interrogate multiple SPARQL endpoints, which are machine-friendly interfaces to RDF knowledge bases that are designed to interact with calling software at other locations on the Web. This is not the only way for SPARQL implementations to access RDF data.

```
<rdf:Description rdf:about="&gene;Catalase">
  <gene:name>Catalase</gene:name>
  <gene:bp rdf:datatype="&xsd;integer">2300</gene:bp>
  <gene:exon_count rdf:datatype="&xsd;integer">13</gene:exon_count>
  <gene:ID>847</gene:ID>
  <gene:KEGG>Peroxisome</gene:KEGG>
  <gene:KEGG>Tryptophan metabolism</gene:KEGG>
</rdf:Description>

<rdf:Description rdf:about="&gene;Acetylcholinesterase">
  <gene:name>Acetylcholinesterase</gene:name>
  <gene:bp rdf:datatype="&xsd;integer">2225</gene:bp>
  <gene:exon_count rdf:datatype="&xsd;integer">6</gene:exon_count>
  <gene:ID>43</gene:ID>
  <gene:KEGG>Glycerophospholipid metabolism</gene:KEGG>
</rdf:Description>

<rdf:Description rdf:about="&gene;Fibrillin-1">
  <gene:name>Fibrillin-1</gene:name>
  <gene:bp rdf:datatype="&xsd;integer">981</gene:bp>
  <gene:exon_count rdf:datatype="&xsd;integer">66</gene:exon_count>
  <gene:ID>2200</gene:ID>
</rdf:Description>

<rdf:Description rdf:about="&gene;SuperoxideDismutase">
  <gene:name>Superoxide dismutase</gene:name>
  <gene:bp rdf:datatype="&xsd;integer">2225</gene:bp>
  <gene:exon_count rdf:datatype="&xsd;integer">5</gene:exon_count>
  <gene:ID>6647</gene:ID>
  <gene:KEGG>Peroxisome</gene:KEGG>
  <gene:KEGG>Huntington's disease</gene:KEGG>
</rdf:Description>

</rdf:RDF>
```

We can combine the RDF graphs represented by proteins.rdf and genes.rdf by equating the entities of both graphs. For this example, we will compare the prot:name and gene:name attributes. In a real application, both RDF graphs would probably refer to one another explicitly using an accession number or other unique ID.

Since both prot:name and gene:name represent properties for the name of the gene and the protein it encodes, we can write a query using the UNION keyword to retrieve all triples with either a prot:name or a gene:name predicate. The [] variable is a *blank node* (bnode) that is used in only one place in the query. A unique blank node will be used to form the triple pattern [201].

```
PREFIX prot:   <http://compbio.charite.de/prot#>
PREFIX gene:   <http://compbio.charite.de/gene#>

SELECT ?n
FROM <proteins.rdf>
FROM <genes.rdf>
WHERE {
        { [] prot:name ?n } UNION { [] gene:name ?n }
      }
```

This returns a list of the names of all triples matching the query.

```
---------------------------
| n                       |
===========================
| "Superoxide dismutase"  |
| "Fibrillin-1"           |
| "Acetylcholinesterase"  |
| "Catalase"              |
| "Superoxide dismutase"  |
| "Fibrillin-1"           |
| "Acetylcholinesterase"  |
| "Catalase"              |
---------------------------
```

In order to retrieve each name only once, the SELECT ?n clause can be replaced by SELECT DISTINCT ?n. An interesting alternative to this query treats the predicate as a variable that can be matched by either prot:name or gene:name.

```
PREFIX prot:   <http://compbio.charite.de/prot#>
PREFIX gene:   <http://compbio.charite.de/gene#>

SELECT DISTINCT ?n
FROM <proteins.rdf>
FROM <genes.rdf>
WHERE {
    [] ?p ?n .
    FILTER( ?p = prot:name || ?p = gene:name )
      }
```

More involved ways of combining two RDF graphs retrieve different information about each RDF graph. The queries are formulated in terms of the *background graph*, which is specified by the FROM clause, and one or more *named graphs* specified by FROM NAMED clauses. In the following query, this is combined with the GRAPH keyword followed by the URI of the RDF graph referenced by the FROM NAMED clause. The graph pattern within the GRAPH clause is applied to the graph identified by the URI. Thus, within the GRAPH clause, a local variable ?x is used to refer to a subject that matches both triples, and within the rest of the query, the variable ?y is used to match triples within the background graph.

```
PREFIX prot:   <http://compbio.charite.de/prot#>
PREFIX gene:   <http://compbio.charite.de/gene#>

SELECT  ?n ?bp ?len
FROM <proteins.rdf>
FROM NAMED <genes.rdf>
WHERE {
        GRAPH <genes.rdf> {
            ?x gene:name ?n .
            ?x gene:bp ?bp .
        } .
        ?y prot:name ?n .
        ?y prot:length ?len .
        }
```

The query returns information from both RDF graphs in a single table, analogous to the way a JOIN clause can combine two tables in an SQL query.

```
-------------------------------------------
| n                        | bp   | len   |
===========================================
| "Superoxide dismutase"   | 2225 | 154   |
| "Fibrillin-1"            | 981  | 2817  |
| "Acetylcholinesterase"   | 2225 | 614   |
| "Catalase"               | 2300 | 527   |
-------------------------------------------
```

16.3   Conclusions

SPARQL is a relatively simple query language for RDF graphs. According to the specifications of the W3C, valid SPARQL implementations must define the results of queries based on RDF simple entailment (see Chapter 13). Other possible entailment regimes include RDF entailment, RDFS entailment, D-entailment, and OWL-DL entailment. Of these, only OWL-DL entailment restricts the set of well-formed graphs [201]. Individual implementations are allowed to support any or all of these types of entailment and still be considered compliant with the W3C standards for SPARQL. At the time of this writing, many implementations do not support more than RDF simple entailment, and thus are not able to perform queries on the basis of subclass relations using the `rdfs:subClassOf` relation.

It is highly likely that intelligent querying of one or more bio-ontologies will become a key use of the Semantic Web for biomedical research during the next decade. Nonetheless, it has to be admitted that at the time of this writing (2011), this kind of application is still relatively rare, and most of the inference in queries involving subclass and part-of relations is performed by graph algorithms rather than general inference engines. However, a number of pioneering works on SPARQL for biomedical queries have appeared. A recent work on the Cell Cycle Ontology showed how to use SPARQL queries as a part of a pipeline to harness Semantic Web technologies to answer complex queries [11]. SPARQL frameworks have been used to organize queries on clinical, demographic and molecular data elements [70], to query a knowledge-base of pseudogenes [119], and to query the thesaurus of the National Cancer Institute [69]. A number of technologies are being developed to query OWL ontologies. At the time of this writing, SWRL, a Semantic Web Rule Language Combining OWL and RuleML, seems particularly promising [127, 190].

It remains to be settled exactly which Semantic Web technologies will become most widely used for biomedical applications. Since many current biomedical ontologies are often mainly subsumption hierarchies with some additional relations, OWL 2 QL may become an attractive option in the future. As mentioned in Chapter 14, OWL 2 has three "profiles" representing subsets of the full OWL 2 language. OWL 2 QL is aimed at applications that use very large volumes of instance data, and where query answering is the most important reasoning task. In OWL 2 QL, quick, polynomial time algorithms can be used to implement the ontology consistency and class expression subsumption reasoning problems. Query answering in OWL 2 QL ontologies can be implemented by rewriting queries into a standard relational query language [170].

16.4 Exercises and Further Reading

The W3C wiki provides a list of SPARQL implementations[4]. In the Exercises, we will use the Jena Java framework, and we have provided some tips on how to get Jena up and running on a linux or Unix system. It is also possible to run Jena in a Windows environment (especially with cygwin, `www.cygwin.com`, which is a linux-like environment for Windows) and Macintosh machines. Jena is highly recommended and has a large user community. Alternatively, many of the applications listed in the W3C wiki can be used

---

[4]`http://esw.w3.org/SparqlImplementations`.

for the Exercises. One recommended application is `Twinkle`, which is a simple GUI interface that act as a wrapper for the ARQ SPARQL query engine.

The individual exercises will ask you to perform or create SPARQL queries. We will use the Jena framework for the Exercises, and you will learn how to set up Jena in the first exercise.

Exercises

16.1 Jena is a Java framework for building Semantic Web applications (http://openjena.org). The framework includes a SPARQL query engine that can be used for the Exercises in this chapter. On linux systems, download the file (currently `arq-2.8.5.zip`) and unzip it, which creates a directory called `ARQ-x.y.z`, where `x.y.z` represents the version number. This directory contains all the necessary libraries in `lib/` as well as a copy of Jena. To use Jena, you will need to set the environment variable `ARQROOT` to the path of this directory and put each and every jar file in `lib/` in your classpath. The classpath is a user-defined environment variable used by Java to determine where predefined classes are located. With a bash shell in linux, you can add the following line to the `.bashrc` startup file (modify the path accordingly):

```
ARQROOT='/path/to/directory/ARQ-2.8.5'
export ARQROOT
export PATH=$PATH:$ARQROOT/bin
X=$(ls -f ${ARQROOT}/lib | grep jar)
for file in $X
do
    CLASSPATH=${file}:${CLASSPATH}
done
export CLASSPATH
```

Right after you have modified your `.bashrc` file, you may need to execute the command `hash -r` to export these definitions into the current shell.

On linux systems, you will need to make the scripts executable to test Jena:

```
chmod u+x $ARQROOT/bin/*
```

If you now add this directory to your path, you can execute the Jena programs from the command line. Follow the instructions on the Jena Web site to install the package on your computer if you have a Windows or Mac OSX system. To check the installation is working correctly, you can enter the command `qtest --all`.

```
qtest --all
```

16.2 We will now execute a first SPARQL query. The following query should be written in a file called q1.rc.

```
PREFIX prot:   <http://compbio.charite.de/prot#>

SELECT ?x
FROM <proteins.rdf>
WHERE { ?x  prot:name  "Catalase" }
```

Now execute the following command from the command line.

```
$ sparql --query=q1.rc
```

It is also possible to specify the RDF data source from the command line. In this case, the query does not have to have a FROM clause. Thus, the following query file and command yield equivalent results as above.

```
PREFIX prot:   <http://compbio.charite.de/prot#>

SELECT ?x
WHERE { ?x  prot:name  "Catalase" }
```

and

```
$ sparql --query=q1.rc --data=proteins.rdf
```

For this exercise, practice using Jena from the command line with the proteins.rdf file and some of the queries from this chapter. Alter the queries to retrieve other data items and to add different restrictions. As you can see, this is an open-ended exercise.

16.3 For the following exercises, you will use an RDF document that contains data obtained from the NCBI Entrez Gene site about a number of human genes, including their gene symbol, their chromosome and chromosome band, and their type. The document is available from the book Web site in the data archive in the SPARQL directory under the name geneinfo.rdf. Here is an excerpt.

```
<?xml version="1.0"?>
<!DOCTYPE rdf:RDF [
  <!ENTITY g "http://compbio.charite.de/gene#">
]>
<rdf:RDF xmlns:rdf="http://www.w3.org/1999/02/22-rdf-syntax-ns#"
         xmlns:g="http://compbio.charite.de/gene#">
  <rdf:Description rdf:about="&g;127">
```

```
      <g:symbol>ADH4</g:symbol>
      <g:chrom>4</g:chrom>
      <g:loc>4q21-q24|4q22</g:loc>
      <g:type>protein-coding</g:type>
    </rdf:Description>
    <rdf:Description rdf:about="&g;32">
      <g:symbol>ACACB</g:symbol>
      <g:chrom>12</g:chrom>
      <g:loc>12q24.11</g:loc>
      <g:type>protein-coding</g:type>
    </rdf:Description>
    <rdf:Description rdf:about="&g;443">
      <g:symbol>ASPA</g:symbol>
      <g:chrom>17</g:chrom>
      <g:loc>17p13.3</g:loc>
      <g:type>protein-coding</g:type>
    </rdf:Description>
    <rdf:Description rdf:about="&g;206">
      <g:symbol>AK4P1</g:symbol>
      <g:chrom>17</g:chrom>
      <g:loc>17q11.2</g:loc>
      <g:type>pseudo</g:type>
    </rdf:Description>
    <rdf:Description rdf:about="&g;118">
      <g:symbol>ADD1</g:symbol>
      <g:chrom>4</g:chrom>
      <g:loc>4p16.3</g:loc>
      <g:type>protein-coding</g:type>
    </rdf:Description>
    <rdf:Description rdf:about="&g;71">
      <g:symbol>ACTG1</g:symbol>
      <g:chrom>17</g:chrom>
      <g:loc>17q25</g:loc>
      <g:type>protein-coding</g:type>
    </rdf:Description>
    <rdf:Description rdf:about="&g;206">
      <g:symbol>AK4P1</g:symbol>
      <g:chrom>17</g:chrom>
      <g:loc>17q11.2</g:loc>
      <g:type>pseudo</g:type>
    </rdf:Description>
  </rdf:RDF>
```

Note the use of the entity declaration. This means that an expression such as rdf:about="&g;798" is equivalent to the longer expression

```
rdf:about="http://compbio.charite.de/gene#798"
```

The number 798 in this example refers to the accession number of the gene CALCP calcitonin pseudogene [Homo sapiens] in the Entrez Gene database of NCBI.[5] For this exercise, write a SPARQL query to get a list of all gene symbols in the RDF document.

16.4 Modify your query from the previous exercise to sort the gene symbols alphabetically using the ORDER BY operator.

16.5 Write a SPARQL query to get a list of all genes whose symbols begin with the letter "z."

16.6 Write a SPARQL query to get a table whose rows contain the gene symbol, chromosome, location, and type for each gene.

16.7 Modify the previous query to restrict the answer to just those genes that are located on chromosome 1.

16.8 Use a SPARQL query to get a list of information for all pseudogenes (those with g:type "pseudo").

16.9 Use a SPARQL query to get a list of genes on the long arm of the X chromosome (for instance, a gene on Xq27.1 is on the long arm, band 2, subband 7, and sub-subband 1 of the X chromosome).

16.10 The geneinfo.rdf file was generated by a Perl script that creates RDF triples based on the tab-separated file format of the gene_info.gz file that can be downloaded from the NCBI Entrez Gene ftp site.[6] For this exercise, you are asked to write a Perl script that generates an RDF file similar to the geneinfo.rdf file from this book. (Of course, you are free to use a script or programming language of your choice, but tasks like this are generally very easy to perform using Perl.) Note that it may be a good idea to limit your RDF file to a single taxon (for instance, the NCBI taxon id for *Homo sapiens* is 9606).

16.11 Adapt the script you made in the previous exercise for a data file from a domain of your choice in order to create an RDF file. Now write several SPARQL queries to interrogate the data.

16.12 Now write a Perl script to create an RDF file for the gene2go.gz file at the Entrez Gene ftp site. Call the file gene2go.rdf. This file should contain the GO annotations for the genes specified by the Entrez IDs. Use the techniques you learned in this chapter to combine the geneinfo.rdf file with the gene2go.rdf file in a query that will return the gene symbol and the GO annotations. Use the Entrez Gene id to join the information from the two files.

---

[5] http://www.ncbi.nlm.nih.gov/gene.

[6] ftp://ftp.ncbi.nih.gov/gene/DATA/.

Appendices

Appendix A

An Overview of R

"Researchers have shown it takes about ten years to develop expertise in any of a wide variety of areas, including chess playing, music composition, telegraph operation, painting, piano playing, swimming, tennis, and research in neuropsychology and topology."

–Peter Norvig, *Teach Yourself Programming in 10 Years*, 2001

Many of the exercises in this book are formulated as problems that can be solved in R. This appendix is intended to provide readers unfamiliar with R enough of a background to work on these exercises, but is by no means a comprehensive account of R or on how to write elegant and correct scripts using R. Consider this a task for the next ten years.

R is a freely available programming language and environment that is primarily designed for statistical and scientific computing and graphics. A number of excellent books are available for learning R [53, 63, 87, 159] and the R homepage offers introductory material and tutorials[1]. Bioconductor is an open-source collection of R packages for computational biology and bioinformatics [88] that has developed into one of the most important resources for bioinformatics. Although Bioconductor is not covered in this book, it is extremely useful for integrative analysis of high-throughput biological data (e.g., microarray or next-generation sequence data) including GO analysis of differentially expressed genes [76, 268].

Installing R

R installation packages can be downloaded from the R homepage or from the Comprehensive R Archive Network (CRAN) site at http://cran.r-project.org. Many linux distributions such as debian allow users to install R directly from the package management system. Read the documentation for your system at the CRAN site.

R Packages

R packages are modular add-ons that typically provide functions and data to implement a specific kind of analysis. Only when a package is loaded are its

---

[1] http://www.r-project.org/.

contents available. There are a number of standard packages that are installed with every R distribution. The `library` command can be used to see which packages are currently installed.

```
> library()
```

Additional packages can be installed directly from R once the R environment is up and running on your system. For Windows users, the R graphical user interface (GUI) has a *Packages* menu with an item for installing packages. Users on all systems can type `help("INSTALL")` or `help("install.packages")` from the R prompt to obtain information on how to install packages on their system.

The `library` command is also used to load a package into the current workspace. For instance, to load the `survival` package, use the following command.

```
> library(survival)
```

R packages come with various kinds of documentation, including a PDF reference manual with explanations of the functions and use examples.

Installing the Robo package

The authors of this book have developed an R package for performing various kinds of analysis on OBO ontologies. The package can be downloaded from the book Web site.[2] Under linux, change to the directory containing the file `Robo.tgz` and enter the command

```
$R CMD INSTALL Robo.tgz
```

The R console window

The console window (or R command line) shows a prompt (>) at which users can type in commands or expressions. The traditional "Hello World" program is particularly easy in R: type in the string "hello world" at the prompt:

```
> "hello world"
[1] "hello world"
```

By default, the R system shows the result of the expression entered at the prompt. In this case, the expression and its result are the same, i.e., the string `"hello world"`. In general, R evaluates the expression and shows the results on the next line (or lines, if there are multiple results):

```
> 1+1
[1] 2
```

---

[2]`http://bio-ontologies-book.org`.

Values can be assigned to variables using either the equals sign ("=") or an arrow ("<" followed by "-").

```
> x <- 1+1
> y = 2+2
> x+y
[1] 6
```

R has a large number of built-in functions and mathematical constants.

```
> pi
[1] 3.141593
> sin(pi/2)
[1] 1
> sqrt(2)
[1] 1.414214
```

The built-in function c combines its arguments to form a vector.

```
> c(1,2,3)
[1] 1 2 3
> c("a","b","c")
[1] "a" "b" "c"
```

Matrices of values can be constructed using the matrix function.

```
> M <- matrix(c("a","b","c","d","e","f","g","h"),nrow=2)
> M
     [,1] [,2] [,3] [,4]
[1,] "a"  "c"  "e"  "g"
[2,] "b"  "d"  "f"  "h"
```

The arguments nrow,ncol, and byrow can be used to determine the number of rows and columns in the matrix.

R is equipped with an extensive help system. For instance, to get more information about the matrix command directly from the R prompt, enter

```
>?matrix
```

This command (or equivalently, help(matrix)), will display an explanation of the function and its arguments and show several examples. If you are not sure exactly where to look for documentation about an R command, using the help.search command will display a list of all potentially relevant help files, e.g.:

```
>help.search("matrix")
```

The command RSiteSearch("matrix") will open a browser window (assuming there is an active network connection) with search results from the R documentation and from other sources including the archives of the R-help mailing list.

Data frames

In a matrix, all elements must have the same datatype (numerical, character, Boolean). A data frame is a useful generalization of a matrix, in which each column needs to be of the same datatype, but different columns can have different datatypes. Data frames can be constructed from vectors using the built-in command data.frame. The following example shows how one can use a data frame to combine several types of data about African mammals.

```
> animal<-c("elephant","giraffe","lion")
> lifespan<-c(70,13,20)
> weight<-c(7400,1200,200)
> carnivore<-c(F,F,T)
> df <- data.frame(name=animal,Lifespan=lifespan,
        Weight=weight,eatsMeat=carnivore)
> df
      name Lifespan Weight eatsMeat
1 elephant       70   7400    FALSE
2  giraffe       13   1200    FALSE
3     lion       20    200     TRUE
```

The built-in function read.table offers an easy way to read data from a file into an R data frame. Arguments to the function control the parse behavior of the function including whether to treat the first line of the input file as a table header and whether the fields of the file are separated by spaces, tabs, or other symbols.

Plotting and Graphics

One of R's major strengths consists of the ease with which publication-quality plots can be generated from data. The basic command for creating a plot is simply plot, but a wealth of arguments exist to modify the basic appearance of plots, add text or legends, create figures with multiple panels, and many other things. Here, we will make a simple plot by creating a vector X with a sequence of numbers from 1 to 20, and using the function runif to create 20 random numbers between 1 and 50. The command plot(X ~ Y), or equivalently plot(X,Y), generates the plot shown in Figure A.1.

```
> X=1:20
> Y=runif(20,min=1,max=50)
> plot(X ~ Y)
```

Among the most important arguments for changing the appearance of plots are pch, which specifies the appearance of the data points, type, which determines what type of plot should be drawn (e.g., points, lines, or both), main (title for the plot), xlab and ylab (titles for the X and Y axis), as well as the many parameter values specified as arguments to par. See the

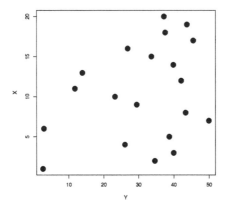

FIGURE A.1: Plotting with R.

R documentation for more details. We will also explain some of the more advanced plotting features later on in this appendix.

Script files

For anything more than casual use, it is recommendable to store R expressions in a text file (called a script file) and execute them all at once by calling the **source** function. For instance, if we write the three commands used to create the above plot in file called **random-plot.R**, we can execute all three commands at once as follows:

```
> source("random-plot.R")
```

It is convention, but not strictly necessary, that script files are stored with the suffix .R. Especially if a long series of commands are required to carry out some analysis, it is helpful to create a script file in order to be able to try out additions and modifications to the commands without having to type everything back in. Comments should be added to script files to document your work and problem solving strategies in case debugging becomes necessary at a future time. Comment lines start with a # and are ignored by R.

The **history()** function will print a list of all commands entered into the current R session that can then be copied and entered into a text file to create a script. The command **savehistory** can be used to save the commands directly to a file.

Finally, if you quit R by the command q(), by default a binary R data file called .RData will be created containing the contents of your current R session. This file will be loaded the next time you start R. This can be a source of errors, because variables can be imported into the current workspace without the user noticing it. Usually it pays to erase all variables from the workspace when starting a new R session. In the following example, a variable called x was assigned the value 2 in one R session, and the variable and its value were stored when the session was ended.

```
> x=2
> q()
Save workspace image? [y/n/c]: y
```

When R is started the next time, this variable and its value have been restored, as we can see using the function ls(). The function rm() is then used to remove all variables from the workspace.

```
> ls()
[1] "x"
> x
[1] 2
> rm(list = ls(all = TRUE))
> ls()
character(0)
```

Useful built-in functions

R has many built-in functions for common operations such as sum, length (number of elements of a vector or many other R objects), sin (sine), cos (cosine), mean, sd (standard deviation), and of course +, -, * (multiplication), and \ (division). The functions work in more or less obvious ways, but it pays to consult the online documentation.

```
> x <- runif(50,min=1,max=100)
> sum(x)
[1] 2482.777
> length(x)
[1] 50
> sum(x)/length(x)
[1] 49.65554
> mean(x)
[1] 49.65554
```

The built-in functions for arithmetic also work in an element-wise fashion on vectors and matrices[3] The following commands provide some examples of how this works.

```
> x=1:3
> 2*x
[1] 2 4 6
```

---

[3]R also has a separate syntax for vector and matrix operations that we will not cover here.

```
> x=1:3
> x
[1] 1 2 3
> 2*x
[1] 2 4 6
> sqrt(x)
[1] 1.000000 1.414214 1.732051
> sin(x)
[1] 0.8414710 0.9092974 0.1411200
```

If two vectors x and y are of the same length, then the result of x*y is element-by-element multiplication (not vector multiplication).

```
> x <- 1:3
> y <- 4:6
> x*y
[1]   4 10 18
```

R for statistics

One of the reasons R has become so important for bioinformaticians is the fact that R offers a comprehensive selection of functions for statistical analysis. For each of the major statistical distributions, R offers a density function, a distribution function, a quantile function, and random generator. For each distribution, there is a help file describing the functions and arguments. We will take some time here to explain how to use these functions, which are important for the exercises about GO overrepresentation analysis (especially the binomial and hypergeometric distributions). For our examples, we will use the functions for the Normal distribution, but the functions for other distributions behave analogously. To read the documentation for the normal distribution, enter ?dnorm (or ?dbinom for the Binomial distribution, and so on).

```
The Normal Distribution
```

```
Description:
```

```
        Density, distribution function, quantile function and random
        generation for the normal distribution with mean equal to
        'mean' and standard deviation equal to 'sd'.
```

```
Usage:
```

```
        dnorm(x, mean = 0, sd = 1, log = FALSE)
        pnorm(q, mean = 0, sd = 1, lower.tail = TRUE, log.p = FALSE)
        qnorm(p, mean = 0, sd = 1, lower.tail = TRUE, log.p = FALSE)
        rnorm(n, mean = 0, sd = 1)
```

The probability density function (pdf) for the normal distribution is defined as

$$f(x) = \frac{e^{-(x-\mu)/2\sigma^2}}{\sqrt{2\pi}} \tag{A.1}$$

The default values of the R functions assume that the mean $\mu = 0$ and the standard deviation $\sigma = 1$ (this is referred to as the standard Normal distribution).

$$f(x) = \frac{e^{-x/2}}{\sqrt{2\pi}} \tag{A.2}$$

Probability distribution functions assign a real number p to each outcome of the sample space such that $p > 0$ and the sum (for discrete probability distributions) or the integral (for continuous distributions) is equal to 1. Thus, for heads and tails, $p_H = p_T = 0.5$ and $\sum_{H,T} p_i = 1$. For a continuous distribution such as the Normal distribution, an integral takes the part of the sum, and we have that

$$\int_{-\infty}^{\infty} \frac{e^{-x/2}}{\sqrt{2\pi}} dx = 1 \tag{A.3}$$

The R function dnorm gives the value of Equation (A.1) for any x. If a mean and standard deviation are not specified, dnorm gives the value of Equation (A.2). To show the values of Equation (A.2) from -3 to 3, we can use the following way of calling the plot function in which the first argument is a function call and the second and third arguments the range of values along the X axis that are to be plotted. Arguments are provided to increase the size of the text on the axis labels (cex.lab) and the axis numbering (cex.axis) by a factor of 2 to improve legibility.

```
> plot(function(x) dnorm(x),-3,3,cex.lab=2,cex.axis=2)
```

This command generates the plot shown in Figure A.2.

The pnorm function corresponds to the cumulative distribution function (cdf), which describes the probability for a random variable to fall in the interval $(-\infty, x]$. For the Normal distribution, the cdf is defined as

$$cdf(x) = \int_{-\infty}^{x} \frac{e^{-(x-\mu)/2\sigma^2}}{\sqrt{2\pi}} dx \tag{A.4}$$

We can plot the cdf in the same way the pdf was plotted.

```
> plot(function(x) pnorm(x),-3,3,cex.lab=2,cex.axis=2)
```

This command generates the plot shown in Figure A.3. The interpretation of this plot is that there is a probability of 50% that x will lie between negative infinity and zero, and a probability of 100% that x will lie between negative and positive infinity. Note that the label of the Y axis is cut off. We will see later on how to adjust the plotting parameters to avoid this.

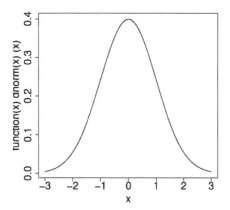

FIGURE A.2: **Normal Distribution: pdf.** The pdf of the standard Normal distribution is shown between the values of -3 and 3.

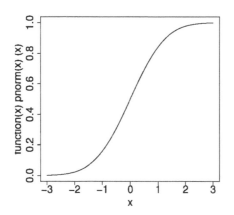

FIGURE A.3: **Normal Distribution: cdf.** The cdf of the standard Normal distribution is shown between the values of -3 and 3.

qnorm represents the quantile function, which is the inverse of the cdf. To demonstrate this, we will first create a series of values ranging from -3 to 3 in increments of 0.01 and then calculate the pdf for the values.

```
> X <- seq(-3,3,by=0.01)
> Y <- dnorm(X)
```

The command plot(X,Y) would create a figure similar to Figure A.2. The variable Y now contains the probability values. We can now use the function qnorm to go back from the probability values to the values of x that correspond to each probability value.

```
> plot(function(x) qnorm(x),Y,cex.lab=2,cex.axis=2)
```

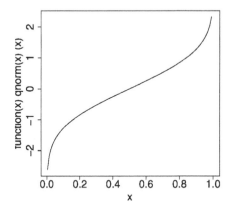

FIGURE A.4: Quantile Function for Normal Distribution. The quantile function of the standard normal distribution is shown. Compare this plot with Figure A.3. The functions pnorm and qnorm are inverse to one another.

Finally, rnorm generates random values according to the Normal distribution. To generate 20 random, Normally distributed values, execute the following command.

```
> r <- rnorm(20)
```

Vectors and Lists

Vectors are the simplest kind of compound data structure in R. They consist of an ordered collection of elements. We have seen examples of numeric vectors above. In contrast to many other programming languages, R uses 1-based numbering, that is, the first element of a vector v is accessed by the expression v[1] rather than by v[0] as in languages such as Java or C. The length command is used to determine how many elements are in a vector. The following example shows how this works for a vector of strings created using built-in constants representing the names of the months.

```
> v<-month.name
> length(v)
[1] 12
> v[4]
[1] "April"
```

Vectors can also be used to store truth values. For instance, if we have a vector of integers $1, 2, \ldots, 10$ and we would like to know which entries of the vector are greater than 7, we can use the following commands.

```
> v <- 1:10
> v > 7
 [1] FALSE FALSE FALSE FALSE FALSE FALSE FALSE  TRUE  TRUE  TRUE
```

Vectors like this are often used to extract a subset of values from other vectors. If we want to create a vector containing just those elements of the vector v that are greater than 7, we can use the logical vector to index the vector v. Just those elements of v for which the index vector has the value TRUE are extracted by the following commands.

```
> i <- v>7
> v[i]
[1]  8  9 10
```

It is also possible to combine the commands using the **which** command.

```
> v[which(v>7)]
[1]  8  9 10
```

An *array* is a generalization of a vector that can have multiple dimensions. A vector is an array with one dimension and n elements. A matrix is an array with two dimensions and $n \times m$ elements. In the R language, a vector is a separate datatype from an array, but is transformed into the array datatype if its dimension attribute (**dim**) is defined.

```
dim(v) <- c(10)
```

A similar command can be used to create a multidimensional array. Say we want to have a $2 \times 3 \times 4$ element three-dimensional array. The array will thus have $2 \times 3 \times 4 = 24$ elements. First, a vector with 24 elements is created, and then its dimension is defined as above. It is now possible to access the elements of the array on the basis of an index in each of the three dimensions.

```
> a <- 1:24
> dim(a) <- c(2,3,4)
> a[1,2,3]
[1] 15
```

A vector or array may only contain a single datatype (numeric, Boolean, or string). Lists are similar to vectors in that the elements of a list are not restricted to a single datatype. A list element can even be a complex datatype such as a matrix or another list. The following code defines an R list representing some aspects of the Beatles.

```
B<-list(name="Beatles",
        type="band",
        members=c("John","Paul","George","Ringo"),
        years=seq(from=1963,to=1970,by=1),
        no.albums=c(2,2,2,1,1,1,2,1))
```

This defines a list with five elements. The first two elements are strings, and the last three elements are vectors. To access an element of a list, the [[indexing operator is used. Thus, $B[[1]]$ retrieves the first list element. Alternatively, one can use a dollar sign followed by the name of the element.

```
> B[[1]]
[1] "Beatles"
> B$name
[1] "Beatles"
```

It is possible to index elements of vectors contained in lists as follows.

```
> B[[3]]
[1] "John"    "Paul"    "George" "Ringo"
> B[[3]][3]
[1] "George"
```

For R lists, there is a subtle difference between B[[1]] and B[1]. B[[1]] extracts the first element of the list. B[1] extracts a sublist from B consisting of the first element of B. '[ ]' is the general subscripting operator. B[1:3] can be used to extract a sublist from B and will return a list consisting of the first three elements of B.

In R, the results of most statistical analyses are returned as a list containing all the relevant values. The following example shows how the results of a *t* test are returned.

```
> x <- rnorm(30)+3
> t <- t.test(x,mu=1)
> t$p.value
[1] 2.283209e-11
> t$method
[1] "One Sample t-test"
> t[[2]]
df
29
```

R as a Programming Language

R has constructs for conditional execution, loops, and functions. The syntax for if-else conditional expressions is

```
if (condition) {
 expression 1
} else {
 expression 2
}
```

For instance, we can write the following program to ask whether P=NP.

```
> P="P"
> NP="NP"
> if (P==NP) {
+ print("!!!")
+ } else {
+ print("thought so")
+ }
[1] "thought so"
```

Any number of expressions can be included within the braces. R has three main constructs for explicitly looping over variables, `for`, `while`, and `repeat`. The syntax is slightly different from the syntax of some other well-known programming languages such as C or Java.

```
> for (i in 1:4) {
+ print(i)
+ }
[1] 1
[1] 2
[1] 3
[1] 4
```

Note in the examples shown above that the continuation of an expression on the R console is shown by a plus sign on the continuation lines.

R functions are defined using the keyword `function` with a list of the argument names in parentheses. In contrast to languages such as C or Java, it is not necessary to stipulate the datatype of the arguments. For example, the following is a function that calculates the square of its argument.

```
sqr <- function(x) {
   x*x
}
```

We can now call the function in our code.

```
> sqr(2)
[1] 4
```

The value of the final expression in the function body is returned to calling code. Functions can have multiple arguments, in which case the order of the arguments is mandatory. An argument can also be assigned a default value. For instance

```
> f <- function(x,y,z=2) {
+   r <- x+y
+   z*r
+ }
> f(2,3)
[1] 10
> f(2,3,1)
[1] 5
```

In general, longer functions are written in files. If some function is stored in a file called foo.R, then the command source("foo.R") will make the function available in the current workspace.

R provides several vectorized options to explicit loops which can save a significant amount of processing time. The lapply function takes a vector or R list as an argument, applies a function to each member of the list and returns the result as a list. Consider the following code.

```
slow <- function(n) {
  x <- 1:n
  y <- list(n)
  for (i in 1:n) {
    y[[i]] <- sqrt(x[i])
  }
 y
}

fast <- function(n) {
  x <- 1:n
  y <- lapply(x,sqrt)
  y
}
```

Each of the functions performs an equivalent calculation. We can compare how long the computer takes to do each calculation using R's built-in system.time function.

```
> system.time(slow(10000))
      User      System     elapsed
     0.588       0.000       0.598
> system.time(fast(10000))
      User      System     elapsed
     0.012       0.000       0.013
```

Users should get familiar with the apply function family, which also includes tapply, sapply, and mapply. Consult the online documentation for details.

Creating Good Plots

As mentioned, it is easy to create plots in R using the `plot` command. However, it usually requires a little more work to create a visually appealing and easy to interpret plot that is suitable for a presentation or publication. To conclude this introduction to R, we present the code used to make Figure 3.2, which demonstrates a few of the more advanced plotting capabilities of R (the line numbers are of course not used in the actual script file).

```
1.  # binomial.R
2.  # Plot the density of the binomial distribution
3.  # for two sets of parameters.
4.
5.  xlab <- expression(italic(k))
6.  margintext <- c("A)","B)")
7.  par(mfrow = c(1,2))
8.  par(pin=c(2,2))   ## set width,height
9.  p <- c(0.5,0.1)
10. n <- 10
11. k <- 0:n
12.
13. for (i in 1:2) {
14.     title <- paste("p=",p[i],sep='')
15.     density <- dbinom(k,n,p[i])
16.     plot(k,density,type='h',lwd=2,xlab=xlab,main=title,ylab='',
17.          ylim=c(0,0.5),cex.lab=1.25,cex.axis=1.25)
18.     points(k,density,type="p",pch=19)
19.     abline(h=0,lwd=2)
20.     mtext(margintext[i],side=3,cex=1.5)
21. }
```

Lines 1–3 are comments. Empty lines in scripts are also ignored by the R interpreter. Line 5 creates an italicized letter (k) using the `expression` command. R has a large number of ways of adding mathematical symbols to plots. Enter > `?plotmath` at the R prompt to see documentation. Line 6 creates a vector of two strings that will be used later on. `par` is a global variable with many individual attributes that can be used to set or query graphical parameters that affect the behavior of the `plot` command. Line 7 sets up a plot with multiple panels. `mfrow` is a vector of the form `c(nrows, ncolumns)`. `pin` sets the current plot dimensions (width,height) in inches.

The for loop in lines 13–21 loops over i=1 and i=2 and is used to create the two panels in the figure, which differ from one another only in the parameters to the binomial distribution. Line 14 creates the title of the panels, whereby `p[i]` refers to the vector created in line 9. Line 15 calculates the probability density function for the binomial distribution (similar to `dnorm` for the Normal distribution as described above).

The plot command plots the vector $k = 0, 1, \ldots, 10$ on the X axis against the binomial densities for these values on the Y axis. The meaning of the arguments is as follows:

- type='h' sets the type of the plot to histogram-like vertical lines.

- lwd=2 sets the line width (of the vertical lines) to '2'.

- xlab=xlab sets the label of the X axis to the value of the variable xlab from line 5.

- main=title sets the overall title of the plot to the value of the variable title from line 14.

- ylim=c(0,0.5) sets the limits of the Y axis to 0–0.5.

- cex.lab=1.25 sets the magnification of the text of the X and Y axis labels to 1.25.

- cex.axis=1.25 sets the magnification of the text of the X and Y axis annotations (numbering, in this case) to 1.25.

The points command in line 18 adds points at the (X,Y) positions given by k and density. The points to be drawn are defined as plain points (type="p") with plotting character pch=19 (a circle). Line 19 draws a horizontal line with line width 2 at Y=0. Finally, line 20 add a text to the margin of the plots at side 3 (top) and user coordinate -2 (at=-2) and magnification 1.5. Often, some amount of trial and error is needed to put the elements of a figure at the desired places. For instance, a first attempt at adding a margin text might be

```
>   mtext(margintext[i],side=3,cex=1.5)
```

This would produce the plot shown in Figure A.5. Usually, the labels for panels A) and B) are shown on the left upper side of the panels. In order to learn how to shift the labels, one would consult the documentation for mtext (> ?mtext), read that the at parameter can be used to give the location of each string in user coordinates, and try various values to adjust the location as desired. In this case, following a little trial and error, at=-2 was found to put the label in the desired place (see Figure 3.2).

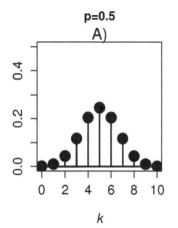
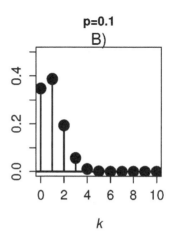

FIGURE A.5: **Binomial Probability Distribution**. This figure was generated using the R code described in the text, except that "at=-2" was not used. Compare Figure 3.2, for which the argument -2 was used for the at parameter.

Appendix B

Information Content and Entropy

This appendix provides additional background about the definition of *information content* used for the calculation of semantic similarity in Chapter 10. The field of information theory was created by Claude Shannon (1916–2001), an American mathematician who published a landmark paper on the mathematical theory of communication in 1948 [231]. Shannon's work led to the concept of information content among many other things.

Basic properties of logarithms

We will begin this appendix with a review of the basic properties of logarithms. A base b logarithm of x is the power to which b must be raised in order to yield x. That is, if $\log_b x = y$, then $b^y = x$. The definition is valid for $b > 1$ and $x > 0$. The most commonly used logarithms are those to the base 2, $\log_2$, natural logarithm to the base e, written simply log or occasionally ln, and logarithms to the base 10, $\log_{10}$. There are a number of familiar identities for logarithms whose proof can be found in basic textbooks of calculus.

$$
\begin{aligned}
\log 1 &= 0 \\
\log xy &= \log x + \log y \\
\log x^r &= r \log x \\
\log \frac{1}{x} &= -\log x \\
\log_a x &= \frac{\log x}{\log a}
\end{aligned}
$$

The first four identities are valid for logarithms to any base, and the final identity provides a method for converting from a logarithm to the base a to a natural logarithm.

Entropy

Let us assume we are analyzing a random variable X, which can take on one of a finite number of specific values $\{x_1, x_2, \ldots, x_n\}$ with a probability $\{p_1, p_2, \ldots, p_n\}$, where the probability of the outcome x_i is p_i and $\sum_{i=1}^{n} p_i = 1$. We will use the notation $p(X = x)$ or simply $p(x)$ to denote the probability that the random variable X takes on some specific value x, where $x \in \{x_1, x_2, \ldots, x_n\}$. We will refer to each of the specific values that X can take on as *outcomes*.

Shannon defined the information content of an outcome x as

$$h(x) = \log_2 \frac{1}{p(x)} \tag{B.1}$$

Note that because of the properties of logarithms, $\log_2 \frac{1}{p(x)} = -\log_2 p(x)$, which is the definition of information content used in Chapter 10. The *entropy* of the random variable X, written as $H(X)$, was defined by Shannon as the average information content of all of the possible outcomes of X.

$$H(X) = \sum_{i=1}^{n} p(x_i) \log_2 \frac{1}{p(x_i)} \tag{B.2}$$

Note that by convention $p(x') \log_2 p(x') = 0$ for $p(x') = 0$. This is necessary because $\log 0$ is undefined, and reasonable because $\lim_{x \to 0} x \log x = 0$.

Entropy and information content are measured in units called *bits* (not to be confused with the definition of a bit as a zero or one in computer science). The entropy of a random variable X can be interpreted as the *uncertainty* of the random variable. For instance, if X has only one outcome (i.e., $p(x') = 1$ for some x'), then the entropy of X is zero.

$$H(X) = p(x') \log_2 \frac{1}{p(x')} = 1 \times (-1) \times \log_2 1 = 0$$

It can easily be shown that the entropy of X is maximized if there is maximal uncertainty about the outcome – this is the case if no outcome is more likely than the others, or stated differently, if X follows a uniform distribution (see Figure B.1).

The function $IC(x) = -\log p(x)$ is a natural one for measuring information content, for several reasons. If there is no uncertainty, i.e., if $p(x = 1)$, then $IC(x) = \log 1 = 0$. On the other hand, the information content can be arbitrarily large as the probability of an outcome goes to arbitrary small quantities: $\lim_{x \to 0} \log x = \infty$. Say we have two random variables, X and Y, that are independent from one another. It is reasonable to assume that the information content of the outcomes x and y must be additive. This is a natural consequence of the logarithm, because

$$IC(x, y) = -\log_2(xy) = -\log_2 x - \log_2 y = IC(x) + IC(y) \tag{B.3}$$

It can be shown that Claude Shannon's definition of entropy as Equation (B.2) can be derived from three postulates.

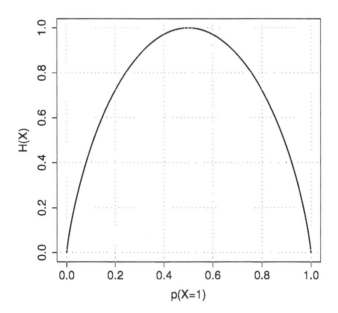

FIGURE B.1: Entropy. The figure shows the entropy of a random variable $X = \{0,1\}$ representing Bernoulli process such as a coin toss for different values of $p(X = 1)$. Maximum entropy is achieved if $p(0) = p(1) = 0.5$, i.e., for a uniform distribution. The minimum entropy of zero occurs if the probability of one of the two outcomes is 1.

Theorem B.1 *Uniqueness Theorem Let p stand for a set of real values $\{p_1, p_2, \ldots, p_n\}$ standing for the probabilities of outcomes $\{x_1, x_2, \ldots, x_n\}$ with $p_i \geq 0$ and $\sum_i p_i = 1$. Consider a function H such that the following three properties hold:*

1. *$H(p_1, p_2, \ldots, p_n)$ takes on its largest value for $p_i = \frac{1}{n}$ for all i.*

2. *$H(AB) = H(A) + H_A(B)$ (if A and B are independent, then $H(AB) = H(A) + H(B)$).*

3. *$H(p_1, p_2, \ldots, p_n) = H(p_1, p_2, \ldots, p_n, 0)$, that is, including an event of zero probability does not change the entropy.*

If for any n the function H is continuous with respect to all its arguments and H has the three properties described above, then H must be of the form

$$H(p_1, p_2, \ldots, p_n) = \lambda \sum_{i=1}^{n} -p_i \log p_i$$

where λ is a positive constant.

A proof of this theorem can be found in Khinchin's book on the mathematical foundations of information theory [140]. It can also be shown that $\lambda = 1$ if $\log_2$ is used.

Thus, we conclude that Equation (B.1) provides a definition of information content that satisfies several important intuitions about the properties that a mathematical definition of information content should have.

Further Reading

An Introduction to Information Theory by Pierce [200] provides a gentle introduction to information theory and its applications. The above-cited book by Khinchin [140] or the book by MacKay [157] offer excellent mathematical treatments of information theory.

Appendix C

W3C Standards: XML, URIs, and RDF

The World Wide Web Consortium (W3C) was founded by Tim Berners-Lee, a computer scientist widely credited with inventing the World Wide Web. The W3C has become the main international standards organization for the World Wide Web, and has developed standards for important technologies such as HTML, CGI, and CSS. The W3C standards for XML, URIs, and RDF are of particular importance for the Semantic Web. However, documentation on the standards is often distributed over multiple Web pages and multiple versions and much of it is written in a highly technical style that can be difficult to understand. This appendix will present enough background on XML, URIs, and the syntax of RDF that readers will be able to understand RDF(S) or OWL ontologies that are serialized in the RDF/XML or OWL/XML formats, rather than to present a comprehensive treatment. Suggestions for further reading are presented at the end of the appendix.

XML

XML, or the eXtensible Markup Language, is a generic markup language for encoding text documents or arbitrary data structures. XML defines a generic syntax to annotate ("mark up") data with simple human- and machine-readable tags. In contrast to HTML, which uses a predefined set of tags such as `<b>some text</b>` to indicate that **some text** should be printed in bold type, XML allows users to define an arbitrary number of arbitrary tags. A tag, together with the data it contains, is called an *element*.

XML documents begin with an XML declaration that specifies the XML syntax version and the encoding. For instance, the following declaration states that the XML document uses XML version 1.0 syntax and is encoded using UTF-16, a 16-bit variable-length character encoding for Unicode.

```
<?xml version='1.0' encoding="UTF-16"?>
```

Tags are markup constructs that begin with < and end with >. A start tag, e.g., `<name>` begins an element, and an end tag, e.g., `</name>` closes it.

An element begins with a start tag, e.g., `<name>` and ends with an end tag, e.g., `</name>`. An element can contain content within the tags as well as an arbitrary number of *child elements*. For instance, the following XML document describes four European countries.

```
<?xml version="1.0" encoding="UTF-8"?>
<europe>
  <country>
    <name>France</name>
    <capital>Paris</capital>
    <language>French</language>
  </country>
  <country>
    <name>Germany</name>
    <capital>Berlin</capital>
    <language>German</language>
  </country>
  <country>
    <name>Spain</name>
    <capital>Madrid</capital>
    <language>Spanish</language>
  </country>
  <country>
    <name>Italy</name>
    <capital>Rome</capital>
    <language>Italian</language>
  </country>
</europe>
```

The element `<europe>` has four child elements called `<country>`, each of which has three child elements called `<name>`, `<capital>`, and `<language>`. Each XML document must have one and only one root element that encloses all other elements (`<europe>`, in the above XML document).

XML elements can also have one or more *attributes*. An attribute is a name-value pair that is put within the start tag of an element. The name is separated from the value by an equals sign, and the value is enclosed in quotation marks. For example, we can use two attributes to indicate the date of the birth and death of Aristotle.

```
<philosopher date-of-birth="384 BC" date-of-death="322 BC">
Aristotle
</philosopher>
```

In many cases, it is possible to convey the same information as child elements or as attributes. For instance, the following XML document conveys the same information about the four European countries.

```
<?xml version="1.0" encoding="UTF-8"?>
<europe>
  <country name="France" capital="Paris" language="French"/>
  <country name="Germany" capital="Berlin" language="German"/>
  <country name="Spain" capital="Madrid" language="Spanish"/>
  <country name="Italy" capital="Rome" language="Italian"/>
</europe>
```

The decision as to whether to use attributes or child elements for certain types of information is left to the developer of the XML document, but as a general rule of thumb, attributes are often used for metadata or annotations about the element and elements (or child elements) for the data. We will see that it is possible to encode a single RDF graph equivalently using both styles, and readers should be familiar with this because they are likely to encounter a number of different styles of encoding RDF graphs about biomedical data.

There are some important syntactical differences between attributes and elements. An attribute name can only be used once within an element. In our example, it would not have been possible to encode the <country> elements as attributes. That is, the following element is *not* valid XML.

```
<?xml version="1.0" encoding="UTF-8"?>
<europe country="France" country="Germany" ... />
```

We will see another important difference between attributes and element names in our discussion of RDF/XML further below.

An XML document is *well formed* if it satisfies the syntactical rules of the XML specification, some of which were explained above. In addition, it is possible to specify the structure of the XML document using a Document Type Definition (DTD) or XML Schema definitions. In brief, DTDs and XML Schema constrain the structure of a document. For instance, a <country> element can be defined such that it must have child elements called <name>, <capital>, and <language>. If a <country> element only has two elements, or if it had an additional element called <national beverage>, the element (and the XML document it is in) would be *invalid*.

XML *entity references* provide a kind of macro that is declared in the DTD and then can be used within the main document. Entities are often used to make XML documents more readable by providing abbreviations for long strings. An entity is declared as an ENTITY element in the DTD and the abbreviation can then be used in the name or within the content of an element. In the following document, entities are used for some long words.

```
<?xml version="1.0" encoding="UTF-8"?>
<!DOCTYPE long-words [
<!ELEMENT long-words (word+)>
<!ELEMENT word (#PCDATA)>
<!ENTITY ade "Antidisestablishmentarianism">
<!ENTITY pph "Pseudopseudohypoparathyroidism">
```

```
<!ENTITY pums "Pneumonoultramicroscopicsilicovolcanoconiosis">
]>
<long-words>
  <word>&ade;</word>
  <word>&pph;</word>
  <word>&pums;</word>
</long-words>
```

In the above XML document, the DTD is specified in the DOCTYPE declaration, which states first that a long-words element must contain one or more word elements and then defines three entities. Each entity is a usually short name or abbreviation followed by a string that is enclosed in quotation marks. The abbreviation can now be used elsewhere in the XML document preceded by an ampersand (&) and followed by a semicolon. Entities can be used within node content or attribute values. We will see that entities are widely used in RDF/XML as abbreviations of URIs to improve legibility of the document.

XML *namespaces* can be used to disambiguate term names. Consider the following well-formed XML document.

```
<?xml version="1.0" encoding="UTF-8"?>
<book>
  <title>The Man Who Mistook His Wife for a Hat</title>
  <author>
    <title>Dr</title>
    <first-name>Oliver</first-name>
    <last-name>Sachs</last-name>
  </author>
</book>
```

Although it is obvious to a human reader that the first title element refers to the title of a book and the second title element to the professional title of its author, there is no way for a computer to "know" this. In larger XML documents, and especially in RDF graphs which combine information from multiple RDF/XML files, confusions such as this can be major problems. We can disambiguate the XML document by declaring the two title elements to belong to different namespaces. The *friend of a friend* (FOAF) vocabulary provides a number of terms for describing people including an element for titles such as Mr, Dr, Professor, and so on [40]. The Dublin Metacore Metadata Initiative provides "core metadata" for simple and generic resource descriptions on the Web. The Dublin Metacore title element can be used for the titles of books and other resources. The following XML document uses both namespaces to disambiguate the use of the English word "title" in the two title elements.

```
<?xml version="1.0" encoding="UTF-8"?>
```

```
<book xmlns:dc="http://purl.org/dc/elements/1.1/"
      xmlns:foaf="http://xmlns.com/foaf/spec/">
  <dc:title>The Man Who Mistook His Wife For a Hat</dc:title>
  <author>
    <foaf:title>Dr</foaf:title>
    <first-name>Oliver</first-name>
    <last-name>Sachs</last-name>
  </author>
</book>
```

Namespaces are used in this way in nearly all RDF/XML documents to bind prefixes (such as dc) to URIs (here, http://purl.org/dc/elements/1.1). We will return to this topic after we have explained the syntax of URLs, URIs, and IRIs.

An XML document can be regarded as a *rooted tree*. In computer science, a tree is a kind of a graph defined by the property that there is exactly one path connecting any two nodes. In a rooted tree, one node is designated as the root. By convention, this node is displayed above all other nodes. Any other node is the root of the *subtree* consisting of the node and all other nodes beneath it. As mentioned, each XML document must have one and only one root element that encloses all other elements. This element is also the root of the tree represented by the XML document, and any element in an XML document is the root of the subtree consisting of the element and any other elements enclosed by the element (Figure C.1).

Uniform Resource Identifier (URI)

A URI is a string that is intended to denote an abstract or concrete resource in an unambiguous fashion. Although URIs are most often used to denote resources in the WWW, such as Web sites, they can be used as a general scheme for identifying a resource in a globally unambiguous way. In the context of RDF; URIs are more often used with the intent of providing globally unique identifiers for entities in the domain being modeled.

URIs can be either Uniform Resource Locators (URLs), Uniform Resource Names (URNs) or both. The usage of these terms can be confusing, and has evolved over time. At the risk of oversimplification, a URL is a subclass of URI that identifies a resource via a representation of its primary access mechanism (e.g., its network "location"), rather than by some other attributes it may have. Thus, http://example.org identifies a resource by giving its address on the WWW. On the other hand, the URN urn:isbn:978-0199219865 unambiguously defines a resource that happens to be a book by its unique ISBN-13 number, but it does not tell us how to gain access to that book.

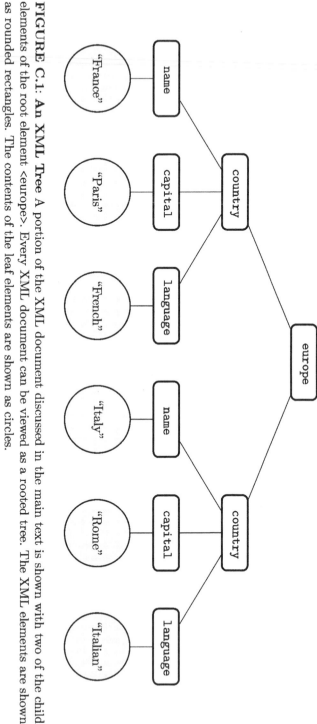

FIGURE C.1: An XML Tree A portion of the XML document discussed in the main text is shown with two of the child elements of the root element <europe>. Every XML document can be viewed as a rooted tree. The XML elements are shown as rounded rectangles. The contents of the leaf elements are shown as circles.

The generic URI syntax consists of a scheme followed by a colon and a hierarchical sequence of components which depends upon the scheme.

```
<scheme>:<scheme-dependent-part>
```

Schemes include HTTP, FTP, mailto, URN, file, and others. Many, but not all schemes use a common basic URI syntax for representing hierarchical relations within the namespace.

```
<scheme>:<authority><path>?<query>
```

Hierarchical URIs use a "/" character for separating hierarchical components. This may be identical to a file path on a computer but does not have to be. Paths over a network are preceded by "//". The following example shows the components of a typical URI.

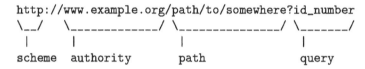

In this example, the `authority` is a server, the *path* indicates the location of the resource according to a hierarchical specification (which may or may not be identical to part of the directory structure on the server), and the `query` is a string of information to be interpreted by the resource (in this example, it might be interpreted as the id number of a Web page or a database entry, but the interpretation is not part of the URI specification – it is left to the resource to decide what to do with the information).

A *URI reference* is a string that represents a URI or the resource identified by that URI. A URI reference may have additional information attached in the form of a fragment identifier. A fragment id can be attached to an absolute or relative URI by appending a # character followed by the fragment id to the end of the URI.

```
http://www.example.org#fragment
```

XML QNames

As we will see, RDF/XML uses XML QNames extensively to provide abbreviations for URIs and URI references. A QName (*qualified name*) is a string of the form `prefix:local_part`. Some examples of QNames are:

```
rdf:about
xsl:template
```

QNames are used to reference elements or attributes within XML or RDF/XML documents. URI references are often rather long and may contain prohibited characters for element/attribute naming. QNames define a mapping between the URI and a namespace prefix that enables abbreviations. For instance, the following namespace declaration associates the prefix `rdf` with the URI reference `http://www.w3.org/1999/02/22-rdf-syntax-ns#`.

```
xmlns:rdf="http://www.w3.org/1999/02/22-rdf-syntax-ns#"
```

The QName `rdf:about` is thus expanded to

```
http://www.w3.org/1999/02/22-rdf-syntax-ns#about
```

The mapping enables the abbreviation of URIs and thus provides a more convenient way to write XML documents. The association between a prefix and a namespace URI is valid within the element in which it is defined. In RDF/XML, the definitions are therefore usually made in the outermost element `rdf:RDF` so that the QNames can be used in the entire RDF/XML document. QNames can be used in XML element or attribute names, but not as a part of the attribute value. Thus, the following is a well-formed XML document. The `rdf:RDF` element has a single child element whose name is the QName `foaf:homepage`. This element has an attribute called `rdf:resource`, whose value is a string literal ("http://www.my-homepage.org/").

```
<?xml version="1.0" encoding="UTF-8"?>
<rdf:RDF xmlns:rdf="http://www.w3.org/1999/02/22-rdf-syntax-ns#"
         xmlns:foaf="http://xmlns.com/foaf/0.1/">
  <foaf:homepage rdf:resource="http://www.my-homepage.org/"/>
</rdf:RDF>
```

It is also possible for a QName to consist entirely of a `local_part`. We will see below in more detail how QNames are used in RDF/XML.

RDF and RDF Serializations

An RDF document describes a directed graph (see chapter 3 for a brief introduction to graph theory). RDF has an abstract syntax that reflects a simple graph-based data model. The semantics of RDF offer a rigorously defined notion of entailment that provides a basis for well-founded deductions with RDF data, which is discussed in chapter 4.

Readers may wonder why XML documents can be represented by a rooted tree, and RDF documents need to be represented as a general graph. XML was designed for the structured representation of data in a way roughly comparable to an SQL database, in that the same bits of information tend to be stored for multiple items. RDF on the other hand follows an open-world assumption: *Anyone is allowed to say anything about anything.* This means we cannot restrict the structure of our data representation *a priori* the way XML does with its predefined tree structure, because the RDF graph may import data from an external source that adds new knowledge. The triple-based syntax of RDF makes it much easier to combine information from multiple sources on the Web, and allows more flexible data structures (graphs) to be constructed from data.

There are multiple ways of representing the same RDF graph as a computer file. The process of converting a data structure into a sequence of bits so that it can be stored as a file is called *serialization*. In the present context, we will use the word serialization to refer to the process of converting an RDF graph into a representation that can be stored as a file, from which the RDF graph can be reconstructed. In this section, we will be concerned with the syntax of the most commonly used serializations for RDF, RDF/XML, N3, N-Triples, and Turtle.

RDF/XML is a syntax that serializes RDF graphs as XML documents. For the most part, the other serialization specifications are intended to be simplified, easily readable forms of the same RDF graph. Therefore, we will begin with a detailed explanation of RDF/XML syntax and then show how to use the simpler non-XML serializations with a few examples.

RDF Triples

As mentioned, RDF graphs represent RDF triples consisting of a subject node, predicate, and object node. The subject can be a URI reference or a bnode and the object can be a URI reference, a bnode, or a literal (see below). The predicate can only be an RDF URI reference which specifies the relationship between the subject node and the object node or defines an attribute value (object node) for the subject node [26]. RDF/XML uses XML QNames to represent RDF URI references. An RDF graph is a collection of directed paths from subject nodes over predicate edges to object nodes. Therefore, each edge in the graph represents a piece of knowledge about the relation between two things.

RDF/XML is flexible, and there are a number of different ways of representing triples that we will now review. Consider first the following RDF graph. We have defined the namespace `http://www.example.org/protein#` for the purpose of this example. The Web site does not exist, which is not relevant for the RDF syntax or for parsers; however, we imagine that the file `protein.rdf` contains some definition of the keywords used just as the W3C Web site for RDF syntax contains definitions of the RDF keywords. Note that we are using ENTITY declarations to provide abbreviations that can be used as part of the attribute values to improve the legibility of the RDF/XML code.

```
<?xml version="1.0" encoding="UTF-8"?>
<!DOCTYPE rdf:RDF [
<!ENTITY uniprot "http://www.uniprot.org/uniprot/">
```

```
<!ENTITY upkeyword "http://www.uniprot.org/keywords/">
]>


<rdf:RDF xmlns:rdf="http://www.w3.org/1999/02/22-rdf-syntax-ns#"
         xmlns:pr="http://www.example.org/protein.rdf#">
  <rdf:Description rdf:about="&uniprot;Q9NUD9">
    <pr:name>GPI mannosyltransferase 2</pr:name>
    <pr:symbol>PIGV</pr:symbol>
    <pr:process rdf:resource="&upkeyword;KW-0337"/>
  </rdf:Description>
</rdf:RDF>
```

In an RDF/XML document, there are two kinds of nodes, those that refer to subjects or objects (*resource nodes*) and those that refer to properties. Resource nodes often (but by no means always) are rdf:Description elements with an rdf:about attribute that specifies the URI of the resource corresponding to the subject of the statement. Resource nodes contain property nodes. In the example, the resource node contains three property nodes, pr:name, pr:symbol, and pr:process. Each property node represents a triple whose subject is defined by the enclosing resource node. Such a statement can have either a resource (URI) or a literal value as its object (or a bnode, as will be discussed below). Thus, the statement defined by the <pr:symbol>PIGV</pr:symbol> element has the literal value "PIGV" as its object, but the statement defined by the <pr:process> element has a resource. In this case, the resource is specified not by the content of the element but rather by the value of the rdf:resource attribute (Figure C.2).

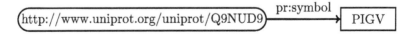

FIGURE C.2: An RDF Triple. This is the graph corresponding to the triple defined by the subject http://www.uniprot.org/uniprot/Q9NUD9, the predicate pr:symbol and the object "PIGV," which is a literal value (symbolized by a rectangular node).

The entire graph represents three triples, all of which have the same subject (Figure C.3).

Note that there are multiple syntactically valid ways of representing the same RDF graph in RDF/XML notation. The following representation encodes two of the triples using attributes of the <rdf:Description> node.

```
<?xml version="1.0" encoding="UTF-8"?>
<!DOCTYPE rdf:RDF [
<!ENTITY uniprot "http://www.uniprot.org/uniprot/">
<!ENTITY upkeyword "http://www.uniprot.org/keywords/">
]>
```

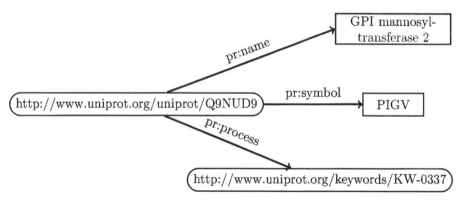

FIGURE C.3: An RDF Graph. This is the graph corresponding to the three triples defined by the subject `http://www.uniprot.org/uniprot/Q9NUD9` as described in the text. Two of the nodes representing objects are literals (symbolized by rectangular nodes), and one represents a resource.

```
<rdf:RDF xmlns:rdf="http://www.w3.org/1999/02/22-rdf-syntax-ns#"
         xmlns:pr="http://www.example.org/protein#">
   <rdf:Description rdf:about="&uniprot;Q9NUD9"
                   pr:name="GPI mannosyltransferase 2"
                   pr:symbol="PIGV">
      <pr:process rdf:resource="&upkeyword;KW-0337"/>
   </rdf:Description>
</rdf:RDF>
```

Another equally valid way of writing the same graph encodes the resource describing the process using a separate `rdf:Description` node.

```
<?xml version="1.0" encoding="UTF-8"?>
<!DOCTYPE rdf:RDF [
<!ENTITY pr "http://www.example.org/protein.rdf#">
<!ENTITY uniprot "http://www.uniprot.org/uniprot/">
<!ENTITY upkeyword "http://www.uniprot.org/keywords/">
]>
<rdf:RDF xmlns:rdf="http://www.w3.org/1999/02/22-rdf-syntax-ns#"
         xmlns:pr="http://www.example.org/protein.rdf#">
   <rdf:Description rdf:about="&uniprot;Q9NUD9">
      <pr:name>GPI mannosyltransferase 2</pr:name>
      <pr:symbol>PIGV</pr:symbol>
      <pr:process>
         <rdf:Description rdf:about="&upkeyword;KW-0337"/>
      </pr:process>
   </rdf:Description>
```

```
</rdf:RDF>
```

This illustrates the fact that an identical RDF graph can be represented by different RDF/XML documents. RDF was designed according to the open-world principle: anyone can say anything about anything. RDF is designed such that additional triples can be added to any graph. If the URIs of a subject or object node match, the new triple is "attached" to one of the nodes in the graph. For instance, in the above example, the object of one of the triples is the resource http://www.uniprot.org/keywords/KW-0337. This is a Web page of UniProt [254] that describes the biological process *GPI-anchor biosynthesis*. We can add an additional RDF triple that describes the name of this resource using the pr:name predicate.

```
<?xml version="1.0" encoding="UTF-8"?>
<!DOCTYPE rdf:RDF [
<!ENTITY uniprot "http://www.uniprot.org/uniprot/">
<!ENTITY upkeyword "http://www.uniprot.org/keywords/">
]>
<rdf:RDF xmlns:rdf="http://www.w3.org/1999/02/22-rdf-syntax-ns#"
         xmlns:pr="http://www.example.org/protein.rdf#">
  <rdf:Description rdf:about="&uniprot;Q9NUD9">
    <pr:name>GPI mannosyltransferase 2</pr:name>
    <pr:symbol>PIGV</pr:symbol>
    <pr:process rdf:resource="&upkeyword;KW-0337"/>
  </rdf:Description>
  <rdf:Description rdf:about="&upkeyword;KW-0337">
    <pr:name>GPI-anchor biosynthesis</pr:name>
  </rdf:Description>
</rdf:RDF>
```

This RDF code makes the object of the triple <Q9NUD9,process,KW-0337> into the subject of the new triple (KW-0337,pr:name,GPI-anchor biosynthesis). Although in this example we have simply added some lines to the RDF/XML file, there is no reason that the new triple could not have come from another Web server. Figure C.4 shows how the additional triple gets integrated into the RDF graph.

In this case again, there are multiple ways of serializing the same RDF graph.

```
<?xml version="1.0" encoding="UTF-8"?>
<!DOCTYPE rdf:RDF [
<!ENTITY uniprot "http://www.uniprot.org/uniprot/">
<!ENTITY upkeyword "http://www.uniprot.org/keywords/">
]>
<rdf:RDF xmlns:rdf="http://www.w3.org/1999/02/22-rdf-syntax-ns#"
         xmlns:pr="http://www.example.org/protein.rdf#">
```

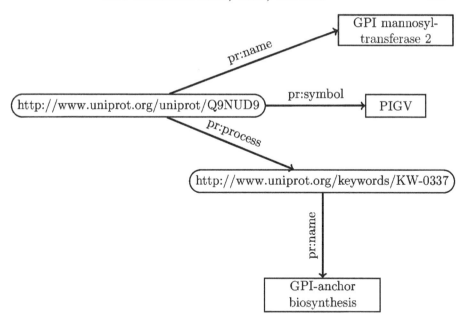

FIGURE C.4: Addition of a Triple to an RDF Graph. This figure shows how the new triple described in the text is integrated into the RDF graph of Figure C.3.

```
<rdf:Description rdf:about="&uniprot;Q9NUD9">
  <pr:name>GPI mannosyltransferase 2</pr:name>
  <pr:symbol>PIGV</pr:symbol>
  <pr:process>
    <rdf:Description rdf:about="&upkeyword;KW-0337">
      <pr:name>GPI-anchor biosynthesis</pr:name>
    </rdf:Description>
  </pr:process>
</rdf:Description>
</rdf:RDF>
```

Recall from Chapter 4 that `rdf:type` is used to state that a resource is an instance of a class. For instance, if we state R `rdf:type` C for some class C, then we are stating that R is an instance of that class. Since this sort of statement is extremely common in RDF and OWL, there is a syntactic shortcut for it that has come to be widely used, whereby the name of the class is used as the name of the corresponding RDF/XML element. In the UniProt RDF document describing entry Q9NUD9, there is a triple defining the entry to be an instance of the class `Protein`. This translates into the following triple.

```
<?xml version="1.0" encoding="UTF-8"?>
<!DOCTYPE rdf:RDF [
```

```
<!ENTITY uniprot "http://www.uniprot.org/uniprot/">
]>
<rdf:RDF xmlns:rdf="http://www.w3.org/1999/02/22-rdf-syntax-ns#">
  <rdf:Description rdf:about="&uniprot;Q9NUD9">
    <rdf:type
        rdf:resource="http://purl.uniprot.org/core/Protein"/>
  </rdf:Description>
</rdf:RDF>
```

This RDF/XML serialization has a completely equivalent meaning to the
following one.

```
<?xml version="1.0" encoding="UTF-8"?>
<!DOCTYPE rdf:RDF [
<!ENTITY uniprot "http://www.uniprot.org/uniprot/">
]>
<rdf:RDF xmlns:rdf="http://www.w3.org/1999/02/22-rdf-syntax-ns#"
        xmlns:uniprot="http://purl.uniprot.org/core/">
  <uniprot:Protein rdf:about="&uniprot;Q9NUD9"/>
</rdf:RDF>
```

Note that we have defined a new namespace and associated it with the
prefix uniprot in order to be able to use uniprot:Protein as a valid QName.
Using http://purl.uniprot.org/core/Protein as the value of the attribute
rdf:about would have led to a parse error, because the http part of the URI
is parsed as the prefix of a QName (see the section on QNames above if you
do not understand this). Both RDF/XML documents encode the graph shown
in Figure C.5.

FIGURE C.5: RDF Triple Representing the rdf:type Relation. Both
the syntactic variants shown in the text correspond to exactly the same RDF
graph.

Literal Values and Blank Nodes

In RDF, URIs are taken to be globally unique names for resources that
exist in the real world. RDF is meant to be published on the Internet, and
URIs are an elegant tool that allows an unambiguous way for different RDF
documents to refer to the same thing. The names for subjects, predicates,
and objects may all be URIs. However, there are statements that require
other kinds of values. The first is the *literal value*, which is simply a string

of characters that can be used instead of objects in an RDF triple. By default, a literal value is interpreted as a string, but other datatypes such as integers and dates can be specified using XML Schema. For instance, `"123"` is an untyped literal, but `"123"^^xsd:integer` will be interpreted as an integer (assuming the prefix `xsd` has been associated with the namespace `http://www.w3.org/2001/XMLSchema#`).

A blank node (also called a bnode or an anonymous node) does not have a URI because it is impossible or not necessary to refer to something by name. Say for instance we know "The P53 protein activates a nuclear protein involved in DNA repair," but we do not know the identity of the nuclear protein. RDF can use a bnode to represent the unknown nuclear protein in a triple (P53,activates,bnode) and can then refer to the same bnode in other triples (bnode,location,nucleus) and (bnode,mediates,DNA repair). We will see how bnodes are used in the following section.

RDF/XML Striping

The notion of "striping" is a very useful conceptual tool for understanding RDF/XML [38]. The RDF/XML syntax is a strategy for encoding the node-edge-node structure of RDF triples using XML elements. Consider the following RDF/XML document, which states that ALK1 phosphorylates ENT2 which in turn interacts with SYP1 (ALK1, ENT2, and SYP1 are proteins in the Baker's yeast *Saccharomyces cerevisiae*).

```
<?xml version="1.0" encoding="UTF-8"?>
<rdf:RDF xmlns:rdf="http://www.w3.org/1999/02/22-rdf-syntax-ns#"
         xmlns:pr="http://www.example.org/protein#">
R1:  <pr:Protein>
P2:    <pr:name>ALK1</pr:name>
P2:    <pr:phosphorylates>
R3:      <pr:Protein>
P4:        <pr:Name>ENT2</pr:Name>
P4:        <pr:interactsWith>
R5:          <pr:Protein>
P6:            <pr:name>SYP1</pr:name>
R5:          </pr:Protein>
P4:        </pr:interactsWith>
R3:      </pr:Protein>
P2:    </pr:phosphorylates>
R1:  </pr:Protein>
</rdf:RDF>
```

We have labeled the lines of the RDF/XML document with an "R" for resource and a "P" for predicate. The "R"s and "P"s form "stripes" that correspond to the nesting level of the XML elements as indicated by the numbers and the indentation level. Figure C.6 shows the RDF graph corresponding to the above RDF/XML serialization. Note that bnodes are used to indicate resources that are mentioned but not explicitly named with a URI.

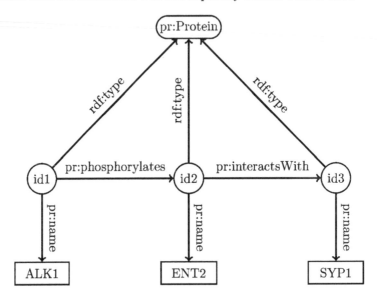

FIGURE C.6: RDF Striping. Compare this figure with the RDF/XML document in the text to understand the "striped" structure of RDF/XML. The blank nodes (bnodes) are shown as circles and have been given arbitrary ids.

An RDF graph can be considered to be a collection of paths formed of node $\xrightarrow{P}$ node triples (where $\xrightarrow{P}$ is a edge representing the predicate). A node that represents the object of one triple can represent the subject of another triple, creating a path of the form $n_1 \xrightarrow{P_1} n_2 \xrightarrow{P_2} n_3$, where n_1, n_2, and n_3 are nodes and $\xrightarrow{P_1}$ and $\xrightarrow{P_2}$ are predicates. In the RDF/XML serialization, these turn into sequences of elements inside elements which alternate between elements for nodes and predicate arcs [38]. In the example shown in Figure C.6, we can see this pattern, whereby n_1=id:1, P_1=pr:phosphorylates, n_2=id:2, P_2=pr:interactsWith, and n_3=id:3. This pattern is called node/arc striping. Note, however, that the syntactic variants of RDF/XML presented above can obscure striping patterns. Nonetheless, looking for patterns such as this can be a helpful way to learn to read RDF/XML documents.

N3 and Turtle

We have already discussed N3 in Chapter 4. Turtle is a subset of N3 that has become widely used for representing RDF graphs in a way that seems more intuitive than RDF/XML to some users. As mentioned many times, RDF is a collection of triples, each with a subject, predicate, and object.

Turtle essentially just writes down the triples in arbitrary order, and provides a number of shortcuts that improve legibility. Turtle was developed from the W3C recommendation *N-Triples*, which essentially just lists all of the triples of an RDF graph one after the other without any of the syntactic shortcuts of Turtle. We will not discuss N-Triples further.

Consider the following triple in N3 notation.

```
<#S> <#P> <#0> .
```

Everything in RDF is identified using a URI (except for objects represented by literal values). The # character identifies the resources #S, #P, #0 in the current document, that is, they are assumed to belong to the default namespace of the document. If the object is a literal value, it is enclosed in quotation marks.

```
<#Q9NUD9> <#symbol> "PIGV" .
```

There is a shortcut for combining triples if there are multiple triples about the same subject:

```
<#Q9NUD9> <#symbol> "PIGV" ;
          <#name> "GPI mannosyltransferase 2" ;
          <#process> <http://www.uniprot.org/keywords/KW-0337> .
```

Thus, there are three triples with the same subject (#Q9NUD). Except for the first triple, the subject is omitted. Except for the last triple, each statement is terminated with a semicolon. Except for the first triple, the subject is omitted. The last triple is terminated with a period. If there are a number of triples with the same subject and predicate but different objects, the objects are separated with a comma:

```
<#TP53> <#interactsWith> <#BAK1>, <#ATM>, <#BAK1>, <#BAX>,
        <#EP300>, <#HIF1A>, <#MDM2>, <#SP1> .
```

Just as QNames and namespaces can be used to abbreviate URIs in RDF/XML documents, prefixes can be used in N3. The following prefix declaration allows the prefix dc to be used in an N3 document the same way as the XML namespace declaration xmlns:dc="http://purl.org/dc/elements/1.1/" would in RDF/XML.

```
@prefix dc:  <http://purl.org/dc/elements/1.1/> .
```

We will now consider again the following RDF/XML document from a previous section of this appendix.

```
<?xml version="1.0" encoding="UTF-8"?>
<rdf:RDF xmlns:rdf="http://www.w3.org/1999/02/22-rdf-syntax-ns#"
         xmlns:pr="http://www.example.org/protein.rdf#">
  <rdf:Description
```

```
       rdf:about="http://www.uniprot.org/uniprot/Q9NUD9">
     <pr:name>GPI mannosyltransferase 2</pr:name>
     <pr:symbol>PIGV</pr:symbol>
     <pr:process
        rdf:resource="http://www.uniprot.org/keywords/KW-0337"/>
   </rdf:Description>
   <rdf:Description
        rdf:about="http://www.uniprot.org/keywords/KW-0337">
     <pr:name>GPI-anchor biosynthesis</pr:name>
   </rdf:Description>
</rdf:RDF>
```

The same graph can be serialized in a much more compact form in N3 (note that lines have been broken to fit on the page):

```
@prefix : <http://www.example.org/protein.rdf#> .

<http://www.uniprot.org/uniprot/Q9NUD9>
     :name "GPI mannosyltransferase 2";
     :process <http://www.uniprot.org/keywords/KW-0337>;
     :symbol "PIGV" .
<http://www.uniprot.org/keywords/KW-0337>
     :name "GPI-anchor biosynthesis" .
```

The namespace http://www.example.org/protein.rdf# has been declared as the default namespace, so that resources prefixed with a semicolon such as :name are understood to belong to it. The two subjects are represented by full URIs, but they could be abbreviated in the same fashion if desired.

In N3 and Turtle, namespaces are declared with the @prefix directive. The RDF namespace http://www.w3.org/1999/02/22-rdf-syntax-ns# is assumed to be bound to rdf.

Since statements with rdf:type are so commonly encountered, N3 provides a convenient abbreviation (a). The following two N3 statements are equivalent.

```
<#S> <rdf:type> <#O> .
<#S> a <#O> .
```

There are several ways in N3 of representing a blank, or unnamed node: the underscore namespace, the square bracket syntax, and the path syntax [30]. We refer to the discussion in Chapter 4 for a description. We conclude this section with a slightly more complicated example. The following is the N3 serialization of the RDF graph of Figure C.6.

```
@prefix : <http://www.example.org/protein#> .

[ a :Protein;
```

```
:name "ALK1";
:phosphorylates  [
    a :Protein;
    :Name "ENT2";
    :interactsWith  [
        a :Protein;
        :name "SYP1" ] ] ].
```

The three blank nodes in this N3 document are enclosed in pairs of square brackets. The outermost bnode states that there is a thing that is an instance of :Protein, and then goes on to state that the thing has the name "ALK1," and so on.

Reification

Consider the following RDF triple: <ex:Kirk,ex:commands,ex:Enterprise>. Reification is a mechanism for expressing information such as who made the statement, when the statement was made, what evidence supports the statement, and so on. The word reification is derived from the Latin words *res* (thing) and *facere* (to make), and thus literally means "to make into a thing." In our context, it refers to making a bnode that refers to an RDF triple representing a statement. The bnode is the "thing" that was made, and can be referred to by other RDF triples to express information such as the author of the statement. To reify a statement, a bnode is created which points to the subject, predicate and object of the statement. Additionally, the bnode is declared to be an instance of rdf:Statement. The following RDF/XML document comprises a reification of the above statement about Captain Kirk.

```
<?xml version="1.0"?>
<rdf:RDF xmlns:rdf="http://www.w3.org/1999/02/22-rdf-syntax-ns#">
  <rdf:Description>
    <rdf:subject
      rdf:resource="http://www.example.org#Kirk"/>
    <rdf:predicate
      rdf:resource="http://www.example.org#commands"/>
    <rdf:object
      rdf:resource="http://www.example.org#Enterprise"/>
    <rdf:type
  rdf:resource="http://www.w3.org/1999/02/22-rdf-syntax-ns#Statement"/>
  </rdf:Description>
</rdf:RDF>
```

The N3 serialization of this RDF graph shows that a single bnode is the subject of all four triples.

```
@prefix : <http://www.example.org#> .
```

```
@prefix rdf: <http://www.w3.org/1999/02/22-rdf-syntax-ns#> .

[  a rdf:Statement;
   rdf:object :Enterprise;
   rdf:predicate :commands;
   rdf:subject :Kirk ].
```

Alternatively, we can use the underscore notation for the bnode.

```
@prefix : <http://www.example.org#> .
@prefix rdf: <http://www.w3.org/1999/02/22-rdf-syntax-ns#> .

_:1 a rdf:Statement .
_:1 rdf:object :Enterprise .
_:1 rdf:predicate :commands .
_:1 rdf:subject :Kirk .
```

Note that a reification of some triple does not entail the triple nor is it entailed by the triple. Rather, the reification only states that the triple exists and what it is about, and does not make any statement about the truth of the triple.

RDF Vocabulary

This section provides a summary of the main elements of the RDF vocabulary.

rdf:RDF: RDF statements are always contained with an rdf:RDF element. As we have seen above, rdf:RDF elements often contain xmlns attributes to define prefixes and namespaces.

rdf:Description: Resource nodes often (but by no means always) are rdf:Description elements with an rdf:about attribute that specifies the URI of the resource corresponding to the subject of the statement. The following rdf:Description tag begins an XML element representing http://www.uniprot.org/uniprot/Q9NUD9 as the subject of one or more RDF triples.

```
<rdf:Description
      rdf:about="http://www.uniprot.org/uniprot/Q9NUD9">
```

rdf:ID: rdf:ID can be used in a similar way to rdf:about, with the difference that rdf:ID always is relative to the base URI of the document, which is specified by the xml:base attribute of the rdf:RDF element. Consider the following RDF/XML document.

```
<?xml version="1.0"?>
<rdf:RDF xmlns:rdf="http://www.w3.org/1999/02/22-rdf-syntax-ns#"
         xmlns:ex="http://example.org/stuff/1.0/"
         xml:base="http://example.org/food.rdf">
  <rdf:Description rdf:ID="snack">
    <ex:prop rdf:resource="nuts/pistachio"/>
  </rdf:Description>
</rdf:RDF>
```

The base URI of this document is `http://example.org/food.rdf`.

The base URI provides a way of abbreviating RDF URI references in XML attributes. The base URI applies to all RDF/XML attributes that deal with RDF URI references which are `rdf:about`, `rdf:resource`, `rdf:ID` and `rdf:datatype`. There is an important difference between `rdf:ID` and the other kinds of attribute. `rdf:about` can be used to indicate a relative URI and `rdf:ID` *must* be used to indicate a relative URI. The attribute `rdf:ID="snack"` is understood to append a "#" followed by the value of the attribute to the base URI. If the attribute `rdf:about` is used to indicate a relative URI, it is understood to append only the value of the attribute to the base URI (without a "#"). Thus, `rdf:ID="snack"` is equivalent to `rdf:about="#snack"`, and both are equivalent to `rdf:about="http://example.org/food.rdf#snack"`.

rdf:type: The subject of a triple with the predicate `rdf:type` is an instance of the class represented by the object of the triple. The following triple states that Q9NUD9 is an instance of the class `Protein`.

```
<?xml version="1.0" encoding="UTF-8"?>
<rdf:RDF xmlns:rdf="http://www.w3.org/1999/02/22-rdf-syntax-ns#"
         xmlns:uniprot="http://purl.uniprot.org/core/">
  <rdf:Description
       rdf:about="http://www.uniprot.org/uniprot/Q9NUD9">
    <rdf:type
         rdf:resource="http://purl.uniprot.org/core/Protein"/>
  </rdf:Description>
</rdf:RDF>
```

Typed Node Elements: RDF allows typed node elements (whose type is indicated by `rdf:type`) to be expressed more concisely by replacing the `rdf:Description` node element name with the namespaced-element corresponding to the RDF URI reference. In the above example, we need to define a namespace prefix in order to create a valid XML element name. The following RDF/XML document is identical in meaning to the document shown above.

```
<?xml version="1.0" encoding="UTF-8"?>
```

```
<rdf:RDF xmlns:rdf="http://www.w3.org/1999/02/22-rdf-syntax-ns#"
         xmlns:uniprot="http://purl.uniprot.org/core/">
  <uniprot:Protein
     rdf:about="http://www.uniprot.org/uniprot/Q9NUD9"/>
</rdf:RDF>
```

RDF Containers: The RDF container vocabulary comprises `rdf:Seq`, `rdf:Bag`, `rdf:Alt`, `rdf:_1`, `rdf:_2`, ... There are three types of container, each of whose members can be enumerated using a fixed set of container membership properties that are indexed by integers (`rdf:_1`, `rdf:_2`, ...). Note that these indices do not necessarily define an ordering of the members (e.g., the index `rdf:_2` does not necessarily "come after" `rdf:_1`). The three container types have different intended modes of use. Items of type `rdf:Bag` are considered to be unordered but allow duplicates; items of type `rdf:Seq` are considered to be ordered; and items of type `rdf:Alt` are considered to represent a collection of alternatives, possibly with a preference ordering. The ordering of items in an ordered container is intended to be indicated by the numerical ordering of the container membership properties (`rdf:_1`, `rdf:_2`, ...). However, these informal interpretations are not reflected in any formal RDF entailments.

```
<?xml version="1.0"?>

<rdf:RDF
xmlns:rdf="http://www.w3.org/1999/02/22-rdf-syntax-ns#"
xmlns:rna="http://www.example.org/nucleicAcid#">

<rdf:Description
rdf:about="http://www.example.org/nucleicAcid#RNA">
  <rna:rna-class>
    <rdf:Bag>
      <rdf:li>tRNA</rdf:li>
      <rdf:li>mRNA</rdf:li>
      <rdf:li>snoRNA</rdf:li>
      <rdf:li>miRNA</rdf:li>
    </rdf:Bag>
  </rna:rna-class>
</rdf:Description>

</rdf:RDF>
```

This RDF/XML document corresponds to the RDF graph in Figure C.7. Note that in the RDF/XML document, the element `rdf:Bag` comes at a place where one normally might expect an `rdf:Description` element (see the discussion of RDF striping above). This has the effect of creating a bnode that has the `rdf:type` of `rdf:Bag`. The four `rdf:li` elements are special property elements that form a kind of shortcut for writing `rdf:_1`, `rdf:_2`, and so on, in order. Replacing the `rdf:Bag` element in the above RDF/XML document with either `rdf:Alt` or `rdf:Seq` would produce an RDF graph with the identical structure whereby only the node for `rdf:Bag` would be replaced by a node for `rdf:Alt` or `rdf:Seq`.

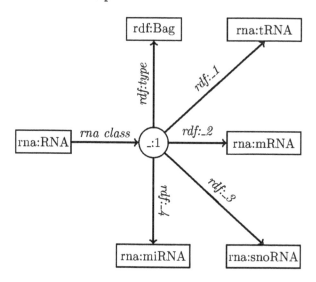

FIGURE C.7: RDF Containers. In this example, the RDF graph for an `rdf:Bag` element with four items is shown.

RDF does not have a mechanism for *closing* a container, that is, of indicating that the container contains no more elements than the ones indicated. This of course goes along with the open-world assumption of RDF. RDF does have constructs for Collections (`rdf:List`, `rdf:first`, `rdf:Last`, `rdf:nil`) that represent lists similar to the way that a linked list data structure in Java or C might be used to represent a list. We refer interested readers to the documentation of the W3C for more information on RDF collections [26].

RDF Reification vocabulary: The keyword `rdf:Statement` is used in triples such as R `rdf:type` `rdf:Statement` to indicate the R represents an RDF statement. A triple such as this is used together with three other triples to describe the subject (`rdf:subject`), predicate (`rdf:predicate`), and object (`rdf:object`) of the statement. Consider for instance, the following sentence: "John claims that the Suriname has the capital city Paramaribo." Here, an individual called *John* is making a statement, "Suriname has the capital city Paramaribo." The subject of the statement is *Suriname*, the predicate is *has capital city*, and the object is *Paramaribo*.
Consider now the following RDF triple.

```
Suriname has_capital Paramaribo .
```

This represents an assertion that it is true that "Suriname has the capital city Paramaribo." The RDF reification vocabulary is used to represent statements such as this that can be referred to by other triples in the RDF graph, without making any assertion whether the triple is true or false. This is most often used to report a statement made by somebody else. Consider the following triples.

```
John claims R .
R rdf:subject Suriname .
R rdf:predicate has_capital .
R rdf:object Paramaribo .
```

These four triples assert that it is true that *John* claimed that "Suriname has the capital city Paramaribo," but themselves make no commitment as to whether John was right or wrong. We can now summarize the meaning of the RDF keywords rdf:subject, rdf:predicate, and rdf:object.

- rdf:subject: The thing the statement is about. May be a URI or a bnode.

- rdf:predicate: The property of the rdf:subject that the statement specifies. Must be a URI.

- rdf:object: Identifies the value of the property specified by the rdf:predicate. May be a URI, blank node, or a Unicode string literal.

rdf:nodeID: Different blank nodes in an RDF graph are distinct but have no URI reference. If it is necessary to refer to the same blank node in multiple places of the RDF/XML document, rdf:nodeID is used to provide the bnode with a document-scope unique identifier. rdf:nodeID can be used in place of rdf:about="RDF URI Reference" on a node element or rdf:resource="RDF URI reference" on a property element.

Recall the RDF graph with three bnodes from our discussion on node/arc striping. We can use the rdf:nodeID attribute to create three unique IDs for these bnodes that can be used anywhere within the document.

```xml
<?xml version="1.0" encoding="UTF-8"?>
<rdf:RDF xmlns:rdf="http://www.w3.org/1999/02/22-rdf-syntax-ns#"
         xmlns:pr="http://www.example.org/protein#">
  <pr:Protein rdf:nodeID="id1">
    <pr:name>ALK1</pr:name>
    <pr:phosphorylates>
      <pr:Protein rdf:nodeID="id2">
        <pr:Name>ENT2</pr:Name>
        <pr:interactsWith>
```

```
  <pr:Protein rdf:nodeID="id3">
    <pr:name>SYP1</pr:name>
  </pr:Protein>
  </pr:interactsWith>
  </pr:Protein>
  </pr:phosphorylates>
</pr:Protein>
<rdf:Description rdf:nodeID="id1">
  <pr:aminoAcids>760</pr:aminoAcids>
</rdf:Description>
</rdf:RDF>
```

The final triple in this document, (bnode=id1,aminoAcids,760), uses the attribute `rdf:nodeID="id1"` to identify the bnode "id1" as the subject of the statement (Figure C.8).

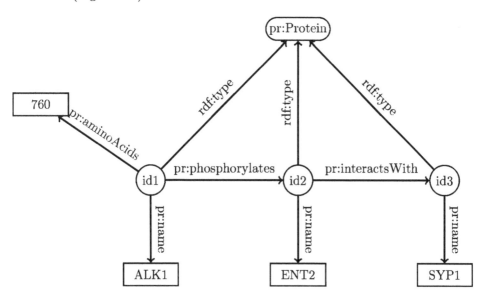

FIGURE C.8: rdf:nodeID. Compare this figure with Figure C.6. An additional triple has been attached to the bnode with the id "id1" using `rdf:nodeID`.

RDF Literals and Datatypes: Property values, such as textual strings, are examples of RDF literals. Literals may be plain or typed, and are usually used to identify values such as numbers and dates by means of a lexical representation. Anything represented by a literal could also be represented by a URI, but it is often more convenient or intuitive to use literals. As mentioned above, a literal may be the object of an RDF statement, but not the subject or the predicate.

A plain literal is a string (with an optional language tag). For instance, "what's up, doc?" and "0.5772156649015328606065" are both literals. RDF does not "know" that the second literal represents a number and the first represents an expression in the English language (that would require a language tag: "what's up, doc?"@en).

An untyped RDF literal is represented in Turtle syntax as

```
ex:a ex:b "whatever" .
```

A typed literal can be declared to be of type xsd:string as follows:

```
ex:a ex:b "whatever"^^xsd:string .
```

If a variable is typed as an XML Literal (e.g., "x"^^rdf:XMLLiteral) then "x" must be a well-formed XML literal (see Section 13.3). The RDF syntax and semantics do not themselves provide special interpretations for data types; this is left to the applications. Datatypes defined in XML Schema [42] are often used in RDF documents. For instance, to specify that the node <pr:aminoAcids> in the RDF/XML document presented on page 422 refers to an integer value, we would modify that document as follows:

```xml
<?xml version="1.0" encoding="UTF-8"?>
<!DOCTYPE rdf:RDF
    [<!ENTITY xsd "http://www.w3.org/2001/XMLSchema#">]>
<rdf:RDF xmlns:rdf="http://www.w3.org/1999/02/22-rdf-syntax-ns#"
              xmlns:pr="http://www.example.org/protein#">
  <pr:Protein rdf:nodeID="id1">
    <pr:name>ALK1</pr:name>
    <pr:phosphorylates>
      <pr:Protein rdf:nodeID="id2">
        <pr:Name>ENT2</pr:Name>
        <pr:interactsWith>
          <pr:Protein rdf:nodeID="id3">
            <pr:name>SYP1</pr:name>
          </pr:Protein>
        </pr:interactsWith>
      </pr:Protein>
    </pr:phosphorylates>
  </pr:Protein>
  <rdf:Description rdf:nodeID="id1">
    <pr:aminoAcids
        rdf:datatype="&xsd;integer">760</pr:aminoAcids>
  </rdf:Description>
</rdf:RDF>
```

We have used a DTD to declare the entity xsd to represent the XML Schema namespace. Within the <pr:aminoAcids> element, we have used an rdf:datatype with the value "&xsd;integer" to indicate that the literal "760" is an integer. Note that "&xsd;integer" expands to "http://www.w3.org/2001/XMLSchema#integer". It is up to application software to use the information that "760" is an integer; this fact is not encoded within RDF semantics itself.

If not otherwise specified, RDF literals are untyped. The treatment of literals in RDF is rather involved, and is not described in detail in this book. Interested readers are referred to the documentation of the W3C for more information [26]. The attribute rdf:datatype is used to indicate the datatype of literals. For instance, the following element declares the literal "614" to be an integer datatype.

```
<prot:length rdf:datatype="&xsd;integer">614</prot:length>
```

Other datatypes defined by XML Schema [42] include string, boolean, decimal, double, and date.

Further Reading

This appendix was intended to familiarize readers with the syntax and vocabulary of RDF rather than to provide a comprehensive treatment of it. Most of the material in this book on RDF and RDFS semantics is presented in Chapter 13. Readers who have studied that chapter and this appendix should be able to follow most of the material in the W3C documentation that is cited below.

XML in a Nutshell by Harold and Means provides a comprehensive and readable account of XML [104]. The W3C provides extensive documentation on XML, URIs, and RDF. An overview is given in Table C.1.

XML	
Extensible Markup Language (XML) 1.0 [264]	Description of the XML language
XML Schema: Formal Description [42]	Description of scope and format of XML Schema
URIs	
URIs, URLs, and URNs [256]	Clarifications and Recommendations of the W3C
RDF	
RDF Primer [160]	Introduction to RDF
RDF Concepts and Abstract Syntax [141]	Defines the abstract syntax on which RDF is based
RDF/XML Syntax [26]	Describes the XML serialization of RDF
RDF Semantics [108]	Describes formal background of RDF entailment
RDF Schema [39]	Describes RDFS

TABLE C.1: Overview about Semantic Web specification documents.

Appendix D

W3C Standards: OWL

Chapter 4 provided an introduction to the Web Ontology Language OWL, and Chapter 14 dealt with inference in OWL. In the body of the book, we have mainly used the Manchester or Turtle syntax for RDF and OWL, but readers are likely to encounter other serializations of OWL ontologies, including especially RDF/XML and OWL/XML serializations. As mentioned in Chapter 4, there are many ways of serializing OWL. For most purposes, the Manchester syntax is the easiest to understand and is the best suited for human consumption. Most OWL 2 files are not written in Manchester syntax, but rather in RDF/XML or OWL/XML format.

This appendix provides a tour of these three serializations based on the documentation of the W3C on this subject [17, 49, 94, 116, 125, 172, 173, 197, 258]. The appendix will deal mainly with the language constructs and syntax of OWL 2, and is intended as a quick reference that is based mainly on the documentation of the W3C on the subject. We will present what is hopefully an intuitive and concise summary of the most important aspects of OWL 2 and the various ways of serializing OWL 2 ontologies, and will provide pointers at the end of the appendix to the relevant documents of the W3C for more information.

An OWL 2 ontology can be defined in abstract terms, but in practice, the ontology needs to be written in some concrete syntax to a file to be stored on a computer or transmitted over a network. The act of writing down a computer data structure to a file is referred to as serialization. There are five main syntaxes that can be used to serialize OWL 2 ontologies (Table D.1).

In this appendix, we will not attempt to explain the full details of the W3C specifications (that is, after all, what the specifications of the W3C are there for). Rather, we will present an intuitive explanation of the most important constructs that should be enough so that readers will understand most OWL 2 constructs encountered in current bio-ontologies written in any of the serializations.

Manchester Syntax

OWL has one normative exchange syntax, RDF/XML, which was discussed in detail in Appendix C, as well as the OWL/XML and the function syntax [173]. None of these formats was designed for ease of use by humans (we hope that the last point was not all too obvious to readers of Appendix C). The

Name	Purpose	Ref.
RDF/XML	Interchange	[26, 197]
OWL/XML	Processing using XML tools	[171]
Functional Syntax	Visualize the formal structure of ontologies	[173]
Manchester Syntax	Easily (human-)comprehensible syntax for reading and writing OWL 2 DL	[125]
Turtle	Easily (human-)comprehensible syntax for reading and writing RDF triples	[27, 197]

TABLE D.1: According to W3C specifications, only the RDF/XML syntax, which was introduced in Appendix C and will also be discussed in this appendix, is mandatory for OWL 2 tools. OWL/XML is an important alternative that can be easily parsed by XML tools, and Manchester syntax is intended to provide a concise and easily comprehensible serialization of OWL 2 DL ontologies for human consumption. These three serializations will be presented in detail in this appendix. Readers are referred to the documentation of the W3C for more information about the functional syntax for OWL 2 and Turtle.

Manchester OWL syntax was created by a number of people at the University of Manchester to meet this need[1]. Manchester syntax uses short intuitive words instead of logical symbols for complex class descriptions, and is based on a frame-like organization principle in which all information about a class is collected [125]. Frames and parts of frames are introduced by keywords that end in colons, and precedence rules are used to avoid an overabundance of parentheses. For instance,

```
Class: Person
  SubClassOf:
      hasAge exactly 1
    and hasGender exactly 1
    and hasGender only {female, male}
```

This fragment of Manchester OWL defines the class of Person to be a subclass of all things that have exactly one value for the predicate `hasAge` as well as exactly one value for the predicate `hasGender`, namely, either `male` or `female`.

OWL RDF/XML Syntax

The vast majority of OWL ontologies published to date have been serialized using the RDF/XML format. Therefore, an important goal of the design of OWL 2 was to ensure backwards compatibility with RDF/XML. Every

---

[1]See the Web page of the group at http://owl.cs.manchester.ac.uk/ as well as the articles in [123, 124, 125].

OWL 1 DL ontology in RDF/XML syntax is also a valid OWL 2 ontology. The RDF/XML syntax for OWL is just the RDF/XML syntax as presented in Appendix C, but with a particular translation for the OWL constructs [197]. The RDF/XML syntax is the only syntax that is mandatory for all OWL 2 tools.

OWL XML Syntax

The W3C has defined a separate XML serialization as an exchange syntax for OWL 2. The XML serialization mirrors the structural specification of OWL 2 and is defined by means of an XML schema and some additional prose constraints [171]. The XML format has some advantages over the more common RDF/XML format in that XML has been much more widely adopted on the Web and is supported by more tools. In addition, the OWL/XML format more closely resembles the underlying specification of OWL 2.

The following sections will present serializations for the most important OWL 2 constructs using the Manchester, RDF/XML, and XML syntaxes. Readers will encounter each of these three syntaxes in publications and on the Web, and it is hoped that by comparing the serializations readers will be able to develop intuition for the meaning of the OWL constructs. There are two other OWL serializations that are commonly used: Turtle [27], which was covered briefly in Appendix C, as well as the functional-style syntax [173] (note that the main purpose of the functional-style syntax is to specify the structure of OWL 2, but it can also be used for serialization).

In this appendix, we will often shorten long IRIs for better legibility. Readers should recognize that in actual ontology files all entities must have either full IRIs or use appropriate entities or prefixes in order that each resource is uniquely identified. We will use references to *Star Trek* for the examples[2].

Classes and Instances

ClassAssertion

`ClassAssertion(i C)` is used to express an assertion that an individual *i* is an instance of class *C*.

As we have seen in Chapters 4 and 14, OWL 2 ontologies describe universes that contain individuals that belong to classes and which can be related to one another by relations (properties/roles). We have seen in Appendix C that there are multiple ways of stating that an individual is an instance of a class

---

[2]With apologies to those readers who do not know much about Star Trek; but the examples should be easy to follow even for non-Trekkies. The alternative would have been yet another example ontology about families or employees.

in RDF/XML syntax. In this appendix, we will employ the following short cut form:

```
<Person rdf:about="McCoy"/>
```

In OWL/XML syntax, the assertion that an individual *i* is an instance of a class *C* is expressed using a `<ClassAssertion>` element.

```
<ClassAssertion>
    <Class IRI="Person"/>
    <NamedIndividual IRI="McCoy"/>
</ClassAssertion>
```

In Manchester syntax, an assertion is made using the keyword `Individual` in order to begin a stanza about the individual. All of the classes which the individual is an instance of are listed after the keyword `Types`.

```
Individual: McCoy
  Types: Person
```

Subclass Relations

SubClassOf

SubClassOf(*C D*) is used to express an assertion that class *C* is a subclass of class *D*, meaning that any instance of *C* is also an instance of *D*.

Using the RDF/XML syntax, we can express the assertion that *Chief Medical Officer* is a subclass of *Star Fleet Officer* using the RDFS keyword `subClassOf`.

```
<owl:Class rdf:about="ChiefMedicalOfficer">
  <rdfs:subClassOf rdf:resource="StarFleetOfficer"/>
</owl:Class>
```

In OWL/XML syntax, a `<SubClassOf>` element is used to express a subclass relation. The order of the two child elements is important. The following XML fragment expresses the assertion that *Chief Medical Officer* is a subclass of *Star Fleet Officer* and not the other way around.

```
<SubClassOf>
  <Class IRI="ChiefMedicalOfficer"/>
  <Class IRI="StarFleetOfficer"/>
</SubClassOf>
```

In OWL Manchester syntax, the subclass assertion is placed in the stanza of the subsumed class.

```
Class: ChiefMedicalOfficer
  SubClassOf: StarFleetOfficer
```

In OWL 2, classes are *reflexive*, i.e., every class is a subset of itself. The subclass relation is *transitive*, i.e., if class A is a subclass of B, and class B is a subclass of C, then class A is also a subclass of class C.

Class Equivalency and Disjointness

EquivalentClasses

`EquivalentClasses(`C D`)` is used to express an assertion that class C is equivalent to class D, meaning that any instance of C is also an instance of D and vice versa.

Classes in an ontology may have different names but effectively refer to the same sets. The OWL keyword `EquivalentClasses` is used to indicate that two classes are semantically equivalent. Two classes are considered equivalent if they contain exactly the same individuals.

Using the RDF/XML syntax, we can express the assertion that the class *Terran* (inhabitants of the planet Earth, or *Terra*) is equivalent to the class *HumanBeing* using the keyword `equivalentClass`.

```
<owl:Class rdf:about="Terran">
  <owl:equivalentClass rdf:resource="HumanBeing"/>
</owl:Class>
```

The same assertion is expressed in OWL/XML syntax with an `<EquivalentClasses>` element.

```
<EquivalentClasses>
  <Class IRI="Terran"/>
  <Class IRI="HumanBeing"/>
</EquivalentClasses>
```

In OWL Manchester syntax, the `EquivalentTo` keyword is used.

```
Class: Terran
  EquivalentTo: HumanBeing
```

There can also be classes that are "incompatible" with one another, meaning that membership in one class excludes membership in another. Failure to specify this is a common modeling error in OWL ontologies. For instance, although it is obvious to us that no single individual can be an instance of both *Cat* and *Dog*, this is human background knowledge that the computer does not possess unless we specify it in our ontologies. If we do not specify it, then the open-world assumption of OWL will mean that inference routines will assume that *Cat* and *Dog* could be different names for the same thing.

DisjointClasses

DisjointClasses(C D) is used to express an assertion that class C is disjoint to class D, meaning that it can be excluded that an instance of C is also an instance of D and vice versa.

If we want to assert in the RDF/XML syntax that inhabitants of the planets Earth, Qo'noS (the Klingon homeworld), and the twin planets Romulus and Remus (the homeworlds of the Romulans) are all different from one another, then we can use the owl:AllDisjointClasses keyword,

```
<owl:AllDisjointClasses>
  <owl:members rdf:parseType="Collection">
    <owl:Class rdf:about="Terran"/>
    <owl:Class rdf:about="Klingon"/>
    <owl:Class rdf:about="Romulan"/>
  </owl:members>
</owl:AllDisjointClasses>
```

Note that this assertion implies that no individual can be an instance of more than one of the three classes. The construct disjointWith can be used for pairwise disjointedness assertions, but the semantics is slightly different.

```
<owl:Class rdf:ID="Terran">
  <owl:disjointWith>
    <rdf:resource="Klingon"/>
    <rdf:resource="Romulan"/>
  </owl:disjointWith>
</owl:Class>
```

The above assertion states that no individual can be an instance of the class *Terran* and of the class *Klingon* or *Romulan*. However, no assertion is made about whether an individual can be an instance of both *Klingon* and *Romulan* or not.

To express the AllDisjointClasses assertion using OWL/XML syntax, a <DisjointClasses> element is used.

```
<DisjointClasses>
    <Class IRI="Terran"/>
    <Class IRI="Klingon"/>
    <Class IRI="Romulan"/>
</DisjointClasses>
```

In OWL Manchester syntax, a statement is made using the `DisjointClasses` keyword. Note that this statement is not part of a class stanza.

```
DisjointClasses: Terran, Klingon, Romulan
```

Object Properties

ObjectPropertyAssertion

`ObjectPropertyAssertion(`P i j`)` is used to express an assertion that individual i is connected to individual j by property P.

In RDF/XML syntax, the object property assertion is made with a standard RDF triple. An individual i is connected by a predicate P to an individual j. For instance, to state that Spock's father is Sarek, we would use the following RDF/XML fragment.

```
<rdf:Description rdf:about="Spock">
  <hasFather rdf:resource="Sarek"/>
</rdf:Description>
```

The corresponding assertion in OWL/XML syntax makes use of an `<ObjectPropertyAssertion>` element.

```
<ObjectPropertyAssertion>
  <ObjectProperty IRI="hasFather"/>
  <NamedIndividual IRI="Spock"/>
  <NamedIndividual IRI="Sarek"/>
</ObjectPropertyAssertion>
```

In OWL Manchester syntax, the object property assertion is listed as one of the `Facts` about the individual who is the subject of the assertion.

```
Individual: Spock
  Facts: hasFather Sarek
```

Note that the entities that describe the way in which individuals are related to one another are called *properties* in OWL (similar entities in RDF and mathematical logic have been called relations, roles, and predicates in other parts of this book).

We can also state that two individuals are not connected by a property. In RDF/XML syntax, we might state that it is not the case that Dr. McCoy gets along very well with Spock as follows.

```
<owl:NegativePropertyAssertion>
  <owl:sourceIndividual rdf:resource="McCoy"/>
  <owl:assertionProperty rdf:resource="getsAlongWith"/>
  <owl:targetIndividual rdf:resource="Spock"/>
</owl:NegativePropertyAssertion>
```

The equivalent assertion in OWL/XML syntax would be:

```
<NegativeObjectPropertyAssertion>
  <ObjectProperty IRI="getsAlongWith"/>
  <NamedIndividual IRI="McCoy"/>
  <NamedIndividual IRI="Spock"/>
</NegativeObjectPropertyAssertion>
```

In OWL Manchester syntax, the keyword not is added to the corresponding Facts statement.

```
Individual: McCoy
  Facts: not getsAlongWith Spock
```

Property Hierarchies

SubObjectPropertyOf

SubObjectPropertyOf$(P\ Q)$ is used to express an assertion that object property P is a subproperty of object property Q.

Properties can form hierarchies much the same way as classes can. In OWL 2, there are two built-in object properties with the IRIs owl:topObjectProperty and owl:bottomObjectProperty. The semantics of these properties are defined as follows:

- owl:topObjectProperty connects all possible pairs of individuals.

- owl:bottomObjectProperty does not connect any pair of individuals.

Thus, in OWL 2, all object properties are subproperties of owl:topObjectProperty.

In RDF/XML syntax, we would express the assertion that hasFather is a subproperty of hasParent as follows.

```
<owl:ObjectProperty rdf:about="hasFather">
  <rdfs:subPropertyOf rdf:resource="hasParent"/>
</owl:ObjectProperty>
```

This means if we make the following assertion

```
<rdf:Description rdf:about="Spock">
  <hasFather rdf:resource="Sarek"/>
</rdf:Description>
```

then we can infer on the basis of the subproperty assertion that the following triple must also be true.

```
<rdf:Description rdf:about="Spock">
  <hasParent rdf:resource="Sarek"/>
</rdf:Description>
```

In OWL/XML syntax, the equivalent assertion is written as follows:

```
<SubObjectPropertyOf>
  <ObjectProperty IRI="hasFather"/>
  <ObjectProperty IRI="hasParent"/>
</SubObjectPropertyOf>
```

In OWL Manchester syntax, the assertion is:

```
ObjectProperty: hasFather
  SubPropertyOf: hasParent
```

Note that the converse of a subproperty assertion is not necessarily true. Spock's (human) mother is called Amanda Grayson, and thus this assertion is true:

```
Individual: Spock
  Facts: hasParent Amanda
```

The following inference (based on an erroneous inference in the "wrong direction" because of the assertion that hasFather is a subproperty of hasParent) is false:

```
Individual: Spock
  Facts: hasFather Amanda
```

Note that the keyword owl:equivalentProperty can be used to state that two (differently named) properties are semantically equivalent. In Manchester syntax, an EquivalentProperties: key-value pair would be used.

Domain and Range Restrictions

ObjectPropertyDomain

ObjectPropertyDomain(P C) is used to express an assertion that the domain of object property P consists of instances of class C. Similarly, ObjectPropertyRange(Q D) is used to express an assertion that the range of object property Q consists of instances of class D.

Frequently, the information that two individuals are interconnected by a certain property allows to draw further conclusions about the individuals themselves. For instance, the information that Sarek `hasWife` Amanda intuitively allows us to infer that Sarek and Amanda are humanoid beings belonging to species that practice the institution of marriage.

Note that the purpose of domain and range restrictions in OWL is not to constrain the kinds of datatypes or variables that are syntactically correct. In contrast, in a programming language such as Java, the following code would not compile because the variable x is declared to be of type `int` but is assigned a string value.

```
int x;
x = "Hello, world!";
```

In OWL, domain and range restriction allow inferences to be made. The following RDF/XML fragment declares that the domain of the property `hasWife` is the class `HumanoidMale` and the range of the property is `HumanoidFemale`.

```
<owl:ObjectProperty rdf:about="hasWife">
  <rdfs:domain rdf:resource="HumanoidMale"/>
  <rdfs:range rdf:resource="HumanoidFemale"/>
</owl:ObjectProperty>
```

If we additionally have the following object property assertion

```
<rdf:Description rdf:about="Sarek">
  <hasWife rdf:resource="Amanda"/>
</rdf:Description>
```

then we can infer from the domain and range restrictions that Sarek is an instance of the class `HumanoidMale` and that Amanda is an instance of the class `HumanoidFemale`.

The equivalent statement about domain and range restrictions in OWL/XML syntax is

```
<ObjectPropertyDomain>
  <ObjectProperty IRI="hasWife"/>
  <Class IRI="HumanoidMale"/>
</ObjectPropertyDomain>
<ObjectPropertyRange>
  <ObjectProperty IRI="hasWife"/>
  <Class IRI="HumanoidFemale"/>
</ObjectPropertyRange>
```

The equivalent assertion in Manchester Syntax is

```
ObjectProperty: hasWife
  Domain: HumanoidMale
  Range: HumanoidFemale
```

Equality and Inequality of Individuals

DifferentIndividuals

`DifferentIndividuals(i j)` is used to express an assertion that individual *i* is different from individual *j*.

In the World Wide Web, different Web sites may use different names to refer to the same entity. The default stance of OWL is not to assume that different names refer to different things unless explicitly stated. The following RDF/XML fragment states that *Jim* and *Leonard* refer to different individuals.

```
<rdf:Description rdf:about="Jim">
  <owl:differentFrom rdf:resource="Leonard"/>
</rdf:Description>
```

OWL/XML:

```
<DifferentIndividuals>
  <NamedIndividual IRI="Jim"/>
  <NamedIndividual IRI="Leonard"/>
</DifferentIndividuals>
```

Manchester syntax:

```
Individual: Jim
  DifferentFrom: Leonard
```

It is also possible to state that two different names refer to the *same* individual.
RDF/XML:

```
RDF/XML:
<rdf:Description rdf:about="James">
  <owl:sameAs rdf:resource="Jim"/>
</rdf:Description>
```

OWL/XML:

```
<SameIndividual>
  <NamedIndividual IRI="James"/>
  <NamedIndividual IRI="Jim"/>
</SameIndividual>
```

Manchester Syntax:

```
Manchester Syntax
Individual: James
  SameAs: Jim
```

A sameAs assertion allows reasoners to infer that any information given about the individual James also holds for the individual Jim.

Data Property Assertions

OWL has two kinds of property assertions. We have already seen that object property assertions are used to connect two individuals. Data property assertions are used to describe how individuals are to be described by data values. For instance, an individual has an age in years, and a property hasAge relates an individual to a data value rather than to another individual. Many of the XML Schema datatypes can be used [42].

For instance, the following fragments state that Sarek's age is 102.

RDF/XML:

```
<Person rdf:about="Sarek">
  <hasAge rdf:datatype="&xsd;integer">102</hasAge>
</Person>
```

OWL/XML:

```
<DataPropertyAssertion>
  <DataProperty IRI="hasAge"/>
  <NamedIndividual IRI="Sarek"/>
  <Literal datatypeIRI="&xsd;integer">102</Literal>
</DataPropertyAssertion>
```

Note that the RDF/XML and OWL/XML fragments assume that an entity has been declared for the XML Schema namespace

```
<!ENTITY xsd "http://www.w3.org/2001/XMLSchema#">
```

The equivalent statement is made in Manchester syntax as follows:

```
Individual: Sarek
  Facts: hasAge "102"^^xsd:integer
```

Negative assertions are possible with syntax that is analogous to that for negative object property assertions.
RDF/XML:

```
<owl:NegativePropertyAssertion>
  <owl:sourceIndividual rdf:resource="Sarek"/>
  <owl:assertionProperty rdf:resource="hasAge"/>
```

```
<owl:targetValue
    rdf:datatype="&xsd;integer">16</owl:targetValue>
</owl:NegativePropertyAssertion>
```

OWL/XML:

```
<NegativeDataPropertyAssertion>
  <DataProperty IRI="hasAge"/>
  <NamedIndividual IRI="Jack"/>
  <Literal datatypeIRI="&xsd;integer">16</Literal>
</NegativeDataPropertyAssertion>
```

Manchester:

```
Individual: Sarek
  Facts: not hasAge "16"^^xsd:integer
```

Domain and range can also be stated for datatype properties. In this case, the domain will be an OWL class and the range will be a datatype. Continuing the above example, we have:

RDF/XML:

```
<owl:DatatypeProperty rdf:about="hasAge">
  <rdfs:domain rdf:resource="Humanoid"/>
  <rdfs:range rdfs:Datatype="&xsd;nonNegativeInteger"/>
</owl:DatatypeProperty>
```

OWL/XML:

```
<DataPropertyDomain>
  <DataProperty IRI="hasAge"/>
    <Class IRI="Humanoid"/>
  </DataPropertyDomain>
  <DataPropertyRange>
    <DataProperty IRI="hasAge"/>
    <Datatype IRI="&xsd;nonNegativeInteger"/>
</DataPropertyRange>
```

Manchester:

```
DataProperty: hasAge
  Domain: Humanoid
  Range:  xsd:nonNegativeInteger
```

Building Blocks for Complex Classes in OWL

Perhaps the most interesting part of the OWL syntax consists of the constructs with which class expressions can be constructed from other classes and property restrictions in order to define new classes. We will begin with three mechanisms for constructing classes based on class intersection, union, and complement (see also the description of the corresponding operators from set theory in Chapter 2). The basic design pattern for creating this kind of class definition is to state that a certain class is *equivalent to* the intersection, union, or complement of other classes.

ObjectUnionOf

`EquivalentClasses(A  ObjectUnionOf(B C))`

This construct is used to express an assertion that class A is equivalent to the union of classes B and C. This means that an individual who is an instance of class A can be inferred to also be an instance of either class B or class C or both.

Let us define the class of *Humanoid* as the union of four species (Terrans, i.e., Human Beings, Vulcans, Klingons, and Romulans).

RDF/XML

```
<owl:Class rdf:about="Humanoid">
  <owl:equivalentClass>
    <owl:Class>
      <owl:unionOf rdf:parseType="Collection">
        <owl:Class rdf:about="Terran"/>
        <owl:Class rdf:about="Vulcan"/>
        <owl:Class rdf:about="Klingon"/>
        <owl:Class rdf:about="Romulan"/>
      </owl:unionOf>
    </owl:Class>
  </owl:equivalentClass>
</owl:Class>
```

OWL/XML

```
<EquivalentClasses>
  <Class IRI="Humanoid"/>
  <ObjectUnionOf>
    <Class IRI="Terran"/>
    <Class IRI="Vulcan"/>
    <Class IRI="Klingon"/>
    <Class IRI="Romulan"/>
  </ObjectUnionOf>
</EquivalentClasses>
```

Manchester

Class: Humanoid
    EquivalentTo: Terran or Vulcan or Klingon or Romulan

ObjectIntersectionOf

EquivalentClasses(A ObjectIntersectionOf(B C))

OWL classes can also be defined as the intersection of two or more other classes. The construct expresses the assertion that any instance of class A is also an instance of class B and class C.

The chief medical officers aboard Federation Starships are both Star Fleet officers and physicians. We can express this using the intersection syntax.

RDF/XML

```
<owl:Class rdf:about="ChiefMedicalOfficer">
  <owl:equivalentClass>
    <owl:Class>
      <owl:intersectionOf rdf:parseType="Collection">
        <owl:Class rdf:about="StarFleetOfficer"/>
        <owl:Class rdf:about="Physician"/>
      </owl:intersectionOf>
    </owl:Class>
  </owl:equivalentClass>
</owl:Class>
```

OWL/XML

```
<EquivalentClasses>
  <Class IRI="ChiefMedicalOfficer"/>
  <ObjectIntersectionOf>
    <Class IRI="StarFleetOfficer"/>
    <Class IRI="Physician"/>
  </ObjectIntersectionOf>
</EquivalentClasses>
```

Manchester

```
Class: ChiefMedicalOfficer
 EquivalentTo: StarFleetOfficer and Physician
```

ObjectComplementOf

```
EquivalentClasses(A
   ObjectIntersectionOf(B ComplementOf(C)))
```

This construct expresses the assertion that class A is equivalent to the intersection of class B and the complement of class C. Any individual that is an instance of class A must be an instance of class B and must *not* be an instance of class C.

The complement of a class corresponds to exactly those objects which are not members of the class itself. A common error is to assume that some background knowledge will restrict the complement of a class. Trekkies among our readers will recall the Star Trek episode "The Omega Glory," which describes a parallel world in which a race of *Yangs*[3] and *Kohms*[4]. On the planet Omega IV, there were just two races of sentient beings: the Yangs and the Kohms. One thus might expect that the complement of a class Yang would be Kohm, but this is not the case. The complement of Yang is actually everything else in the universe, perhaps including *Automobile*, *Spaceship*, and *Supernova*. The correct definition would be that the class Kohm is defined as the intersection of OmegaIVSentientBeing with the class *Yang*:

```
EquivalentClasses(Kohm
   ObjectIntersectionOf(OmegaIVSentientBeing ComplementOf(Yang)))
```

The corresponding class definitions in the three OWL serializations are as follows.

RDF/XML

```
<owl:Class rdf:about="Kohm">
  <owl:equivalentClass>
    <owl:Class>
      <owl:intersectionOf rdf:parseType="Collection">
        <owl:Class rdf:about="OmegaIVSentientBeing"/>
        <owl:Class>
          <owl:complementOf rdf:resource="Yang"/>
        </owl:Class>
      </owl:intersectionOf>
```

---

[3]The Yangs presumably correspond to "Yankees," or Americans. In the episode, the Yangs worship the American flag and Constitution without understanding the meaning of democracy.

[4]The Kohms presumably correspond to "Communists," who fought a bacterial war with the Yangs over 1,000 years ago. The episode exemplifies Star Trek's "Hodgkin's law of parallel planetary development," and provides an interesting commentary to some aspects of the Cold War, which existed in full force at the time the episode was originally aired in 1968.

```
      </owl:Class>
    </owl:equivalentClass>
  </owl:Class>
```

OWL/XML

```
<EquivalentClasses>
  <Class IRI="Kohm"/>
  <ObjectIntersectionOf>
    <Class IRI="OmegaIVSentientBeing"/>
    <ObjectComplementOf>
      <Class IRI="Yang"/>
    </ObjectComplementOf>
  </ObjectIntersectionOf>
</EquivalentClasses>
```

Manchester

```
Class: Kohm
  EquivalentTo: OmegaIVSentientBeing and not Yang
```

These definitions completely specify the class Kohm. That is, the definition using the equivalentClass syntax provides necessary and sufficient conditions for being an instance of the class Kohm. It is also possible to specify necessary, but not sufficient conditions for a class using subclass assertions.

SubClassOf

SubClassOf(A ObjectIntersectionOf(B C)

This construct expresses the assertion that class A is a subclass of the intersection of class B and class C. Any individual that is an instance of class A must be an instance of class B and also an instance of class C. However, no assertion is made as to whether an individual that is an instance of class B and C must also be an instance of class A.

For instance, we can state that a Terran (i.e., a human being) is a subclass of all entities that are both sentient beings and humanoids. The converse is not true, because, for instance, Vulcans are also sentient beings and humanoids, but they are not Terrans[5].

RDF/XML

```
<owl:Class rdf:about="Terran">
  <rdfs:subClassOf>
```

---

[5]We will ignore borderline cases such as Spock, whose father was a Vulcan and whose mother was a Terran.

```
  <owl:Class>
    <owl:intersectionOf rdf:parseType="Collection">
      <owl:Class rdf:about="SentientBeing"/>
      <owl:Class rdf:about="Humanoid"/>
    </owl:intersectionOf>
  </owl:Class>
</rdfs:subClassOf>
</owl:Class>
```

OWL/XML

```
<SubClassOf>
  <Class IRI="Terran"/>
  <ObjectIntersectionOf>
    <Class IRI="SentientBeing"/>
    <Class IRI="Humanoid"/>
  </ObjectIntersectionOf>
</SubClassOf>
```

Manchester

```
Class: Terran
  SubClassOf: SentientBeing and Humanoid
```

Complex classes can be used wherever named classes can be used. For instance, to declare that Kirk is an instance of the class *Terran* using the above definitions, we could use the following statements.

RDF/XML

```
<rdf:Description rdf:about="Kirk">
  <rdf:type>
    <owl:Class>
      <owl:intersectionOf rdf:parseType="Collection">
        <owl:Class rdf:about="SentientBeing"/>
        <owl:Class rdf:about="Humanoid"/>
      </owl:intersectionOf>
    </owl:Class>
  </rdf:type>
</rdf:Description>
```

OWL/XML

```
<ClassAssertion>
  <ObjectIntersectionOf>
    <Class IRI="SentientBeing"/>
    <Class IRI="Humanoid"/>
  </ObjectIntersectionOf>
  <NamedIndividual IRI="Kirk"/>
</ClassAssertion>
```

Manchester

```
Individual: Kirk
  Types: SentientBeing and Humanoid
```

Class Definitions with Property Restrictions

OWL provides a syntax for defining classes based on property restrictions. The general pattern is that a class is defined to consist of all entities that satisfy a certain restriction on some property. Two of the most important kinds of restriction are the *existential* and *universal* restrictions that were discussed in the section on first-order logic in Chapter 2 on page 21.

ObjectSomeValuesFrom

```
EquivalentClasses(A ObjectSomeValuesFrom(R B))
```

This construct expresses the assertion that class A is equivalent to the class of all entities that are linked to an instance of class B by the R property.

For instance, we can define the class of starship captain as being equivalent to all entities that command a starship.

RDF/XML

```
<owl:Class rdf:about="Captain">
  <owl:equivalentClass>
    <owl:Restriction>
      <owl:onProperty rdf:resource="commands"/>
      <owl:someValuesFrom rdf:resource="Starship"/>
    </owl:Restriction>
  </owl:equivalentClass>
</owl:Class>
```

OWL/XML

```
<EquivalentClasses>
  <Class IRI="Captain"/>
  <ObjectSomeValuesFrom>
    <ObjectProperty IRI="commands"/>
    <Class IRI="Starship"/>
  </ObjectSomeValuesFrom>
</EquivalentClasses>
```

Manchester

```
Class: Captain
  EquivalentTo: commands some Starship
```

This construct means that for every instance of the class *Captain* there exists at least one ("some") instance of the class *Starship* such that the *Captain* instance commands the instance of *Starship*. Note that the keyword some is used to represent the someValuesFrom relation in Manchester syntax.

ObjectAllValuesFrom

EquivalentClasses(A ObjectAllValuesFrom(R B))

This construct expresses the assertion that class A is equivalent to the class of all entities for which all individuals who are related to an instance of A by the R property must be instances of the class B. Note that there may be an instance of A that is not related to anything else by the R property.

The syntax is essentially the same as for the ObjectSomeValuesFrom restriction. Here, we show only the Manchester version.

Manchester

```
Class: StarshipCaptain
    EquivalentTo: commands only Starship
```

This construct means that if a StarshipCaptain commands anything, that anything must be a Starship. It is, however, possible that there is a StarshipCaptain who does not command anything[6].

ObjectHasValue

EquivalentClasses(A ObjectHasValue(R i))

This construct expresses the assertion that class A is equivalent to the class of all entities that are linked to an individual i by the R property.

Property restrictions can also be used to describe classes of individuals that are related to one particular individual. For instance, the following constructs express the assertion that the class *KirksCrew* is equivalent to the set of things that have Kirk as captain.

RDF/XML

```
<owl:Class rdf:about="KirksCrew">
  <owl:equivalentClass>
    <owl:Restriction>
      <owl:onProperty rdf:resource="hasCaptain"/>
```

---

[6]One example might be Commodore Mathew Decker, who lost his ship called the USS Constellation to the doomsday machine.

```
          <owl:hasValue rdf:resource="Kirk"/>
        </owl:Restriction>
      </owl:equivalentClass>
</owl:Class>
```

OWL/XML

```
<EquivalentClasses>
  <Class IRI="KirksCrew"/>
  <ObjectHasValue>
    <ObjectProperty IRI="hasCaptain"/>
    <NamedIndividual IRI="Kirk"/>
  </ObjectHasValue>
</EquivalentClasses>
```

Manchester

```
Class: KirksCrew
  EquivalentTo: hasCaptain value Kirk
```

A new feature of OWL 2 is the ability to specify local reflexivity in class definitions. For instance, a narcissist is someone who loves himself, and an "autophosphorylating kinase" is a kinase that phosphorylates itself. Recall that the Tribbles are small, furry creatures that are born pregnant. Because of their prodigious reproductive rate, the Tribbles essentially overrun a space station in a well-known Trek episode called "The Trouble with Tribbles." Although the physiological mechanism for the Tribbles breeding capacity is not specified in the original episode, assuming that the Tribbles can undergo asexual reproduction, it is fair to state that a Tribble reproducesWith *self*.

RDF/XML

```
<owl:Class rdf:about="Tribble">
  <rdfs:subClassOf>
    <owl:Restriction>
      <owl:onProperty rdf:resource="reproducesWith"/>
      <owl:hasSelf rdf:datatype="&xsd;boolean">
        true
      </owl:hasSelf>
    </owl:Restriction>
  </rdfs:subClassOf>
</owl:Class>
```

OWL/XML

```
<SubClassOf>
  <Class IRI="Tribble"/>
  <ObjectHasSelf>
```

```
    <ObjectProperty IRI="reproducesWith"/>
   </ObjectHasSelf>
 </SubClassOf>
```

Manchester

```
Class: Tribble
  SubClassOf: reproducesWith Self
```

Property Cardinality Restrictions

ObjectSomeValuesOf

```
ClassAssertion(ObjectMinCardinality(n R [C] ) i)
```

This construct expresses the assertion individual i has at least *n* items of class C to which i is related by the relation R. Note that the class expression C is optional.

For instance, in order to express the assertion that the individual Enterprise has at least two shuttlecraft (small, short-range starships possessing only impulse drive capability), we would use the following statements.
RDF/XML

```
<rdf:Description rdf:about="Enterprise">
  <rdf:type>
    <owl:Restriction>
      <owl:minQualifiedCardinality
          rdf:datatype="&xsd;nonNegativeInteger">
          2
       </owl:minQualifiedCardinality>
      <owl:onProperty rdf:resource="hasAuxiliaryCraft"/>
      <owl:onClass rdf:resource="Shuttle"/>
    </owl:Restriction>
  </rdf:type>
</rdf:Description>
```

OWL/XML

```
<ClassAssertion>
  <ObjectMinCardinality cardinality="2">
    <ObjectProperty IRI="hasAuxiliaryCraft"/>
    <Class IRI="Shuttle"/>
  </ObjectMinCardinality>
  <NamedIndividual IRI="Enterprise"/>
</ClassAssertion>
```

Manchester

```
Individual: Enterprise
  Types: hasAuxiliaryCraft min 2 Shuttle
```

We can use similar statements about the maximum number of Shuttlecraft.
Manchester

```
Individual: Enterprise
  Types: hasAuxiliaryCraft max 8 Shuttle
```

(The RDF/XML and OWL/XML assertions are similar, with `max` replaced by `min`). If we know that the exact number of shuttlecraft is 4, then we can assert the following.

RDF/XML

```
<rdf:Description rdf:about="Enterprise">
  <rdf:type>
    <owl:Restriction>
      <owl:qualifiedCardinality
        rdf:datatype="&xsd;nonNegativeInteger">
        4
      </owl:qualifiedCardinality>
      <owl:onProperty rdf:resource="hasAuxiliaryCraft"/>
      <owl:onClass rdf:resource="Shuttle"/>
    </owl:Restriction>
  </rdf:type>
</rdf:Description>
```

OWL/XML

```
<ClassAssertion>
  <ObjectExactCardinality cardinality="4">
    <ObjectProperty IRI="hasAuxiliaryCraft"/>
    <Class IRI="Shuttle"/>
  </ObjectMinCardinality>
  <NamedIndividual IRI="Enterprise"/>
</ClassAssertion>
```

Manchester

```
Individual: Enterprise
  Types: hasAuxiliaryCraft exactly 4 Shuttle
```

Finally, it is optional to indicate the class involved in the cardinality restriction. If we just want to indicate that the Enterprise has 4 auxiliary crafts and not that the craft are shuttles, we can use the following assertion.

Manchester

```
Individual: Enterprise
  Types: hasAuxiliaryCraft exactly 4
```

Enumerated Classes

ObjectOneOf

```
EquivalentClasses(C  ObjectOneOf(i j k))
```

This construct expresses the assertion that class C comprises exactly the individuals i, j, and k.

We can define the class of EnterpriseOfficer as follows (for simplicity, we will limit ourselves to five officers, but there is no limit on the number of individuals that can be named in enumerated class definitions).

RDF/XML

```
<owl:Class rdf:about="EnterpriseOfficer">
   <owl:equivalentClass>
    <owl:Class>
      <owl:oneOf rdf:parseType="Collection">
        <rdf:Description rdf:about="Kirk"/>
        <rdf:Description rdf:about="Spock"/>
        <rdf:Description rdf:about="McCoy"/>
        <rdf:Description rdf:about="Uhura"/>
        <rdf:Description rdf:about="Scotty"/>
      </owl:oneOf>
    </owl:Class>
   </owl:equivalentClass>
 </owl:Class>
```

OWL/XML

```
 <EquivalentClasses>
   <Class IRI="EnterpriseOfficer"/>
   <ObjectOneOf>
     <NamedIndividual IRI="Kirk"/>
     <NamedIndividual IRI="Spock"/>
     <NamedIndividual IRI="McCoy"/>
     <NamedIndividual IRI="Uhura"/>
     <NamedIndividual IRI="Scotty"/>
   </ObjectOneOf>
 </EquivalentClasses>
```

Manchester

```
Class: EnterpriseOfficer
  EquivalentTo: {Kirk, Spock, McCoy, Uhura, Scotty}
```

The semantics of this kind of axiom implies not only that the class *Enterprise* includes the individuals listed in the definitions, but also that they are the *only* members of the class. Classes defined by enumeration are sometimes referred to as closed classes for this reason. Axioms about enumerated classes allow the inference that individuals not named in the definition do not belong to the class.

Property Characteristics

InverseObjectProperties

```
InverseObjectProperties(R S)
```

This construct expresses the assertion that property S is the inverse of property R.

Recalling that the Borg are a race of humanoid beings that are part organic, part machine. Individual Borg are implanted with bio-chips that link their consciousness with a central collective consciousness. In the course of time, the Borg have conquered thousands of worlds in the Delta Quadrant of the galaxy by means of forced assimilation of the individuals and technology of the alien races into the Borg collective. We can assert that `assimilate` and `joinTheCollectiveOf` are inverse relations.

RDF/XML

```
<owl:ObjectProperty rdf:about="assimilate">
   <owl:inverseOf rdf:resource="joinCollectiveOf"/>
 </owl:ObjectProperty>
```

OWL/XML

```
<InverseObjectProperties>
  <ObjectProperty IRI="assimilate"/>
  <ObjectProperty IRI="joinCollectiveOf"/>
</InverseObjectProperties>
```

Manchester

```
ObjectProperty: assimilate
 InverseOf: joinCollectiveOf
```

This would allow us to infer from some assertion `<Borg,assimilate,Picard>` that Picard has joined the collective of the Borg.

Symmetric and Asymmetric Properties

A property is symmetric if whenever $X \xrightarrow{P} Y$ holds, then it is also true that $Y \xrightarrow{P} X$. Properties such as hasFriend, hasSpouse, isAdjacentTo are common examples of symmetric properties. In OWL, we can define a symmetric property with the following constructs.

RDF/XML

```
<owl:SymmetricProperty rdf:about="hasCrewmate"/>
```

OWL/XML

```
<SymmetricObjectProperty>
  <ObjectProperty IRI="hasCrewmate"/>
</SymmetricObjectProperty>
```

Manchester

```
ObjectProperty: hasCrewmate
  Characteristics: Symmetric
```

For instance, if we assert that Sulu hasCrewmate Chekov, then we can infer that Chekov hasCrewmate Sulu.

Similarly, there are properties for which if the relation $X \xrightarrow{P} Y$ holds then it cannot also be true that $Y \xrightarrow{P} X$. Common examples of this are hasChild and part_of and others.

It is possible to declare relations to be disjoint, which is the case if there are no two individuals that are interlinked by both properties.

DisjointObjectProperties

```
DisjointObjectProperties(R S)
```

This construct expresses the assertion that no two individuals can be interlinked by both the property R and the property R.

RDF/XML

```
<rdf:Description rdf:about="hasAlly">
  <owl:propertyDisjointWith rdf:resource="hasEnemy"/>
</rdf:Description>
```

OWL/XML

```
<DisjointObjectProperties>
  <ObjectProperty IRI="hasAlly"/>
  <ObjectProperty IRI="hasEnemy"/>
</DisjointObjectProperties>
```

Manchester

```
DisjointProperties: hasAlly, hasEnemy
```

On the basis of this assertion, it would be an error to relate the `Federation` to the `KlingonEmpire` using both the relation `hasEnemy` and `hasAlly`.

Properties can be *irreflexive*, meaning that no individual can be related to itself by the property. Typical irreflexive properties are `hasParent` or `isChildOf`.

RDF/XML

```
<owl:IrreflexiveProperty rdf:about="parentOf"/>
```

OWL/XML

```
<IrreflexiveObjectProperty>
  <ObjectProperty IRI="parentOf"/>
</IrreflexiveObjectProperty>
```

Manchester

```
ObjectProperty: parentOf
  Characteristics: Irreflexive
```

Properties can also be *reflexive*, meaning that such a property relates everything to itself. Recall that by definition, every class is also a subclass of itself. If desired, we can specify that other relations are reflexive as well. The serialization syntax is analogous to that for irreflexive properties.

Functional Object Properties

```
FunctionalObjectProperty(R)
```

This construct expresses the assertion that an individual can be linked to at most one other individual by the property R. The converse assertion is made with the InverseFunctionalObjectProperty construct.

```
InverseFunctionalObjectProperty(R)
```

This construct expresses the assertion that no more than one individual can be linked to a particular individual by the property R.

Under the assumption that a Captain commands at most one Starship, and a Starship is commanded by at most one *Captain*, we might want to declare that the relation `commandsShip` is functional.

RDF/XML

```
<owl:FunctionalProperty rdf:about="commandsShip"/>
```

OWL/XML

```
<FunctionalObjectProperty>
  <ObjectProperty IRI="commandsShip"/>
</FunctionalObjectProperty>
```

Manchester

```
ObjectProperty: commandsShip
  Characteristics: Functional
```

If we make an assertion that *Kirk* commandsShip *Enterprise*, then we can infer that *Kirk* commands the Enterprise and no other ship.

RDF/XML

```
<owl:OnverseFunctionalProperty rdf:about="commandsShip"/>
```

OWL/XML

```
<InverseFunctionalObjectProperty>
  <ObjectProperty IRI="commandsShip"/>
</InverseFunctionalObjectProperty>
```

Manchester

```
ObjectProperty: commandsShip
  Characteristics: InverseFunctional
```

If we have also asserted that commandsShip is an inverse functional property, then we can infer from the assertion *Kirk* commandsShip *Enterprise* that Kirk is the only commander of the Enterprise.

Transitive Properties and Property Chains

TransitiveObjectProperty

TransitiveObjectProperty(R)

This construct expresses the assertion that the property R is transitive. That is, if $X \xrightarrow{R} Y$ and $Y \xrightarrow{R} Z$ then we can infer that $X \xrightarrow{R} Z$.

We define isCommandedBy as a transitive property as follows.

RDF/XML

```
<owl:TransitiveProperty rdf:about="isCommandedBy"/>
```

OWL/XML

```
<TransitiveObjectProperty>
 <ObjectProperty IRI="isCommandedBy"/>
</TransitiveObjectProperty>
```

Manchester

```
ObjectProperty: isCommandedBy
 Characteristics: Transitive
```

Thus, if we assert that Nurse Chapel is commanded by Dr. McCoy and Dr. McCoy is commanded by Captain Kirk, we can infer that Nurse Chapel is also under the command of Captain Kirk.

We may also want to state that some property is equivalent to a chain of two or more other properties. For instance, `isGrandparentOf` can be considered to be equivalent to two `isParentOf` properties.

Further Reading

This appendix was intended to familiarize readers with the three most important serializations for OWL ontologies and to explain the most important language constructs. It is hoped that readers will now be able to understand the documentation of the W3C. The following table lists the most important documents about OWL 2 at the time of writing.

OWL 2 Introduction	
Document overview	Introduction to OWL 2 and description of W3C documents about OWL 2 [258].
OWL 2 Serializations	
RDF/XML	Serialization of OWL 2 ontology as RDF graph [197].
OWL/XML	Serialization of OWL 2 ontology as XML [171].
Manchester syntax	Serialization using Manchester syntax [125].
Turtle	Optional serialization as Turtle graph [27].
OWL 2 Core Specifications	
Functional syntax	Defines the constructs of OWL 2 ontologies, and defines OWL 2 DL ontologies in terms of global restrictions on OWL 2 ontologies [173].
Direct semantics	Model-theoretic semantics of OWL 2 ontologies [172]
RDF semantics	Defines the meaning of OWL 2 ontologies via an extension of RDF Semantics [49].
OWL 2 for Users	
Primer	OWL 2 Primer with introductory material [116].
New features	New features of OWL 2 and their rationale [94].
Reference	Quick reference guide to OWL 2 [17].

Glossary

cardinality The cardinality of a set refers to the number of items in the set.

class extension The set of individuals associated with an RDFS or OWL class is called the *class extension*. The individuals in the class extension are called the *instances* of the class.

commutative An operation is commutative if changing the order of the operands does not change the result: $A * B = B * A$ for a commutative operator $*$.

conjunction In logic, conjunction is an operation on two logical propositions that is true if both of the propositions are true. A conjunction of the propositions P and Q is written as $P \wedge Q$. This is equivalent to the Boolean logical AND operator in C and many other computer languages: P && Q.

differential diagnosis The process by which a physician estimates the probability of various possible causes (diagnoses) that could account for a patient's illness. For instance, the differential diagnosis of cough and fever includes the common cold, pneumonia, tuberculosis, and many other things.

directed acyclic graph (DAG) A directed graph is *acyclic* if it contains no directed cycle.

disjoint Two sets are *disjoint* if they have no elements in common: $A \cap B = \varnothing$.

disjunction In logic, disjunction is an operation on two logical propositions that is true if one or both of the propositions is true. A disjunction of the propositions P and Q is written as $P \vee Q$. This is equivalent to the Boolen OR operator in C and many other computer languages: P || Q.

domain A portion of reality that forms the subject matter of a branch of science, e.g., glycobiology or genomics.

family In graph theory, the *family* of a vertex is defined as the set consisting of the vertex and all of its parents.

FDR The FDR (short for *false-discovery rate*) is the proportion of statistical tests called significant that are type I errors (false-positive).

FWER The FWER (short for *family-wise error rate*) is the probability of making at least one type I error (i.e., false-positive result) when multiple statistical tests are performed.

homology Two proteins are homologous if they have a common evolutionary ancestor. Homology is generally recognized on the basis of a high degree of sequence similarity.

joint probability distribution The joint probability distribution (JPD) of random variables $X_1, X_2, \ldots, X_n$ defines the probability of events that are defined in terms of $X_1, X_2, \ldots, X_n$. If $n = 2$, the JPD is called a bivariate distribution. For a discrete JPD, the joint probability distribution function describes the probability of outcomes corresponding to combinations of the possible values of the variables $X_1, X_2, \ldots, X_n$.

MTC Multiple testing correction, a procedure used when multiple hypothesis tests are performed simultaneously in order to reduce the chance of incorrectly rejecting the null hypothesis (i.e., of incorrectly inferring association simply because a large number of tests are performed).

NP complete In computational complexity theory, the class of NP complete problems are verifiable in polynomial time (i.e., if a solution is given, one can confirm the solution in polynomial time). However, there is no quick (polynomial time) way of finding a solution if none is known. Informally, NP complete problems are difficult to solve computationally and generally require approximation algorithms to be addressed at all.

ontology An *ontology* is what this book is about.

orthology Two proteins are homologous if they have a common evolutionary ancestor. If the proteins were separated by a speciation event, they are additionally said to be orthologous. For instance, the PTPN11 protein in humans is orthologous to the PTPN11 protein in mice.

pmf Probability mass function, a function that is used to describe a discrete probability distribution by giving the probability that a discrete random variable assumes some value.

qname Qualified name, a simplified way of expressing a URI using a namespace and an identifier separated by a colon. The namespace is a mapping to globally unique identifiers such as a URL (e.g., http://www.example.com).

resource description framework (RDF) The fundamental framework for the distribution of semantically coded data on the Web, and one of the components of OWL.

RDF schema A vocabulary description language for RDF that defines classes and properties that may be used to describe classes, properties and other resources.

RDF store A system for storing and managing RDF data (roughly speaking, an RDF database).

term-for-term approach An analysis method for GO term overrepresentation analysis in which each term is analyzed separately by the exact Fisher test.

triple A statement in RDF consisting of a subject, predicate, and object.

type I error In statistical hypothesis testing, a type I error occurs if the null hypothesis is falsely rejected (i.e., a false-positive). The probability of a type I error is symbolized by the Greek letter alpha (α).

type II error In statistical hypothesis testing, a type II error occurs if the null hypothesis is not rejected although the alternative hypothesis is true (i.e., a false-negative). The probability of a type II error is symbolized by the Greek letter beta (β).

W3C The W3C (short for World-Wide Web Consortium), an international organization that is responsible for Web and Semantic Web standards such as HTML, XML, RDF, and OWL.

Bibliography

[1] Abbott A (2010) Mouse project to find each gene's role. *Nature*, **465**:410.

[2] Abdi H (2007) Bonferroni and Sidak corrections for multiple comparisons. In *Encyclopedia of Measurement and Statistics*, pages 103–107. Sage, Thousand Oaks, CA.

[3] Aickin M, Gensler H (1996) Adjusting for multiple testing when reporting research results: the Bonferroni vs Holm methods. *Am J Public Health*, **86**:726–728.

[4] Alexa A, Rahnenführer J, Lengauer T (2006) Improved scoring of functional groups from gene expression data by decorrelating GO graph structure. *Bioinformatics*, **22**:1600–1607.

[5] Allanson JE, Biesecker LG, Carey JC, Hennekam RCM (2009) Elements of morphology: introduction. *Am J Med Genet A*, **149A**:2–5.

[6] Allemang D, Hendler JA (2008) *Semantic Web for the Working Ontologist: Effective Modeling in RDFS and OWL*. Morgan Kaufmann, Burlington, MA.

[7] Allison DB, Cui X, Page GP, Sabripour M (2006) Microarray data analysis: from disarray to consolidation and consensus. *Nat Rev Genet*, **7**:55–65.

[8] Altschul SF, Madden TL, Schffer AA, Zhang J, Zhang Z, Miller W, Lipman DJ (1997) Gapped BLAST and PSI-BLAST: a new generation of protein database search programs. *Nucleic Acids Res*, **25**:3389–3402.

[9] Amberger J, Bocchini CA, Alan F Scott AF, Hamosh A (2009) Mckusick's Online Mendelian Inheritance in Man (OMIM). *Nucleic Acids Res*, **37(Database issue)**:D793–D796.

[10] Andrieu C, de Freita N, Doucet A, Jordan MI (2003) An introduction to MCMC for machine learning. *Machine Learning*, **50**:5–43.

[11] Antezana E, Egaña M, Blondé W, Illarramendi A, Bilbao I, De Baets B, Stevens R, Mironov V, Kuiper M (2009). The Cell Cycle Ontology: an application ontology for the representation and integrated analysis of the cell cycle process. *Genome Biol*, **10**:R58.

[12] Arighi CN, Liu H, Natale DA, Barker WC, Drabkin H, Blake JA, Smith B, Wu CH (2009) TGF-beta signaling proteins and the Protein Ontology. *BMC Bioinformatics*, **10 Suppl 5**:S3.

[13] Arp R, Smith B (2008) Function, Role, and Disposition in Basic Formal Ontology. *Nature Precedings*.

[14] Ashburner M, Ball CA, Blake JA, Botstein D, Butler H, Cherry JM, Davis AP, Dolinski K, Dwight SS, Eppig JT, Harris MA, Hill DP, Issel-Tarver L, Kasarskis A, Lewis S, Matese JC, Richardson JE, Ringwald M, Rubin GM, Sherlock G (2000) Gene Ontology: tool for the unification of biology. The Gene Ontology Consortium. *Nat Genet*, **25**:25–29.

[15] Baader F, Calvanese D, McGuinness DL, Nardi D (editors) (2003) *The Description Logic Handbook: Theory, Implementation, and Applications.* Cambridge University Press.

[16] Baader F, Ulrike Sattler U (2001) An overview of tableau algorithms for description logics. *Studia Logica*, **69**:5–40.

[17] Bao J, Kendall EF, McGuinness DL, Patel-Schneider PF (2009) OWL 2 Web Ontology Language: Quick Reference Guide. W3C Recommendation, October 2009. http://www.w3.org/TR/owl-quick-reference/.

[18] Bar-Hillel M (1980) The base-rate fallacy in probability judgments. *Acta Psychologica*, **44**:211–233.

[19] Bard J, Rhee SY, Ashburner M (2005) An ontology for cell types. *Genome Biol*, **6**:R21.

[20] Bard JBL (2005) Anatomics: the intersection of anatomy and bioinformatics. *J Anat*, **206**:1–16.

[21] Bard JBL, Rhee SY (2004) Ontologies in biology: design, applications and future challenges. *Nat Rev Genet*, **5**:213–222.

[22] Barrell D, Dimmer E, Huntley RP, Binns D, O'Donovan C, Apweiler R (2009) The GOA database in 2009–an integrated Gene Ontology Annotation resource. *Nucleic Acids Res*, **37(Database issue)**:D396–D403.

[23] Bauer S, Gagneur J, Robinson PN (2010) GOing Bayesian: model-based gene set analysis of genome-scale data. *Nucleic Acids Res*, **38**:3523–3532.

[24] Bauer S, Grossmann S, Vingron M, Robinson (2008) Ontologizer 2.0–a multifunctional tool for GO term enrichment analysis and data exploration. *Bioinformatics*, **24**:1650–1651.

[25] Beck T, Morgan H, Blake A, Wells S, Hancock JM, Mallon AM (2009) Practical application of ontologies to annotate and analyse large scale raw mouse phenotype data. *BMC Bioinformatics*, **10 Suppl 5**:S2.

[26] Beckett D (2004) RDF/XML Syntax Specification (Revised). W3C Recommendation, 10 February 2004. http://www.w3.org/TR/REC-rdf-syntax/.

[27] Beckett D, Berners-Lee T (2008) Turtle - Terse RDF Triple Language. W3C Team Submission, 14 January 2008. http://www.w3.org/TeamSubmission/turtle/.

[28] Benjamini Y, Hochberg Y (1995) Controlling the false discovery rate: A practical and powerful approach to multiple testing. *Journal of the Royal Statistical Society. Series B*, **57**:289–300.

[29] Benjamini Y, Yekutieli D (2001) The control of the false discovery rate in multiple testing under dependency. *The Annals of Statistics*, **29**:1165–1188.

[30] Berners-Lee T (2006) Notation 3, March 2006. http://www.w3.org/DesignIssues/Notation3.

[31] Berners-Lee T, Connolly D (2008) Notation3 (N3): A readable RDF syntax. Technical report, W3C, January 2008.

[32] Berners-Lee T, Hendler J, Lassila O (2001) The Semantic Web. *Scientific American*, **284**:34–43.

[33] Biesecker LG (2005) Mapping phenotypes to language: a proposal to organize and standardize the clinical descriptions of malformations. *Clin Genet*, **68**:320–326.

[34] Blake JA, Bult CJ (2006) Beyond the data deluge: data integration and bio-ontologies. *J Biomed Inform*, **39**:314–320.

[35] Bland JM, Altman DG (1995) Multiple significance tests: the Bonferroni method. *BMJ*, **310**:170.

[36] Bodenreider O (2008) Biomedical ontologies in action: role in knowledge management, data integration and decision support. *Yearb Med Inform*, pages 67–79.

[37] Brazma A, Hingamp P, Quackenbush J, Sherlock G, Spellman P, Stoeckert C, Aach J, Ansorge W, Ball CA, Causton HC, Gaasterland T, Glenisson P, Holstege FC, Kim IF, Markowitz V, Matese JC, Parkinson H, Robinson A, Sarkans U, Schulze-Kremer S, Stewart J, Taylor R, Vilo J, Vingron M (2001) Minimum information about a microarray experiment (MIAME)-toward standards for microarray data. *Nat Genet*, **29**:365–371.

[38] Brickley D (2002) RDF: Understanding the striped RDF/XML syntax, August 2002. http://www.w3.org/2001/10/stripes/.

[39] Brickley D, Guha RV (2004) RDF Vocabulary Description Language 1.0: RDF Schema. W3C Recommendation, 10 February 2004. http://www.w3.org/TR/rdf-schema/.

[40] Brickley D, Miller L (2010) FOAF vocabulary specification 0.98, August 2010. http://xmlns.com/foaf/spec/.

[41] Brinkman RR, Courtot M, Derom D, Fostel JM, He Y, Lord P, Malone J, Parkinson H, Peters B, Rocca-Serra P, Ruttenberg A, Sansone SA, Soldatova LN, Stoeckert CJ, Turner JA, Zheng J, and the O.B.I. consortium (2010) Modeling biomedical experimental processes with OBI. *J Biomed Semantics*, 1 **Suppl** 1:S7.

[42] Brown A, Fuchs M, Robie J, Wadler P (2001) XML Schema: Formal Description. W3C Working Draft, 25 September 2001. http://www.w3.org/TR/xmlschema-formal/.

[43] Bulmer MG (1979) *Principles of Statistics*. Dover.

[44] Bult CJ, Eppig JT, Kadin JA, Richardson JE, Blake JA, and Mouse Genome Database Group. The Mouse Genome Database (MGD): Mouse biology and model systems. *Nucleic Acids Res*, **36(Database issue)**:D724–D728.

[45] Bult CJ, Kadin JA, Richardson JE, Blake JA, Eppig JT, and Mouse Genome Database Group (2010) The Mouse Genome Database: Enhancements and updates. *Nucleic Acids Res*, **38(Database issue)**:D586–D592.

[46] Burgun A (2006) Desiderata for domain reference ontologies in biomedicine. *J Biomed Inform*, **39**:307–313.

[47] Camon E, Magrane M, Barrell D, Lee V, Dimmer E, Maslen J, Binns D, Harte N, Lopez R, Apweiler R (2004) The Gene Ontology Annotation (GOA) database: Sharing knowledge in Uniprot with Gene Ontology. *Nucleic Acids Res*, **32(Database issue)**:D262–D266.

[48] Cannata N, Schröder M, Marangoni R, Romano P (2008) A Semantic Web for bioinformatics: Goals, tools, systems, applications. *BMC Bioinformatics*, **9 Suppl** 4:S1.

[49] Carroll J, Herman I, Patel-Schneider PF (2009) OWL 2 Web Ontology Language: RDF-Based Semantics. W3C Recommendation, 27 October 2009. http://www.w3.org/TR/owl2-rdf-based-semantics/.

[50] Carroll JJ, Dickinson I, Dollin C, Reynolds D, Seaborne A, Wilkinson K (2003) Jena: Implementing the semantic web recommendations. In *WWW Alt. 04: Proceedings of the 13th international World Wide Web conference on Alternate track papers & posters*, pages 74–83, New York, NY, USA, 2004. ACM.

[51] Castillo-Davis CI, Hartl DL (2003) Genemerge–post-genomic analysis, data mining, and hypothesis testing. *Bioinformatics*, **19**:891–892.

[52] Cimino JJ, Zhu X (2006) The practical impact of ontologies on biomedical informatics. *Yearb Med Inform*, pages 124–135.

[53] Cohen Y, Cohen JY (2008) *Statistics and Data with R. An Applied Approach Through Examples*. Wiley.

[54] Coletti MH, Bleich HL (2001) Medical subject headings used to search the biomedical literature. *J Am Med Inform Assoc*, **8**:317–323.

[55] Cook DL, Mejino JLV, Rosse C (2004) The foundational model of anatomy: a template for the symbolic representation of multi-scale physiological functions. *Conf Proc IEEE Eng Med Biol Soc*, **7**:5415–5418.

[56] Cook DL, Mejino JLV, Rosse C (2004) Evolution of a foundational model of physiology: symbolic representation for functional bioinformatics. *Stud Health Technol Inform*, **107**:336–340.

[57] Cormen TH, Leiserson CE, Ronald L Rivest RL, Stein C (2001) *Introduction to Algorithms*. MIT Press, Cambridge, MA, second edition.

[58] Cornet R, de Keizer N (2008). Forty years of SNOMED: a literature review. *BMC Med Inform Decis Mak*, **8 Suppl 1**:S2.

[59] Côté RG, Jones P, Lennart Martens L, Apweiler R, Hermjakob H (2008) The ontology lookup service: More data and better tools for controlled vocabulary queries. *Nucleic Acids Res*, **36(Web Server issue)**:W372–W376.

[60] Couto F, Silva MJ, Coutinho PM (2007) Measuring semantic similarity between Gene Ontology terms. *Data Knowledge & Engineering*, **61**:137–152.

[61] Cuenca B, Horrocks I, Motik B, Parsia B, Patel-Schneider P, Sattler U (2008) OWL 2: The next step for OWL. *Web Semantics*, **6**:309–322.

[62] Dahdul WM, Lundberg JG, Midford PE, Balhoff JP, Lapp H, Vision TJ, Haendel MA, Westerfield M, Mabee PM (2010) The teleost anatomy ontology: Anatomical representation for the genomics age. *Syst Biol*, **59**:369–383.

[63] Dalgaard P (2002) *Introductory Statistics With R*. Springer.

[64] Day-Richter J, Harris MA, Haendel M, Gene Ontology OBO-Edit Working Group, Lewis S (2007) OBO-Edit–an ontology editor for biologists. *Bioinformatics*, **23**:2198–2200.

[65] de Coronado S, Wright LW, Fragoso G, Haber MW, Hahn-Dantona EA, Hartel FW, Quan SL, Safran T, Thomas N, Whiteman L (2009) The NCI Thesaurus quality assurance life cycle. *J Biomed Inform*, **42**:530–539.

[66] de Matos P, Alcántara R, Dekker A, Ennis M, Hastings J, Haug K, Spiteri I, Turner S, Steinbeck C (2010) Chemical entities of biological interest: An update. *Nucleic Acids Res*, **38(Database issue)**:D249–D254.

[67] Degtyarenko K, de Matos P, Ennis M, Hastings J, Zbinden M, McNaught A, Alcántara R, Darsow M, Guedj M, Ashburner M (2008) ChEBI: A database and ontology for chemical entities of biological interest. *Nucleic Acids Res*, **36(Database issue)**:D344–D350.

[68] Delaurier A, Burton N, Bennett M, Baldock R, Davidson D, Mohun TJ, Logan MPO (2008) The Mouse Limb Anatomy Atlas: An interactive 3D tool for studying embryonic limb patterning. *BMC Dev Biol*, **8**:83.

[69] Landon T Detwiler LT, Dan Suciu D, and James F Brinkley JF (2008) Regular paths in SparQL: Querying the NCI Thesaurus. *AMIA Annu Symp Proc*, pages 161–165.

[70] Deus HF, Veiga DF, Freire PR, Weinstein JN, Mills GB, Almeida JS (2010) Exposing the cancer genome atlas as a SPARQL endpoint. *J Biomed Inform*, **43**:998–1008.

[71] Diaconis P (2009) The Markov chain Monte Carlo revolution. *Bull. Amer. Math. Soc.*, **46**:179–205.

[72] Diaconis P, Saloff-Coste L (1995) What do we know about the Metropolis algorithm? In *STOC '95: Proceedings of the twenty-seventh annual ACM symposium on Theory of computing*, pages 112–129, New York, NY, USA, ACM.

[73] Doms A, Schroeder M (2005) GoPubMed: Exploring PubMed with the Gene Ontology. *Nucleic Acids Res*, **33(Web Server issue)**:W783–W786.

[74] Dudoit S, van der Laan MJ, Pollard KS (2004) Multiple testing. Part I. Single-step procedures for control of general type I error rates. *Stat Appl Genet Mol Biol*, **3**:Article13.

[75] Ewens WJ, Grant GR (2005) *Statistical Methods in Bioinformatics: An Introduction.* Springer, 2nd edition.

[76] Falcon S, Gentleman R (2007) Using GOstats to test gene lists for GO term association. *Bioinformatics*, **23**:257–258.

[77] Federative Committee on Anatomical Termi (1998) *Terminologia Anatomica: International Anatomical Terminology.* Thieme Stuttgart.

[78] Ferreira JD, Couto FM (2010) Semantic similarity for automatic classification of chemical compounds. *PLoS Comput Biol,* **6**:e1000937.

[79] Fog A (2008) Calculation methods for Wallenius' noncentral hypergeometric distribution. *Communications in Statistics - Simulation and Computation,* **37**:258–273.

[80] Fragoso G, de Coronado S, Haber M, Hartel F, Wright L (2004) Overview and utilization of the NCI Thesaurus. *Comp Funct Genomics,* **5**:648–654.

[81] Fredman D, Munns G, Rios D, Sjöholm F, Siegfried M, Lenhard B, Lehväslaiho H, Brookes AJ (2004) HGVbase: a curated resource describing human DNA variation and phenotype relationships. *Nucleic Acids Res,* **32(Database issue)**:D516–D519.

[82] Freimer N, Sabatti C (2003) The human phenome project. *Nat Genet,* **34**:15–21.

[83] Gangemi A, Guarino N, Masolo C, Oltramari A, Schneider L (2002) Sweetening Ontologies with DOLCE. In Gomez Perez A, Benjamins VR (editors) *Knowledge Engineering and Knowledge Management. Ontologies and the Semantic Web. Proceedings of 13th International Conference on Knowledge Engineering and Knowledge Management (EKAW),* Berlin: Springer Verlag, p.166–181.

[84] Ge Y, Dudoit S, Speed TP (2003) Resampling-based multiple testing for microarray data analysis. *TEST,* **12**:1–44.

[85] Gene Ontology Consortium (2001) Creating the Gene Ontology resource: design and implementation. *Genome Res,* **11**:1425–1433.

[86] Gene Ontology Consortium (2010) The Gene Ontology in 2010: extensions and refinements. *Nucleic Acids Res,* **38(Database issue)**:D331–D335.

[87] Gentleman R (2008) *R Programming for Bioinformatics.* Chapman & Hall/CRC.

[88] Gentleman RC, Carey VJ, Bates DM, Bolstad B, Dettling M, Dudoit S, Ellis B, Gautier L, Ge Y, Gentry J, Hornik K, Hothorn T, Huber W, Iacus S, Irizarry R, Leisch F, Li C, Maechler M, Rossini AJ, Sawitzki G, Smith C, Smyth G, Tierney L, Yang JYH, Zhang J (2004) Bioconductor: open software development for computational biology and bioinformatics. *Genome Biol,* **5**:R80.

[89] George RA, Smith TD, Callaghan S, Hardman L, Pierides C, Horaitis O, Wouters MA, Cotton RGH (2008) General mutation databases: Analysis and review. *J Med Genet*, **45**:65–70.

[90] Gkoutos GV, Green ECJ, Mallon AM, Hancock JM, Davidson D (2004) Building mouse phenotype ontologies. *Pac Symp Biocomput*, pages 178–189.

[91] Gkoutos GV, Green ECJ, Mallon AM, Hancock JM, Davidson D (2005) Using ontologies to describe mouse phenotypes. *Genome Biol*, **6**:R8.

[92] Gkoutos GV, Mungall C, Dölken S, Ashburner M, Lewis S, Hancock J, Schofield P, Köhler S, Robinson PN (2009) Entity/quality-based logical definitions for the human skeletal phenome using PATO. *Conf Proc IEEE Eng Med Biol Soc*, **2009**:7069–7072.

[93] Golbreich C, Horridge M, Horrocks I, Motik B, Shearer R, OBO and OWL: Leveraging Semantic Web technologies for the life sciences (2007) In Aberer K, Choi KS, Noy N, Allemang D, Lee KI, Nixon LJB, Golbeck J, Mika P, Maynard D, Schreiber G, Cudré-Mauroux P (editors) *Proceedings of the 6th International Semantic Web Conference and 2nd Asian Semantic Web Conference (ISWC/ASWC2007), Busan, South Korea*, volume 4825 of *LNCS*, pages 169–182, Berlin, Heidelberg. Springer Verlag.

[94] Golbreich C, Wallace EK (2009) OWL 2 Web Ontology Language: New Features and Rationale. W3C Recommendation, 27 October 2009. http://www.w3.org/TR/owl2-new-features/.

[95] Grenon P (2003) Spatio-temporality in Basic Formal Ontology: SNAP and SPAN, upper-level ontology, and framework for formalization, part I. Technical report, Institute for Formal Ontology and Medical Information Science (IFOMIS).

[96] Grenon P, Smith B, Goldberg L (2004) Biodynamic ontology: Applying BFO in the biomedical domain. *Stud Health Technol Inform*, **102**:20–38.

[97] Grossmann S, Bauer S, Robinson PN, Vingron M (2006) An improved statistic for detecting over-represented Gene Ontology annotations in gene sets. In *RECOMB*, pages 85–98.

[98] Grossmann S, Bauer S, Robinson PN, Vingron M (2007) Improved detection of overrepresentation of Gene-Ontology annotations with parent child analysis. *Bioinformatics*, **23**:3024–3031.

[99] Gruber T (2007) Ontology of folksonomy: A mash-up of apples and oranges. *International Journal on Semantic Web & Information Systems*, **3**:1–11.

[100] Gruber TR (1993) A translation approach to portable ontology specifications. *Knowledge Acquisition*, **5**:199–220.

[101] Grumbling G, Strelets V (2006) FlyBase: Anatomical data, images and queries. *Nucleic Acids Res*, **34(Database issue)**:D484–D488.

[102] Haendel MA, Gkoutos GV, Lewis SE, Mungall C (2009) Uberon: Towards a comprehensive multi-species anatomy ontology. *Available from Nature Precedings <http://dx.doi.org/10.1038/npre.2009.3592.1>*.

[103] Haendel MA, Neuhaus F, Osumi-Sutherland D, Mabee PM, Mejino JLV, Mungall CJ, Smith B (2007) CARO - The Common Anatomy Reference Ontology. In Burger A, Davidson D, Baldock R (editors) *Anatomy Ontologies for Bioinformatics: Principles and Practice*, pages 327–350. Springer.

[104] Harold ER, Means WS (2004) *XML in a Nutshell*. A Nutshell handbook. O'Reilly, Beijing, 3rd rev. ed. edition.

[105] Harris TW, Antoshechkin I, Bieri T, Blasiar D, Chan J, Chen WJ, De La Cruz N, Davis P, Duesbury M, Fang R, Fernandes J, Han M, Kishore R, Lee R, Müller HM, Nakamura C, Ozersky P, Petcherski A, Rangarajan A, Rogers A, Schindelman G, Schwarz EM, Tuli MA, Van Auken K, Wang D, Wang X, Williams G, Yook K, Durbin R, Stein LD, Spieth J, Sternberg PW (2010) WormBase: A comprehensive resource for nematode research. *Nucleic Acids Res*, **38(Database issue)**:D463–D467.

[106] Hawking SW (2005) *God created the integers: The mathematical breakthroughs that changed history*. Running Press, Philadelphia, PA.

[107] Hayamizu TF, Mangan M, Corradi JP, Kadin JA, Ringwald M (2005) The adult mouse anatomical dictionary: A tool for annotating and integrating data. *Genome Biol*, **6**:R29.

[108] Hayes P (2004) RDF Semantics. W3C Recommendation, 10 February 2004. http://www.w3.org/TR/rdf-mt/.

[109] Hecht J, Seitz V, Urban M, Wagner F, Robinson PN, Stiege A, Dieterich C, Kornak U, Wilkening U, Brieske N, Zwingman C, Kidess A, Stricker S, Mundlos S (2007) Detection of novel skeletogenesis target genes by comprehensive analysis of a Runx2$^{(-/-)}$ mouse model. *Gene Expr Patterns*, **7**:102–112.

[110] Hecht J, Kuhl H, Haas SA, Bauer S, Poustka AJ, Lienau J, Schell H, Stiege AC, Seitz V, Reinhardt R, Duda GN, Mundlos S, Robinson PN (2006) Gene identification and analysis of transcripts differentially regulated in fracture healing by EST sequencing in the domestic sheep. *BMC Genomics*, 7:172, 2006.

[111] Heckerman D (1995) A tutorial on learning with Bayesian networks. Technical report, MSR-TR-95-06, Microsoft Research.

[112] Herre H (2010) General Formal Ontology (GFO): A foundational ontology for conceptual modelling. In Roberto Poli and Leo Obrst (editors) *Theory and Applications of Ontology*, volume 2. Springer, Berlin.

[113] Herre H, Heller B, Burek P, Hoehndorf R, Loebe F, Michalek H (2006) General Formal Ontology (GFO): A foundational ontology integrating objects and processes. part I: Basic principles (version 1.0). Onto-Med Report 8, Research Group Ontologies in Medicine (Onto-Med), University of Leipzig.

[114] Hill DP, Blake JA, Richardson JE, Ringwald M (2002) Extension and integration of the Gene Ontology (GO): Combining GO vocabularies with external vocabularies. *Genome Res*, **12**:1982–1991.

[115] Hill DP, Smith B, McAndrews-Hill MS, Blake JA (2008) Gene Ontology annotations: What they mean and where they come from. *BMC Bioinformatics*, **9 Suppl 5**:S2.

[116] Hitzler P, Krötzsch M, Parsia B, Patel-Schneider PF, Rudolph S (2009) OWL 2 Web Ontology Language: Primer. W3C Recommendation, 27 October 2009. http://www.w3.org/TR/owl2-features/.

[117] Hitzler P, Krötzsch M, Rudolph S (2009) *Foundations of Semantic Web Technologies*. CRC Press, Boca Raton, FL.

[118] Hitzler P, Krötzsch M, Rudolph S, Sure Y (2008) *Semantic Web: Grundlagen*. Springer Verlag, Berlin.

[119] Holford ME, Khurana E, Cheung KH, Gerstein M (2010) Using semantic web rules to reason on an ontology of pseudogenes. *Bioinformatics*, **26**:i71–i78.

[120] Holland BS, Copenhaver MD (1988) Improved Bonferroni-type multiple testing procedures. *Psychological Bulletin*, **104**:145–149.

[121] Horridge M (2009) *A Practical Guide To Building OWL Ontologies Using Protégé 4 and CO-ODE Tools Edition 1.2*. University of Manchester, 1.2 edition.

[122] Horridge M, Bechhofer S (2008) The OWL API: A Java API for Working with OWL 2 Ontologies. In Hoekstra R, Patel-Schneider PF (editors) *OWLED*, volume 529 of *CEUR Workshop Proceedings*. CEUR-WS.org.

[123] Horridge M, Drummond N, Goodwin J, Rector AL, Stevens R, Wang H (2006) The Manchester OWL syntax. In Cuenca Grau B, Hitzler P, Shankey C, Wallace E (editors) *OWLED*, volume 216 of *CEUR Workshop Proceedings*. CEUR-WS.org.

[124] Horridge M, Patel-Schneider PF (2008) Manchester syntax for OWL 1.1. In *OWLED 2008, 4th international workshop OWL: Experiences and Directions (2008) Live Extraction 1223*.

[125] Horridge M, Patel-Schneider PF (2009) OWL 2 Web Ontology Language: Manchester Syntax. W3C Working Group Note, 27 October 2009. http://www.w3.org/TR/owl2-manchester-syntax/.

[126] Horrocks I, Kutz O, Sattler U (2005) The irresistible $\mathcal{SHIQ}$. In *In Proc. of OWL: Experiences and Directions*, 2005.

[127] Horrocks I, Patel-Schneider PF, Boley H, Tabet S, Grosofand B, Dean M (2004) SWRL: A semantic web rule language combining OWL and RuleML. W3C Member Submission, May 2004. Last accessed on Dec 2008 at: http://www.w3.org/Submission/SWRL/.

[128] Horrocks I, Sattler U (2004) Decidability of $\mathcal{SHIQ}$ with complex role inclusion axioms. *Artificial Intelligence*, **160**:79–104.

[129] Howe D, Costanzo M, Fey P, Gojobori T, Hannick L, Hide W, Hill DP, Kania R, Schaeffer M, St Pierre S, Twigger S, White O, Rhee SY (2008) Big data: The future of biocuration. *Nature*, **455**:47–50.

[130] Hubbard TJP, Aken BL, Beal K, Ballester B, Caccamo M, Chen Y, Clarke L, Coates G, Cunningham F, Cutts T, Down T, Dyer SC, Fitzgerald S, Fernandez-Banet J, Graf S, Haider S, Hammond M, Herrero J, Holland R, Howe K, Howe K, Johnson N, Kahari A, Keefe D, Kokocinski F, Kulesha E, Lawson D, Longden I, Melsopp C, Megy K, Meidl P, Ouverdin B, Parker A, Prlic A, Rice S, Rios D, Schuster M, Sealy I, Severin J, Slater G, Smedley D, Spudich G, Trevanion S, Vilella A, Vogel J, White S, Wood M, Cox T, Curwen V, Durbin R, Fernandez-Suarez XM, Flicek P, Kasprzyk A, Proctor G, Searle S, Smith J, Ureta-Vidal A, Birney E (2007) Ensembl 2007. *Nucleic Acids Res*, **35(Database issue)**:D610–D617.

[131] Huntley RP, Binns D, Dimmer E, Barrell D, O'Donovan C, Apweiler R (2009) QuickGO: A user tutorial for the web-based Gene Ontology browser. *Database (Oxford)*, **2009**:bap010.

[132] Husmeier D (2003) Sensitivity and specificity of inferring genetic regulatory interactions from microarray experiments with dynamic Bayesian networks. *Bioinformatics*, **19**:2271–2282.

[133] Irizarry RA, Wang C, Zhou Y, Speed TP (2009) Gene set enrichment analysis made simple. *Statistical Methods in Medical Research*, **18**:565–575.

[134] Jiang JJ, Conrath DW (1997) Semantic similarity based on corpus statistics and lexical taxonomy. In *International Conference Research on Computational Linguistics*, p. 19–33.

[135] Johansson I (2009) Four kinds of 'is_a' relations. In Munn K, Smith B (editors) *Applied Ontology: An Introduction*, p. 235–253. Ontos Verlag.

[136] Jonquet C, Musen MA, Shah NH (2010) Building a biomedical ontology recommender web service. *J Biomed Semantics*, 1 **Suppl** 1:S1.

[137] Kalyanpur A, Parsia B, Sirin E, Hendler J (2005) Debugging unsatisfiable classes in OWL ontologies. *Web Semantics: Science, Services and Agents on the World Wide Web*, **3**:268–293.

[138] Kanehisa M, Goto S (2000) KEGG: Kyoto Encyclopedia of Genes and Genomes. *Nucleic Acids Res*, **28**:27–30.

[139] Khatri P, Drăghici S (2005) Ontological analysis of gene expression data: current tools, limitations, and open problems. *Bioinformatics*, **21**:3587–3595.

[140] Khinchin AI (1957) *Mathematical Foundations of Information Theory*. Dover.

[141] Klyne G, Carroll JJ (2004) Resource Description Framework (RDF): Concepts and Abstract Syntax. W3C Recommendation, 10 February 2004. http://www.w3.org/TR/rdf-primer/.

[142] Knublauch H, Dameron O, Musen MA (2004) Weaving the biomedical semantic web with the Protégé OWL plugin. In *International Workshop on Formal Biomedical Knowledge Representation*.

[143] Knublauch H, Fergerson RW, Noy NF, Musen MA (2004) The Protégé OWL plugin: An open development environment for semantic web applications. In *Third International Semantic Web Conference - ISWC 2004*, Hiroshima, Japan.

[144] Köhler S, Bauer S, Horn D, Robinson PN (2008) Walking the interactome for prioritization of candidate disease genes. *Am J Hum Genet*, **82**:949–958.

[145] Köhler S, Schulz MH, Krawitz P, Bauer S, Dölken S, Ott CE, Mundlos C, Horn D, Mundlos S, Robinson PN (2009) Clinical diagnostics in human genetics with semantic similarity searches in ontologies. *Am J Hum Genet*, **85**:457–464.

[146] Kume T, Deng KY, Winfrey V, Gould DB, Walter MA, Hogan BL (1998) The forkhead/winged helix gene Mf1 is disrupted in the pleiotropic mouse mutation congenital hydrocephalus. *Cell*, **93**:985–996.

[147] Labrou Y, Finin T (1999) Yahoo! as an ontology: Using yahoo! categories to describe documents. In *CIKM '99: Proceedings of the eighth international conference on Information and knowledge management*, pages 180–187, New York, NY, USA, ACM.

[148] Lage K, Karlberg EO, Størling ZM, Olason PI, Pedersen AG, Rigina O, Hinsby AM, Tümer Z, Pociot F, Tommerup N, Moreau Y, Brunak S (2007) A human phenome-interactome network of protein complexes implicated in genetic disorders. *Nat Biotechnol*, **25**:309–316.

[149] Lee RYN, Sternberg PW (2003) Building a cell and anatomy ontology of *Caenorhabditis elegans*. *Comp Funct Genomics*, 4:121 126.

[150] Lenffer J, Nicholas FW, Castle K, Rao A, Gregory S, Poidinger M, Mailman MD, Ranganathan S (2006) OMIA (Online Mendelian Inheritance in Animals): An enhanced platform and integration into the Entrez search interface at NCBI. *Nucleic Acids Res*, **34(Database issue)**:D599–D601.

[151] Lin D (1998) An information-theoretic definition of similarity. In *Proc. of the 15th International Conference on Machine Learning*, p. 296–304, Morgan Kaufmann.

[152] Lomax J (2005) Get ready to GO! A biologist's guide to the Gene Ontology. *Brief Bioinform*, **6**:298–304.

[153] Lord PW, Stevens RD, Brass A, Goble CA (2003) Investigating semantic similarity measures across the Gene Ontology: the relationship between sequence and annotation. *Bioinformatics*, **19**:1275–1283.

[154] Lord P, Stevens R (2010) Adding a little reality to building ontologies for biology. *PLoS ONE*, **5**:e12258.

[155] Lu Y, Rosenfeld R, Simon I, Nau GJ, Bar-Joseph Z (2008) A probabilistic generative model for GO enrichment analysis. *Nucleic Acids Res*, **36**:e109.

[156] Lutz C (2003) Description logics with concrete domains—a survey. In *Advances in Modal Logics Volume 4*. King's College Publications.

[157] MacKay DJC (2005) *Information Theory, Inference, and Learning Algorithms*. Cambridge University Press, 7.2 edition.

[158] Mailman MD, Feolo M, Jin Y, Kimura M, Tryka K, Bagoutdinov R, Hao L, Kiang A, Paschall J, Phan L, Popova N, Pretel S, Ziyabari L, Lee M, Shao Y, Wang ZY, Sirotkin K, Ward M, Kholodov M, Zbicz K, Beck J, Kimelman M, Shevelev S, Preuss D, Yaschenko E, Graeff A, Ostell J, Sherry ST (2007) The NCBI dbGaP database of genotypes and phenotypes. *Nat Genet*, **39**:1181 1186.

[159] Maindonald J, Braun J (2007) *Data Analysis and Graphics Using R: An Example-Based Approach, Second Edition*. Cambridge University Press.

[160] Manola F, Miller E (2004) RDF Primer. W3C Recommendation, 10 February 2004. http://www.w3.org/TR/rdf-primer/.

[161] Mascardi V, Cordì V, Rosso P (2006) A comparison of upper ontologies. Technical Report DISI-TR-06-21, Dipartimento di Informatica e Scienze dell'Informazione (DISI), Universitá degli Studi di Genova, Italy.

[162] Mason DM, Schuenemeyer JH (1983) A modified Kolmogorov-Smirnov test sensitive to tail alternatives. *Ann. Statist.*, 11:933–946.

[163] McGuinness DL (2003) Ontologies come of age. In Fensel D, Hendler J, Lieberman H, Wahlster W (editors) *Spinning the Semantic Web: Bringing the World Wide Web to Its Full Potential*, pages 171–194. MIT Press.

[164] Michael J, Mejino JL, Rosse C (2001) The role of definitions in biomedical concept representation. *Proc AMIA Symp*, pages 463–467.

[165] Miller GA (1995) WordNet: A lexical database for English. *Communications of the ACM*, 38:39–41.

[166] Mootha VK, Lindgren CM, Eriksson KF, Subramanian A, Sihag S, Lehar J, Puigserver P, Carlsson E, Ridderstråle M, Laurila E, Houstis N, Daly MJ, Patterson N, Mesirov JP, Golub TR, Tamayo P, Spiegelman B, Lander ES, Hirschhorn JN, Altshuler D, Groop LC (2003) PGC-1α-responsive genes involved in oxidative phosphorylation are coordinately downregulated in human diabetes. *Nat Genet*, 34:267–273.

[167] Morgan H, Beck T, Blake A, Gates H, Adams N, Debouzy G, Leblanc S, Lengger C, Maier H, Melvin D, Meziane H, Richardson D, Wells S, White J, Wood J, EUMODIC Consortium, Hrabé de Angelis M, Brown SDM, Hancock JM, Mallon AM (2010). EuroPhenome: a repository for high-throughput mouse phenotyping data. *Nucleic Acids Res*, 38(Database issue):D577–D585.

[168] Morin R, Bainbridge M, Fejes A, Hirst M, Krzywinski M, Pugh T, McDonald H, Varhol R, Jones S, Marra M (2008) Profiling the HeLa S3 transcriptome using randomly primed cDNA and massively parallel short-read sequencing. *BioTechniques*, 45:81–94.

[169] Mortazavi A, Williams BA, McCue K, Schaeffer L, Wold B (2008) Mapping and quantifying mammalian transcriptomes by RNA-Seq. *Nature Methods*, 5:621–628.

[170] Motik B, Cuenca Grau B, Ian Horrocks I, Wu Z, Fokoue A, Lutz C (2009) OWL 2 Web Ontology Language: Profiles. W3C Recommendation, 27 October 2009. http://www.w3.org/TR/owl2-profiles/.

[171] Motik B, Parsia B, Patel-Schneider PF (2009) OWL 2 Web Ontology Language: XML Serialization. W3C Recommendation, 27 October 2009. http://www.w3.org/TR/owl2-xml-serialization/.

[172] Motik B, Patel-Schneider PF, Cuenca Grau B (2009) OWL 2 Web Ontology Language: Direct Semantics. W3C Recommendation, 27 October 2009. http://www.w3.org/TR/owl2-direct-semantics/.

[173] Motik B, Patel-Schneider PF, Parsia B (2009) OWL 2 Web Ontology Language: Structural Specification and Functional-Style Syntax. W3C Recommendation, 27 October 2009. http://www.w3.org/TR/owl2-syntax/.

[174] Motik B, Shearer R, Horrocks I (2007) A Hypertableau Calculus for $\mathcal{SHIQ}$. In Calvanese D, Franconi E, Haarslev V, Lembo D, Motik B, Tessaris S, Turhan AY (editors), *Proc. of the 20th Int. Workshop on Description Logics (DL 2007)*, pages 419–426, Brixen/Bressanone, Italy, June 8–10 2007. Bozen/Bolzano University Press.

[175] Motik B, Shearer R, Horrocks I (2009) Hypertableau reasoning for description logics. *J. Artif. Intell. Res. (JAIR)*, **36**:165–228.

[176] Mundlos S (2009) The brachydactylies: a molecular disease family. *Clin Genet*, **76**:123–136.

[177] Mungall CJ, Bada M, Berardini TZ, Deegan J, Ireland A, Harris MA, Hill DP, Lomax J (2010) Cross-product extensions of the Gene Ontology. *J Biomed Inform*, **44**:80–86.

[178] Mungall CJ, Batchelor C, Eilbeck K (2011) Evolution of the sequence ontology terms and relationships. *J Biomed Inform*, **44**:87-93..

[179] Mungall CJ, Gkoutos GV, Smith CL, Haendel MA, Lewis SE, Ashburner M (2010) Integrating phenotype ontologies across multiple species. *Genome Biol*, **11**:R2.

[180] Natale DA, Arighi CN, Barker WC, Blake J, Chang TC, Hu Z, Liu H, Smith B, Wu CH (2007) Framework for a protein ontology. *BMC Bioinformatics*, **8 Suppl 9**:S1.

[181] Natale DA, Arighi CN, Barker WC, Blake JA, Bult CJ, Caudy M, Drabkin HJ, D'Eustachio P, Evsikov AV, Huang H, Nchoutmboube J, Roberts NV, Smith B, Zhang J, Wu CH (2011) The Protein Ontology: a structured representation of protein forms and complexes. *Nucleic Acids Res*, **39(Database issue)**:D539–D545.

[182] Neapolitan RE (2003) *Learning Bayesian Networks*. Prentice Hall.

[183] Neches R, Fikes RE, Finin T, Gruber T, Patil R, Senator T, Swartout WR (1991) Enabling technology for knowledge sharing. *AI Magazine*, **3**:16–36.

[184] Niles I, Pease A (2001) Towards a standard upper ontology. In *Proceedings of the international conference on Formal Ontology in Information Systems - Volume 2001*, pages 2–9. ACM Press, New York.

[185] Nishiwaki-Yasuda K, Suzuki A, Kakita A, Sekiguchi S, Asano S, Nishii K, Nagao S, Oiso Y, Itoh M (2007) Vasopressin stimulates Na-dependent phosphate transport and calcification in rat aortic smooth muscle cells. *Endocr J*, **54**:103–112.

[186] Noy NF, McGuinness D (2001) Ontology Development 101: A Guide to Creating your First Ontology. Technical Report SMI-2001-0880, Stanford Medical Informatics, Stanford.

[187] Noy NF, Crubezy M, Fergerson RW, Knublauch H, Tu SW, Vendetti J, Musen MA (2003) Protégé-2000: An open-source ontology-development and knowledge-acquisition environment. *AMIA Annu Symp Proc*, **2003**:953.

[188] Noy NF, Rubin DL (2008) Translating the foundational model of anatomy into OWL. *Web Semant*, **6**:133–136.

[189] Noy NF, Shah NH, Whetzel PL, Dai B, Dorf M, Griffith G, Jonquet C, Rubin DL, Storey MA, Chute CG, Musen MA (2009) BioPortal: Ontologies and integrated data resources at the click of a mouse. *Nucleic Acids Res.*, **37**:W170–173.

[190] O'Connor MJ, Shankar RD, Tu SW, Nyulas C, Das AK, Musen MA (2007) Efficiently querying relational databases using OWL and SWRL. In Massimo Marchiori, Jeff Z. Pan, and Christian de Sainte Marie, (editors) *RR*, volume 4524 of *Lecture Notes in Computer Science*, pages 361–363. Springer.

[191] Oshlack A, Wakefield M (2009) Transcript length bias in RNA-seq data confounds systems biology. *Biology Direct*, **4**:14.

[192] Oti M, Huynen MA, Brunner HG (2008) Phenome connections. *Trends Genet*, **24**:103–106.

[193] Ott CE, Grünhagen J, Jäger M, Horbelt D, Schwill S, Kallenbach K, Guo G, Manke T, Knaus P, Mundlos S, Robinson PN (2011) MicroRNAs Differentially Expressed in Postnatal Aortic Development Downregulate Elastin via 3' UTR and Coding-Sequence Binding Sites. *PLoS ONE*, **6**:e16250.

[194] Ott CE, Bauer S, Manke T, Ahrens S, Rödelsperger C, Grünhagen J, Kornak U, Duda G, Mundlos S, Robinson PN (2009). Promiscuous and depolarization-induced immediate-early response genes are induced by mechanical strain of osteoblasts. *J Bone Miner Res*, **24**:1247–1262.

[195] Pandey G, Myers CL, Kumar V (2009) Incorporating functional inter-relationships into protein function prediction algorithms. *BMC Bioinformatics*, **10**:142.

[196] Parkinson H, Kapushesky M, Kolesnikov N, Rustici G, Shojatalab M, Abeygunawardena N, Berube H, Dylag M, Emam I, Farne A, Holloway E, Lukk M, Malone J, Mani R, Pilicheva E, Rayner TF, Rezwan F, Sharma A, Williams E, Bradley XZ, Adamusiak T, Brandizi M, Burdett T, Coulson R, Krestyaninova M, Kurnosov P, Maguire E, Neogi SG, Rocca-Serra P, Sansone SA, Sklyar N, Zhao M, Sarkans U, Brazma A (2009) ArrayExpress update–from an archive of functional genomics experiments to the atlas of gene expression. *Nucleic Acids Res*, **37**(**Database issue**):D868–D872.

[197] Patel-Schneider PF, Motik B (2009) OWL 2 Web Ontology Language: Mapping to RDF Graphs. W3C Recommendation, 27 October 2009. http://www.w3.org/TR/owl-mapping-to-rdf/.

[198] Pesquita C, Faria D, Falcão AO, Lord P, Couto FM (2009) Semantic similarity in biomedical ontologies. *PLoS Comput Biol*, **5**:e1000443.

[199] Pesquita C, Faria D, Bastos H, E N Ferreira A, Falcão AO, Couto FM (2008) Metrics for GO based protein semantic similarity: a systematic evaluation. *BMC Bioinformatics*, **9 Suppl 5**:S4.

[200] Pierce JR (1980) *An Introduction to Information Theory*. Dover Publications.

[201] Prud'hommeaux E, Seaborne A (2008) SPARQL Query Language for RDF. W3C Recommendation, 15 January 2008. http://www.w3.org/TR/rdf-sparql-query/.

[202] Reese MG, Moore B, Batchelor C, Salas F, Cunningham F, Marth GT, Stein L, Flicek P, Yandell M, Eilbeck K (2010) A standard variation file format for human genome sequences. *Genome Biol*, **11**:R88.

[203] Reiner A, Yekutieli D, Benjamini Y (2003) Identifying differentially expressed genes using false discovery rate controlling procedures. *Bioinformatics*, **19**:368–375.

[204] Resnik P (1995) Using information content to evaluate semantic similarity in a taxonomy. In *Proceedings of the 14th International Joint Conference on Artificial Intelligence*, pages 448–453.

[205] Resnik P (1999) Semantic similarity in a taxonomy: An information-based measure and its application to problems of ambiguity in natural language. *Journal of Artificial Intelligence Research*, **11**:95–130.

[206] Rhee SJ, Wood V, Dolinski K, Drăghici S (2008) Use and misuse of the Gene Ontology annotations. *Nat Rev Genet*, **9**:509–515.

[207] Richardson L, Venkataraman S, Stevenson P, Yang Y, Burton N, Rao J, Fisher M, Baldock RA, Davidson DR, Christiansen JH (2010) EMAGE mouse embryo spatial gene expression database: 2010 update. *Nucleic Acids Res*, **38(Database issue)**:D703–D709.

[208] Robinson MD, Smyth GK (2007) Moderated statistical tests for assessing differences in tag abundance. *Bioinformatics*, **23**:2881–2887.

[209] Robinson PN, Mundlos S (2010) The Human Phenotype Ontology. *Clin Genet*, **77**:525–534.

[210] Robinson PN, Böhme U, Lopez R, Mundlos S, Nürnberg P (2004) Gene-Ontology analysis reveals association of tissue-specific 5' CpG-island genes with development and embryogenesis. *Hum Mol Genet*, **13**:1969–1978.

[211] Robinson PN, Köhler S, Bauer S, Seelow D, Horn D, Mundlos S (2008) The Human Phenotype Ontology: a tool for annotating and analyzing human hereditary disease. *Am J Hum Genet*, **83**:610–615.

[212] Robinson PN, Wollstein A, Böhme U, Beattie B (2004) Ontologizing gene-expression microarray data: characterizing clusters with gene ontology. *Bioinformatics*, **20**:979–981.

[213] Rosenthal N, Brown S (2007) The mouse ascending: perspectives for human-disease models. *Nat Cell Biol*, **9**:993–999.

[214] Rosse C, Mejino JLV (2003) A reference ontology for biomedical informatics: the foundational model of anatomy. *J Biomed Inform*, **36**:478–500.

[215] Rubin DL, Lewis SE, Mungall CJ, Misra S, Westerfield M, Ashburner M, Sim I, Chute CG, Solbrig H, Storey MA, Smith B, Day-Richter J, Noy NF, Musen MA (2006) National Center for Biomedical Ontology: Advancing biomedicine through structured organization of scientific knowledge. *OMICS*, **10**:185–198.

[216] Rubin DL, Shah NH, Noy NF (2008) Biomedical ontologies: A functional perspective. *Brief Bioinform*, **9**:75–90.

[217] Russell S, Norvig P (1995) *Artificial Intelligence: A Modern Approach*. Prentice Hall.

[218] Schild K (1991) A correspondence theory for terminological logics. In *Proceedings of the 12th IJCAI*, pages 466–471.

[219] Schlicker A, Domingues FS, Rahnenführer J, Lengauer T (2006) A new measure for functional similarity of gene products based on gene ontology. *BMC Bioinformatics*, **7**:302.

[220] Schlicker A, Lengauer T, Albrecht M (2010) Improving disease gene prioritization using the semantic similarity of Gene Ontology terms. *Bioinformatics*, **26**:i561–i567.

[221] Schmidt-Strauss M, Smolka G (1990) Attributive concept descriptions with complements. *Artificial Intelligence*, **48**:1–26.

[222] Schober D, Smith B, Lewis SE, Kusnierczyk W, Lomax J, Mungall C, Taylor CF, Rocca-Serra P, Sansone SA (2009) Survey-based naming conventions for use in OBO foundry ontology development. *BMC Bioinformatics*, **10**:125.

[223] Schofield PN, Gruenberger M, Sundberg JP (2010) Pathbase and the MPATH ontology: Community resources for mouse histopathology. *Vet Pathol*, **47**:1016–1020.

[224] Schofield PN, Bard JBL, Booth C, Boniver J, Covelli V, Delvenne P, Ellender M, Engstrom W, Goessner W, Gruenberger M, Hoefler H, Hopewell J, Mancuso M, Mothersill C, Potten CS, Quintanilla-Fend L, Rozell B, Sariola H, Sundberg JP, Ward A (2004) Pathbase: A database of mutant mouse pathology. *Nucleic Acids Res*, **32(Database issue)**:D512–D515.

[225] Schofield PN, Gkoutos GV, Gruenberger M, Sundberg JP, Hancock JM (2010) Phenotype ontologies for mouse and man: Bridging the semantic gap. *Dis Model Mech*, **3**:281–289.

[226] Schofield PN, Sundberg JP (2009) One medicine, one pathology and the one health concept. *J Am Vet Med Ass*, **234**:1530–1531.

[227] Schulz MH, Köhler S, Bauer S, Vingron M, Robinson PN (2009) Exact score distribution computation for similarity searches in ontologies. In *WABI'09: Proceedings of the 9th international conference on Algorithms in bioinformatics*, pages 298–309, Springer-Verlag.

[228] Segerdell E, Bowes JB, Pollet N, Vize PD (2008) An ontology for *Xenopus* anatomy and development. *BMC Dev Biol*, **8**:92.

[229] Sevilla JL, Segura V, Podhorski A, Guruceaga E, Mato JM, Martínez-Cruz LA, Corrales FJ, Rubio A (2005) Correlation between gene expression and GO semantic similarity. *IEEE/ACM Trans Comput Biol Bioinform*, **2**:330–338.

[230] Shah NH, Bhatia N, Jonquet C, Rubin D, Chiang AP, Musen MA (2009) Comparison of concept recognizers for building the open biomedical annotator. *BMC Bioinformatics*, **10 Suppl 9**:S14.

[231] Shannon CE (1948) A mathematical theory of communication. *Bell Systems Technical Journal*, **27**:379–423, 623–656.

[232] Sirin E, Parsia B, Cuenca Grau B, Kalyanpur A, Katz Y (2007) Pellet: A practical OWL-DL reasoner. *Web Semantics*, **5**:51–53.

[233] Smith B, Brochhausen M (2010) Putting biomedical ontologies to work. *Methods Inf Med*, **49**:135–40.

[234] Smith B (1998) The basic tools of formal ontology. In Nicola Guarino (editor) *Formal Ontology in Information Systems*, pages 19–28. IOS Press.

[235] Smith B (2004) Beyond concepts: Ontology as reality representation. pages 73–84. IOS Press.

[236] Smith B, Ashburner M, Rosse C, Bard J, Bug W, Ceusters W, Goldberg LJ, Eilbeck K, Ireland A, Mungall CJ, Leontis N, Rocca-Serra P, Ruttenberg A, Sansone SA, Scheuermann RH, Shah N, Whetzel PL, Lewis S (2007) The OBO Foundry: Coordinated evolution of ontologies to support biomedical data integration. *Nat Biotechnol*, **25**:1251–1255.

[237] Smith B, Brogaard B (2003) Sixteen days. *The Journal of Medicine and Philosophy*, **28**:45–78.

[238] Smith B, Ceusters W, Klagges B, Köhler J, Kumar A, Lomax J, Mungall C, Neuhaus F, Rector AL, Rosse C (2005) Relations in biomedical ontologies. *Genome Biol*, **6**:R46.

[239] Smith B, Rosse C (2004) The role of foundational relations in the alignment of biomedical ontologies. *Stud Health Technol Inform*, **107**:444–448.

[240] Smith B, Williams J, Schulze-Kremer S (2003) The ontology of the Gene Ontology. *AMIA Annu Symp Proc*, pages 609–613.

[241] Smith CL, Goldsmith CAW, Eppig JT (2005). The Mammalian Phenotype Ontology as a tool for annotating, analyzing and comparing phenotypic information. *Genome Biol*, **6**:R7.

[242] Smith MK, Welty C, McGuinness DL (2004) OWL Web Ontology Language Guide. W3C Recommendation, 10 February 2004. http://www.w3.org/TR/owl-guide/.

[243] Soldatova LN, King RD (2005) Are the current ontologies in biology good ontologies? *Nat Biotechnol*, **23**:1095–1098.

[244] Spellman PT, Miller M, Stewart J, Troup C, Sarkans U, Chervitz S, Bernhart D, Sherlock G, Ball C, Lepage M, Swiatek M, Marks WL, Goncalves J, Markel S, Iordan D, Shojatalab M, Pizarro A, White J, Hubley R, Deutsch E, Senger M, Aronow BJ, Robinson A, Bassett D, Stoeckert CJ, Brazma A (2002) Design and implementation of microarray gene expression markup language (MAGE-ML). *Genome Biol*, **3**:RESEARCH0046.

[245] Stenson PD, Mort M, Ball EV, Howells K, Phillips AD, St Thomas N, Cooper DN (2009) The Human Gene Mutation Database: 2008 update. *Genome Med*, 1:13.

[246] Stewart WJ (2009) *Probability, Markov Chains, Queues, and Simulation: The Mathematical Basis of Performance Modeling.* Princeton University Press, Princeton, NJ, USA.

[247] Subramanian A, Tamayo P, Mootha VK, Mukherjee S, Ebert BL, Gillette MA, Paulovich A, Pomeroy SL, Golub TR, Lander ES, Mesirov JP (2005) Gene set enrichment analysis: A knowledge-based approach for interpreting genome-wide expression profiles. *Proc Natl Acad Sci USA*, **102**:15545–15550.

[248] ter Horst HJ (2005) Completeness, decidability and complexity of entailment for RDF Schema and a semantic extension involving the OWL vocabulary. *Web Semantics: Science, Services and Agents on the World Wide Web*, 3(2-3):79–115.

[249] Thomas E, Pan JZ, Ren Y (2010) TrOWL: Tractable OWL 2 Reasoning Infrastructure. In *the Proc. of the Extended Semantic Web Conference (ESWC2010)*.

[250] Thomas PD, Campbell MJ, Kejariwal A, Mi H, Karlak B, Daverman R, Diemer K, Muruganujan A, Narechania A (2003) PANTHER: a library of protein families and subfamilies indexed by function. *Genome Res*, **13**:2129–2141.

[251] Tirmizi SH, Miranker DP (2006) OBO2OWL: Roundtrip between OBO and OWL. Technical Report TR-06-47, The University of Texas at Austin, Department of Computer Sciences.

[252] Tsarkov D, Horrocks I (2006) FaCT++ description logic reasoner: System description. In *Proc. of the Int. Joint Conf. on Automated Reasoning (IJCAR 2006)*, volume 4130 of *Lecture Notes in Artificial Intelligence*, pages 292–297. Springer.

[253] Twigger SN, Shimoyama M, Bromberg S, Kwitek AE, Jacob HJ, RGD Team (2007) The Rat Genome Database, update 2007–easing the path from disease to data and back again. *Nucleic Acids Res*, **35(Database issue)**:D658–D662.

[254] UniProt Consortium (2009) The Universal Protein Resource (UniProt) 2009. *Nucleic Acids Res*, **37(Database issue)**:D169–D174.

[255] UniProt Consortium (2010) The Universal Protein Resource (UniProt) in 2010. *Nucleic Acids Res*, **38(Database issue)**:D142–D148.

[256] URI Planning Interest Group, W3C/IETF (2001) URIs, URLs, and URNs: Clarifications and Recommendations 1.0. W3C Note, 21 September 2001. http://www.w3.org/TR/uri-clarification/.

[257] van der Laan MJ, Dudoit S, Pollard KS (2004) Multiple testing. Part II. Step-down procedures for control of the family-wise error rate. *Stat Appl Genet Mol Biol*, **3**:Article14.

[258] W3C OWL Working Group (2009) OWL 2 Web Ontology Language: Document Overview. W3C Recommendation, 27 October 2009. http://www.w3.org/TR/owl-overview/.

[259] Wang Z, Gerstein M, Snyder M (2009) RNA-Seq: A revolutionary tool for transcriptomics. *Nature Reviews Genetics*, **10**:57–63.

[260] Washington NL, Haendel MA, Mungall CJ, Ashburner M, Westerfield M, Lewis SE (2009) Linking human diseases to animal models using ontology-based phenotype annotation. *PLoS Biol*, **7**:e1000247.

[261] Wit E, McClure J (2004) *Statistics for Microarrays: Design, Analysis and Inference*. Wiley.

[262] World Wide Web Consortium (2004) OWL Web Ontology Language Reference, February 2004. http://www.w3.org/TR/owl-ref/.

[263] World Wide Web Consortium (2004) OWL Web Ontology Language. Semantics and Abstract Syntax, February 2004. http://www.w3.org/TR/owl-semantics/.

[264] World Wide Web Consortium (2008) Extensible Markup Language (XML) 1.0 (Fifth Edition), November 2008. http://www.w3.org/TR/REC-xml/.

[265] Wu CH, Nikolskaya A, Huang H, Yeh LSL, Natale DA, Vinayaka CR, Hu ZZ, Mazumder R, Kumar S, Kourtesis P, Ledley RS, Suzek BE, Arminski L, Chen Y, Zhang J, Cardenas JL, Chung S, Castro-Alvear J, Dinkov G, Barker WC (2004) PIRSF: Family classification system at the protein information resource. *Nucleic Acids Res*, **32(Database issue)**:D112–D114.

[266] Xu T, Gu J, Zhou Y, Du L (2009) Improving detection of differentially expressed gene sets by applying cluster enrichment analysis to Gene Ontology. *BMC Bioinformatics*, **10**:240.

[267] Young M, Wakefield M, Smyth G, Oshlack A (2010) Gene ontology analysis for RNA-seq: Accounting for selection bias. *Genome Biology*, **11**:R14.

[268] Yu G, Li F, Qin Y, Bo X, Wu Y, Wang S (2010) GOSemSim: An R package for measuring semantic similarity among GO terms and gene products. *Bioinformatics*, **26**:976–978.

[269] Yu H, Jansen R, Stolovitzky G, Gerstein M (2007) Total ancestry measure: Quantifying the similarity in tree-like classification, with genomic applications. *Bioinformatics*, **23**:2163–2173.

Index